SYSTEM ENGINEERING MANAGEMENT

FOURTH EDITION

Benjamin S. Blanchard

Professor—Emeritus
Department of Industrial and Systems Engineering
Virginia Polytechnic Institute and State University
Blacksburg, Virginia

WILEY

JOHN WILEY & SONS, INC.

Library of Congress Cataloging-in-Publication Data:

Blanchard, Benjamin S.
    System engineering management / Benjamin S. Blanchard. – 4th ed.
        p. cm. – (Wiley series in systems engineering and management)
    Includes bibliographical references and index.
    ISBN 978-0-470-16735-9 (cloth : alk. paper)
  1. Systems engineering.   I. Title.
    TA168.B53 2008
    620.001'171–dc22

                                                    2008019000

CONTENTS

6 System Engineering Program Planning

FOREWORD

The first three editions of Ben Blanchard's *System Engineering Management* text are exceptionally fine works and reflect very well the continued evolutionary and emerging developments in the important subject of system engineering and closely related management of system engineering programs and projects. It is hard to believe that the third edition could be made better. However, and much to the credit of the dedication of the author, the fourth edition of the work reflects well and improves upon the exemplary work reflected in the previous three editions through continued evolutionary enhancements.

The author is one of the most respected system engineering and management educators in the world today, and this work reflects his continuing dedication to the profession. The third edition of the book is not, in any sense, obsolete. It is quite useful and relevant. This fourth edition presents an increased emphasis on such subjects as system engineering processes, design, systems architecting, and organization of programs and projects for system engineering. This new edition is not revolutionary in any sense; it is indeed a significant evolutionary improvement over the present excellent third edition and is even more relevent to the world of system engineering management practice.

This text continues to be most useful as a senior, or very beginning graduate level, text in the management of system engineering programs and projects. It could well be used as a text for all engineering disciplines who definitely need, and who often do not get, some exposure in their curricula to the systems approach and systems-level thinking. The continued inclusion of more basic system engineering and analysis material not in the previous editions to the same extent makes the work even more useful for this latter group than the previous editions. The work appears especially useful as a single-source introduction to system engineering and engineering management for undergraduate engineers in all disciplines. It will also be useful as a text in system engineering programs, and could well be used as the basic text for a single semester course in engineering management taught at the senior or introductory graduate level.

System engineering management needs abound, as evidenced by the continual unrelenting pressure for enhanced productivity and responsiveness. There is increasing

ABET pressure for inclusion of system engineering material in engineering curricula. Thus, the work fills several needs today. It is eminently suitable for use by system engineering majors in an introductory course in system engineering management, as well as for all engineering undergraduate students who wish a definitive presentation of the role of system engineering in enhancing productivity in the public and private sectors.

It is a special pleasure to note that this edition of the work has found a most appropriate home in the Wiley Series in *Systems Engineering and Management.*

ANDREW P. SAGE

PREFACE

Current trends indicate that, in general, the complexity of *systems* is increasing, and the challenges associated with bringing new systems into being are greater than ever! Requirements are constantly changing with the introduction of new technologies on a continuing and evolutionary basis; the life cycles of many systems are being extended while, at the same time, the life cycles of individual and specific technologies are becoming shorter; and systems are being viewed more in terms of interoperability requirements and within a system of systems (SOS) context. There is a greater degree of outsourcing and the utilization of suppliers throughout the world, and international competition is increasing in a global environment. Available resources are dwindling worldwide; and many of the systems (products) in use today are not meeting the needs of the customer/user in terms of performance, quality, and overall cost-effectiveness.

Given today's environment, there is an ever-increasing need to develop and produce systems that are robust, reliable and of high quality, supportable, cost-effective from a total life-cycle perspective, and responsive to the needs of the customer/user in a satisfactory manner. Further, future systems must be designed with an open-architecture approach in mind in order to facilitate the incorporation of quick configuration changes and new technology insertions, and to be able to respond to system interoperability requirements on an expedited basis.

From past experience, the majority of the problems noted have been the direct result of not applying a total *systems* approach, from the beginning, in meeting the desired objectives. That is, the overall requirements for the system in question were not very well defined initially; a bottom-up (versus top-down) approach was followed in the system development process; the overall perspective pertaining to meeting the customer's need was relatively short-term; and, in many instances, the philosophy has been to "design-it-now-and-fix-it-later." In essence, the system design and development process has suffered somewhat from the lack of good early planning and the definition of *requirements* in a complete and methodical manner, and total *life-cycle* considerations have basically been addressed after the fact! This approach has turned out to be quite costly in the long term, particularly in assessing, the risks associated with the decision-making processes during the early stages of system development.

The combination of these and related factors has created a critical need—that is, the requirements for developing and producing (or constructing) well-integrated, high-quality, reliable, supportable, and cost-effective systems with complete customer (user) satisfaction in mind. In this highly competitive resource-constrained environment, it is now more important than ever to ensure that the principles and concepts of *system engineering* are properly implemented, both in the design and development of new systems and/or in the reengineering and modification of existing systems. System requirements must be well-defined from the beginning. The system must be viewed in terms of *all* of its components on a totally integrated basis—to include prime equipment, software, operating personnel, facilities, associated data and information, its associated production and distribution process, and the elements of maintenance and support (and not limited to just those elements utilized to accomplish a given mission scenario). A top-down (and bottom-up) integrated approach must be assumed, with the appropriate allocation of requirements from the system level and down to its various elements. The system must be addressed within a higher-level system of systems (SOS) context, as appropriate, and considering applicable interoperability requirements. Further, the system must be viewed in terms of its entire life cycle; that is, from conceptual through preliminary and detailed design, production and/or construction, system utilization, maintenance and support, and system retirement and material recycling and/or disposal. Decisions made in any one phase of the system life cycle will likely have a significant impact on the activities in the other phases. Thus, a total *system's life-cycle approach* must be assumed.

These concepts are not necessarily new or novel. System engineering, in its current context, has been a subject of interest since the late 1950s and early 1960s (and perhaps even earlier). The principles have been successfully applied in a few programs. However, in most instances, although we may believe that we utilize these methods successfully, we really do not implement them very well (if at all). The successful implementation of system engineering requires not only a *technical* thrust, but a *management* thrust as well. It is essential that one select the appropriate technologies, utilize the proper analytical tools, and apply the necessary resources to enhance the system engineering process. In addition, the proper *organizational environment* must be established to allow for the proper implementation of this process. Thus, it is necessary, first, to understand and believe in the process and, second, to establish the proper management and organizational structure that will allow it to happen! This approach, in turn, provides a *cultural* challenge for the future.

This text was developed with the preceding objectives in mind. The basic principles and concepts, the need for system engineering and its applications, and introduction to some key terms and definitions are covered in Chapter 1. This leads to a comprehensive presentation of the *system engineering process* in Chapter 2. This process commences with the identification of a consumer need and extends though the definition of system operational requirements and the maintenance and support concept; the identification and prioritization of technical performance measure (TPMs); a description of system architecture and the elements of the system in *functional* terms; the allocation of top system-level requirements to the various components of the system in form of input design-to criteria; system synthesis, analysis, and design optimization; test, evaluation, and validation; production and/or construction; distribution, installation, and system utilization in the user's environment; system maintenance and sustaining life-cycle support; and system retirement and material recycling and/or disposal. Key areas of emphasis for system engineering are noted throughout. A thorough understanding of this process is fundamental in dealing with the overall subject area, and the material in Chapter 2 serves as a baseline for discussion in subsequent chapters.

Given the preceding overview, it is appropriate to delve further into some of the objectives of system engineering. One goal includes the *integration* of a wide variety of key design support disciplines into the total mainstream system design effort. Chapter 3 provides an introduction to some of these disciplines to include software engineering, reliability, and maintainability engineering, human factors and safety engineering, manufacturing and production, logistics and supportability, disposability, quality, environmental and value/cost engineering. Chapter 4 follows with a discussion pertaining to the application of design methods and tools, utilized in such a manner as to enhance the fulfillment of system engineering objectives. The appropriate application of electronic commerce (EC), information technology (IT), electronic data interchange (EDI), and computer-aided design (CAD) methods allows for "front-end" analysis, leading to a better system definition at an earlier stage in the life cycle. Chapter 5 discusses the checks and balances in the design process, provided through the accomplishment of formal design review, evaluation, feedback and control, and the initiation of changes for corrective action as necessary. An objective of system engineering is to provide a strong engineering leadership role relative to the initial definition of system requirements, the necessary *integration* of design activities to ensure effective and efficient results, and the follow-on measurement and evaluation functions to ensure that the initially specified requirements have been met.

The next step addresses the *management* issues pertaining to the application of system engineering requirements to different projects. Chapter 6 leads off with an in-depth discussion of planning and the development of the *System Engineering Management Plan* (*SEMP*). System engineering tasks, the development of a work breakdown structure (WBS), program task schedules, and the preparation of cost projections are included. Customer, producer (prime contractor), and supplier activities are covered. Of particular note is the identification, selection, and contracting with key suppliers. Chapter 7 addresses system engineering in a typical project organizational structure, highlighting the differences between functional, product-line, project, and matrix structures. The many interfaces between the customer (consumer), the producer (contractor), and suppliers are discussed, as well as the human resources requirements pertaining to the staffing and management of a system engineering department/group. Having covered the planning, organization, and implementation of a system engineering program, it is essential that one consider a formal *evaluation* to properly measure and assess the degree to which the organization is performing in accomplishing its overall objectives. Chapter 8 introduces organizational *benchmarking* and the application of several different models for the purposes of evaluation and feedback (e.g., the *SECM* and the *CMMI* models). Dealing with the issues of planning and organization only, without the benefit of evaluation and feedback, constitutes only part of the process and tends to inhibit future growth.

The six appendixes provide excellent supplemental material in support of the various topics covered throughout the eight chapters in the text. Appendix A includes case-study illustrations of the functional analysis; Appendix B describes in detail the steps involved in performing a *life-cycle cost analysis (LCCA)*; Appendix C includes seven different case-study examples of various types of design analysis; Appendix D includes an extensive design-review checklist; Appendix E contains a supplier evaluation checklist; and Appendix F constitutes an extensive bibliography.

In summary, the intent herein is to describe system engineering in terms of its objectives and applications and the steps in the system engineering process, and to provide a management perspective for the implementation of programs with a strong system engineering thrust. It is believed that this text can be effectively utilized in the academic classroom (both at the undergraduate and graduate levels), in support of a continuing education seminar or workshop, and as an on-the-job reference guide. Questions and problem exercises are included at the end of each chapter to provide the necessary emphasis where required, and an instructor's guide is available for academic classroom support.

Finally, I wish to express my thanks and appreciation to Dr. Andrew P. Sage, editor, *Wiley Series in Systems Engineering and Management*, for his support relative to including this text within the series, for preparation of the Foreword, and for his continuing friendship through the years.

BENJAMIN S. BLANCHARD

1

INTRODUCTION TO SYSTEM ENGINEERING

This text deals with *system engineering,* or the orderly process of bringing a system into being and the subsequent effective and efficient operation and support of that system throughout its projected life cycle. It constitutes an interdisciplinary approach and means for enabling the realization and follow-on deployment of a successful system.

A *system* comprises a complex combination of resources (in the form of human beings, materials, equipment, software, facilities, data, information, services, etc.), integrated in such a manner as to fulfill a designated need. A system is developed to accomplish a specific function, or a series of functions, with the objective of responding to some identified need. The various elements of a system must be directly tied to and supportive of the accomplishment of some given mission scenario, or series of scenarios.

A system may be classified as a *natural system, human-made system, physical system, conceptual system, closed-loop system, open-loop system, static system, dynamic system,* and so on. This text addresses primarily human-made systems that are physical, dynamic, and open-loop in structure. Further, the objective is to address the system in the context of its *whole* versus dealing with its components only. Of significant importance is the realization that ultimate system performance is dependent not only on the complete and timely integration of its various components, but on establishing the proper interrelationships among these components. By accomplishing such through the application of system engineering principles and concepts, a value-added component can be realized.

A system may vary in form, fit, and/or function. One may be dealing with a group of aircraft accomplishing a mission at a specific geographical location, a communication network for the processing of information on a worldwide basis, a power

1

distribution capability involving waterways and electrical power-generating units, a healthcare capability including a group of hospitals and mobile units serving a given community, a manufacturing facility that produces x products in a designated time frame, or a small vehicle providing the transportation of certain cargo from one location to another. A system may also be contained within some overall hierarchy such as an aircraft within an airline system, which is within a larger regional transportation system, which is within a worldwide transportation capability, and so on. In this context, we may be dealing with *system of systems (SOS)*, a popular term currently being applied in describing highly complex systems within some higher-level structure. The objective is to be able to adequately define and describe the overall boundaries of the particular system being addressed, and its interfaces (and interrelationships) across the board.

The objectives of this chapter are: to address the subject of systems in general; to define some key terms and the characteristics of systems; to identify the need for and the basic requirements for bringing systems into being and for later evaluating systems in terms of their effectiveness in a user's environment; and to provide an introduction to *system engineering* and the associated management activities inherent in the system engineering process.

1.1   DEFINITION OF A SYSTEM

In order to ensure a good and common understanding of the material throughout this text, it seems appropriate to commence with a few definitions. As a start, one should first establish a basic definition for a *system*. Although this may appear to be overly simplistic, experience has indicated that people throughout the world tend to utilize the term rather loosely to describe many different situations and configurations. Further, there is a lack of consistency in the application of system engineering principles and concepts. Thus, it is important to first review a few terms to establish somewhat of a baseline for further discussion.

1.1.1   The Characteristics of a System

The term *system* stems from the Greek *systēma*, meaning an "organized whole." *Merriam-Webster's Collegiate Dictionary* defines a system as "a regularly interacting or interdependent group of items forming a unified whole."[1] One of the early Military Standards on the subject, MIL-STD-499, defines a system as "a composite of equipment, skills, and techniques capable of performing and/or supporting an operational role. A complete system includes all equipment, related facilities, material, software, services, and personnel required for its operation and support to the degree that it can be considered a self-sufficient unit in its intended environment."[2]

---

[1]*Merriam-Webster's Collegiate Dictionary*. 10th ed. (Springfield, MA: Merriam-Webster, Inc., 1998).
[2]Military Standard, MIL-STD-499. *System Engineering Management*. (Department of Defense, July 17, 1969).

A more recent document, EIA/IS-632, defines a system as "an integrated composite of people, products, and processes that provide a capability to satisfy a stated need or objective."[3]

Given the variations in the basic definition of a "system," the leadership of INCOSE (International Council on Systems Engineering) assigned the current Fellows of the Council to develop a consensus definition. After a few iterations the following definition evolved:

> A "system" is a construct or collection of different elements that together produce results not obtainable by the elements alone. The elements, or parts, can include people, hardware, software, facilities, policies, and documents; that is, all things required to produce system-level results. The results include system-level qualities, properties, characteristics, functions, behavior, and performance. The value added by the system as a whole, beyond that contributed independently by the parts, is primarily created by the relationship among the parts; that is, how they are interconnected.[4]

In essence, a system constitutes a set of interrelated components working together with the common objective of fulfilling some designated need.

Although the preceding definitions reflect a good initial overview, a greater degree of detail and precision is required to provide a good working definition acceptable for describing the principles and concepts of system engineering. To facilitate this objective, a system may be defined further in terms of the following general characteristics:

1. A system constitutes a *complex combination of resources* in the form of human beings, materials, equipment, software, facilities, data, money, and so on. To accomplish many functions often requires large amounts of personnel, equipment, facilities, and data (e.g., an airline or a manufacturing capability). Such resources must be combined in an *effective* manner, as it is too risky to leave this to chance alone.

2. A system is contained within some form of *hierarchy.* An airplane may be included within an airline, which is part of an overall transportation capability, which is operated in a specific geographic environment, which is part of the world, and so on. As such, the system being addressed is highly influenced by the performance of the higher-level system, and these external factors must be evaluated.

3. A system may be broken down into *subsystems* and related components, the extent of which depends on complexity and the function(s) being performed. Dividing the system into smaller units allows for a simpler approach relative to the initial allocation of requirements and the subsequent analysis of the system and its functional interfaces. A system is made up of many different

---

[3]EIA/IS-632, *Processes for Engineering a System*, Electronic Industries Association (EIA), Washington, DC, December 1994.

[4]INCOSE, 2150 N. 107th Street, Suite 205, Seattle, WA 98133. This definition was developed in the fall of 2001.

components, these components *interact* with each other, and these interactions must be thoroughly understood by the system designer and/or analyst. Because of these interactions among components, it is impossible to produce an effective design by considering each component separately. One must view the system as a whole, break down the system into components, study the components and their interrelationships, and then put the system back together.

4. A system must have a *purpose*. It must be functional, able to respond to some identified need, and able to achieve its overall objective in a cost-effective manner. There may be a conflict of objectives, influenced by the higher-level system in the hierarchy, and the system must be capable of meeting its stated purpose in the best way possible.

As a point of emphasis, a system must respond to an identified *functional need*. Thus, the elements of a system must include not only those items that relate directly to the accomplishment of a given scenario or mission profile, but must also include those elements of logistics and the maintenance and support infrastructure that have to be available and in place should a failure of a prime element(s) occur. In other words, if one is to ensure the successful completion of a mission, all of the supporting elements must be available, in place, and ready to respond to a given need.[5]

1.1.2 Categories of Systems

In defining systems in terms of the general characteristics presented, it readily becomes apparent that some degree of further classification is desirable. There are many different types of systems, and there are some variations in terms of similarities and dissimilarities. To provide some insight into the variety of systems in existence, a partial listing of categories follows:[6]

1. *Natural and man-made systems.* Natural systems include those that came into being through natural processes. Examples include a river system and an energy system. Man-made systems are those that have been developed by human beings, the results of which include a wide variety of capabilities. As all man-made systems are embedded in the natural world, there are numerous interfaces that must be addressed. For instance, the development and construction of a hydroelectric power system located on a river system creates impacts on both sides of the spectrum, and it is essential that the systems approach involving both the natural and man-made segments of this overall capability be implemented.

---

[5]In many instances, the logistics and maintenance support infrastructure is not addressed or included as an element of the system, or as a major subsystem, but treated separately and "after the fact." The approach assumed throughout this text is that this is included in the context of a major subsystem and addressed as an inherent part of the whole.

[6]This categorization follows the general form presented in B. Blanchard and W. Fabrycky. *Systems Engineering and Analysis*, 4th ed. (Upper Saddle River, NJ: Pearson Prentice-Hall, 2006). These categories represent only a few of those that could be described.

2. *Physical and conceptual systems.* Physical systems are those made up of real components occupying space. By contrast, conceptual systems can be an organization of ideas, a set of specifications and plans, a series of abstract concepts, and so on. Conceptual systems often lead directly into the development of physical systems, and there is a certain degree of commonality in terms of the type of processes employed. Again, the interfaces may be many, and there is a need to address these elements in the context of a higher-level system in the overall hierarchy.

3. *Static and dynamic systems.* Static systems include those having structure, but without activity (as viewed in a relatively short period of time). A highway bridge and a warehouse are examples. A dynamic system is one that combines structural components with activity. An example is a production capability combining a manufacturing facility, capital equipment, utilities, conveyors, workers, transportation vehicles, data, software, managers, and so on. Although there may be specific points in time when all system components are static in nature, the successful accomplishment of system objectives does require activity and the dynamic aspects of system operation do prevail throughout a given scenario.

4. *Closed and open-loop systems.* A closed system is one that is relatively self-contained and does not significantly interact with its environment. The environment provides the medium in which the system operates; however, the impact is minimal. A chemical equilibrium process and an electrical circuit (with a built-in feedback and control loop) are examples. Conversely, open-loop systems interact with their environments. Boundaries are crossed (through the flow of information, energy, and/or matter), and there are numerous interactions both among the various system components and up and down the overall system hierarchical structure. A system/product logistic support capability is an example.

These categories are presented to stimulate further thought relative to the definition of a system. It is not easy to classify a system as being either closed or open, and the precise relationships between natural and man-made systems may not be well defined. However, the objective here is to gain a greater appreciation for the many different considerations required in dealing with system engineering and its process. This text tends to deal mainly with *man-made* systems that are *physical* by nature, *dynamic* in operation, and of the *open-loop* variety.

The systems addressed herein may include a wide variety of *functional* entities. There are transportation systems, communication systems, manufacturing systems, information processing systems, and so on, as indicated in Figure 1.1. In each instance, there are *inputs*, there are *outputs*, there are external *constraints* imposed on a system, and there are the required *mechanisms* necessary to realize the desired results. Within the framework of the system, there are products and processes.

A system is composed of many different elements, including those that are directly utilized in the actual accomplishment of a mission (e.g., prime equipment,

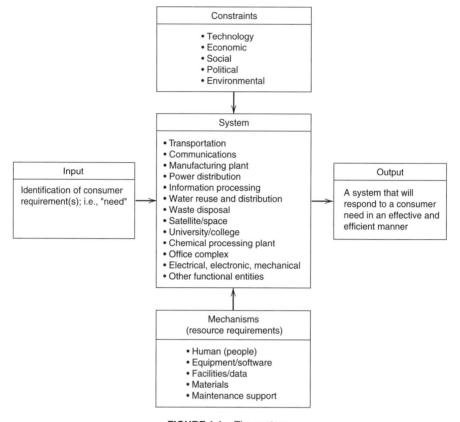

FIGURE 1.1   The system.

operating software, operating personnel) and the elements of maintenance support (e.g., maintenance personnel, test equipment, facilities, spares and repair parts). Although the support infrastructure is not often considered an element of a system per se, the system may not be able to complete its designated function in its absence. Thus, the support infrastructure is addressed as a major system element, presented in the context of the system life cycle. Figure 1.2 identifies the major elements of a system.

1.1.3 System of Systems (SOS)

A system may be contained within some form of *hierarchy*, as shown in Figure 1.3, where there are different layers of systems within an overall configuration. For example, there is an aircraft system, within an air transportation system (e.g., commercial airline), within an overall regional transportation capability, and so on. Often, and

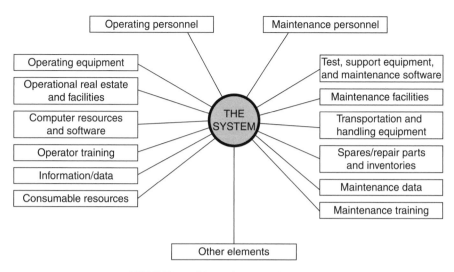

FIGURE 1.2 Major elements of a system.

particularly in dealing with large-scale defense systems, such as configuration may be referred to as a *system of systems (SOS)*. Basically, an SOS may be defined as:[7]

> a collection of component systems that produce results unachievable by the individual systems alone. Each system in the SOS structure is likely to be operational in its own right, as well be contributing in the accomplishment of some higher-level mission requirement. The life cycles of the individual systems may vary somewhat as there will be additions and deletions at different times, as long as the mission requirements for any given system are met. Thus, there may be some new developments in progress at the same time as other elements are being retired for disposal.

Referring to Figure 1.3, the question is—are we addressing a transportation system including many different type of air and ground vehicles, or an airline including many different aircraft, or a specific aircraft with its crew and all of its support? It is not uncommon for a group of individuals to get together and discuss a particular issue, each having a different perception as to the *system* being addressed. One person's system of interest can be viewed as an element in another person's system of interest.

In defining the requirements for a system, one must be careful in relating such to a specific functional objective, establishing the appropriate hierarchical relationships, defining the boundaries for each system in the hierarchy, and identifying some of the interrelationships that exist throughout. In regard to the systems shown in Figure 1.3, both upward and downward impacts must be considered. Decisions

[7]INCOSE-TP-2003-002-03, *Systems Engineering Handbook*, Version 3, International Council on Systems Engineering, 2150 N. 107th Street, Suite 205, Seattle, WA 98133, 2006.

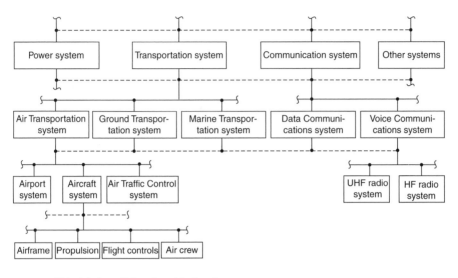

------ Major interfaces (Integration of Systems)

FIGURE 1.3 Multiple systems (system of systems).

pertaining to the aircraft system may have an upward impact on the air transportation system (e.g., the airline), and certainly will have a downward impact on the aircraft airframe, propulsion and so on. For example, the maintenance support infrastructure for the aircraft system may have to be compatible with the maintenance concept specified for the higher-level air transportation system (e.g., the airline). In addition, this concept may also be imposed as a constraint in the design of the airframe and its components. In any event, these interaction effects may be significant and must be addressed.

1.2 THE CURRENT ENVIRONMENT: SOME CHALLENGES

Having a good understanding of the overall *environment*, and some of the challenges ahead, is certainly a prerequisite to the successful implementation of system engineering principles and concepts. Although individual perceptions will differ, depending on what various individuals observe, there are a few trends that appear to be significant. These trends, as summarized and illustrated in Figure 1.4, are all interrelated and need to be addressed in total, and as an integrated set in determining the ultimate requirements for systems and in the implementation of the system engineering process:

1. *Constantly changing requirements.* The requirements for new systems are frequently changing because of the dynamic conditions worldwide, changes in mission thrusts and priorities, and the continuous introduction of new technologies. Further, it is often difficult to define the "real" requirements for new

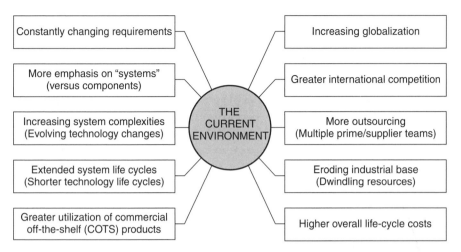

FIGURE 1.4 The current environment.

systems because of the lack of a good definition of the problem(s) to be solved and the subsequent lack of good communications between the ultimate user and the system developer from the beginning.

2. *More emphasis on systems*. There is a greater degree of emphasis on total *systems* versus the *components* of a system. One must look at the system in total, and throughout its entire life cycle, to ensure that the functions that need to be performed are being accomplished in an effective and efficient manner. At the same time, components need to be addressed within the context of some overall system configuration.

3. *Increasing system complexities*. It appears that the structures of many systems are becoming more complex with the introduction of evolving new technologies. Further, the interaction effects between different systems, within a higher-level system of systems (SOS) configuration, often lead to added complexities. It will be necessary to design systems so that changes can be incorporated quickly, efficiently, and without causing a significant impact on the overall configuration of the system.

4. *Extended system life cycles—shorter technology life cycles*. The life cycles of many of the systems in use today are being extended for one reason or another while, at the same time, the life cycles of most technologies are relatively much shorter. It will be necessary to design systems with an *open-architecture* approach in mind so that the incorporation of a new technology can be accomplished easily and efficiently (this trend, of course, closely relates to item 3).

5. *Greater utilization of commercial off-the-shelf (COTS) products*. With current goals pertaining to lower initial costs and shorter and more efficient procurement and acquisition cycles, there has been a greater emphasis on the

utilization of best commercial practices, processes, and commercial off-the-shelf (COTS) equipment and software. As a result, there is a greater need for a good definition of requirements from the beginning, and there is a greater emphasis on the design of systems (and their major subsystems) versus the design of components.

6. *Increasing globalization.* The "world is becoming smaller" (as they say), and there is more trading and dependency on different countries (and manufacturers) throughout the world than ever before. This trend, of course, is being facilitated through the introduction of rapid and improved communications practices, the availability of quicker and more efficient packaging and transportation methods, the application of electronic commerce (EC) methods for expediting procurement and related processes, and so on.

7. *Greater international competition.* Along with the noted trends toward increasing globalization, there is more international competition than ever before. This, of course, is facilitated not only through improvements in communications and transportation methods, but through the greater utilization of COTS items and the establishment of effective partnerships worldwide.

8. *More outsourcing.* There is more outsourcing and procurement of COTS items (equipment, software, processes) from external sources of supply than ever before. Thus, there are more suppliers associated with any given program. This trend, in turn, requires greater emphasis on the early definition and allocation of system-level requirements, the development of a good and complete set of specifications, and a closely coordinated and integrated level of activity throughout the system development and acquisition process.

9. *Eroding industrial base.* The aforementioned trends (increasing globalization, more outsourcing, and greater international competition), combined with some decline in available resources worldwide, have resulted in a decrease in the number of available manufacturers of many products. In the design of systems, it is necessary to take care to select and utilize components for which there are stable and reliable sources of supply for at least the duration of the life cycle for the system in question.

10. *Higher overall life-cycle costs.* In general, experience indicates that the life-cycle costs of many of the systems in use today are increasing. Although a great deal of emphasis has been placed on minimizing the costs associated with the procurement and acquisition of systems, little attention has been paid to the costs of system operation and support. In the design of systems, it is important to view *all* decisions in the context of *total cost* if one is to properly assess the risks associated with the decision in question.

Although these and related trends have evolved over time and have had a direct impact on our day-to-day activities, we often tend to ignore some of the changes that have taken place and continue with a business-as-usual approach by implementing some past practices that ultimately have had a negative impact on the systems we

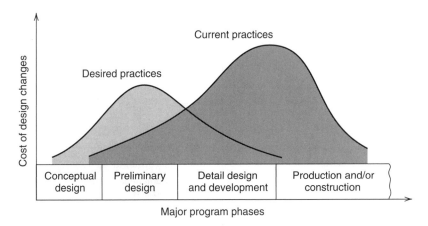

FIGURE 1.5   The cost impact due to changes.

have developed. From past experience, it is clear that many of the problems noted have been the direct result of not applying a *disciplined* "systems approach" to meet the desired objectives. The overall requirements for the system in question were not well defined from the beginning; the perspective in terms of responding to a consumer (user) need was a relatively "short-term" focus; and, in many instances, the approach followed was to *design it now and fix it later!* In essence, the system design and development process has suffered somewhat from a lack of good early planning and the subsequent definition and allocation of requirements in a complete and methodical manner.

In regard to *requirements*, the trend has been to keep things "loose" in the beginning by developing a system-level specification that is very general in content, providing an opportunity for the introduction of the "latest and greatest" in technology developments just prior to going into the construction/production stage. Traditionally, many engineers do not want to be forced into design-related commitments any earlier than necessary, and the basis for defining lower-level requirements is often very "fluid" from the beginning. Thus, there are a lot of last-minute changes in design, and many of these late changes are introduced in haste and without concern for any form of configuration management. Furthermore, sometimes these changes are actually incorporated at a later stage. In any event, the introduction of late changes and the lack of good configuration control can be rather costly. Figure 1.5 provides a comparison of the cost impact due to the incorporation of changes early in the design process versus those incorporated later.[8]

These and related past practices have had a great impact on the overall costs of systems. In fact, in recent years and for many systems, there has been an *imbalance*

---

[8]Referring to Figure 1.5, it should be noted that the curves show the relative costs of actually incorporating the changes in design and not the "downstream" costs resulting from the impact of these changes over the life cycle.

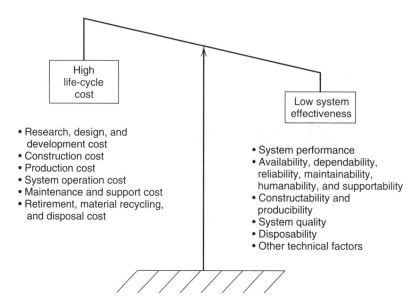

FIGURE 1.6 The imbalance between system cost and effectiveness factors.

between the "cost" side of the spectrum and the "effectiveness" side, as illustrated in Figure 1.6. Many systems have grown in complexity, and although there has been an increase in emphasis on some *performance* factors, the resultant reliability and quality have been decreasing. At the same time, the overall long-term costs have been increasing. Thus, there is a need to provide the proper balance in the development of systems in the future, as any specific design decision will have an impact on both sides of the balance and the interaction effects can be significant.

In addressing the aspect of economics, one often finds that there is a lack of total cost visibility, as illustrated by the "iceberg" in Figure 1.7. For many systems, design and development costs (and production costs) are relatively well known; however, the costs associated with system operation and maintenance are somewhat hidden. In essence, the design community has been successful in dealing with the short-term aspects of cost, but has not been very responsive to the long-term effects. At the same time, experience has indicated that a large segment of the life-cycle cost for a given system is associated with the operational and maintenance support activities accomplished downstream in the life cycle (e.g., up to 75% of the total cost in some instances). Thus, although our budgeting and current practices tend to heed the short-term cost impacts, we cannot adequately assess the risks associated with the ongoing decision-making process without projecting these decisions in the context of the entire system life cycle. In other words, we may wish to make a design decision based on some short-term aspect of cost, but it is important to address the life-cycle implications prior to finalizing the decision.

Moreover, in considering cause-and-effect relationships, it has been determined that a major portion of the projected life-cycle cost for a given system stems from

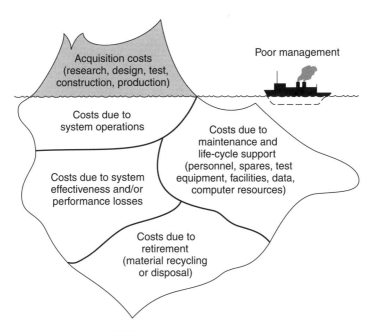

FIGURE 1.7   Total cost visibility.

the consequences of decisions made during the early stages of advance planning and system conceptual design. Such decisions, which can have a significant impact on downstream costs, relate to the definition of operational requirements (the number of consumer sites assumed, the selection of a given mission profile, specified utilization factors, the assumed life cycle), maintenance and support policies (two versus three levels of maintenance, levels of repair, in-house versus third-party maintenance support), allocations associated with manual versus automation applications, equipment packaging schemes and diagnostic routines, hardware versus software applications, the selection of materials, the selection of a manufacturing process, whether a COTS item should be selected versus the pursuit of a new design approach, and so on. In Figure 1.8 it can be seen that the greatest opportunity for influencing life-cycle cost can be realized in the early stages of system design and development. In other words, early design decisions should be evaluated on the basis of *total life-cycle cost.*

Given the environment of constantly changing requirements, greater utilization of COTS items, increased globalization and more outsourcing, and so forth, there is an ever-increasing need to review our current practices for bringing new systems into being. A highly disciplined approach must be pursued in the design and development of new systems, with the objective of providing the consumer (user) with a high-quality system that is cost-effective, considering the proper balance among the factors identified in Figure 1.6. In addition, there must be more emphasis on *systems,* from a life-cycle perspective, which must be established from the beginning, as illustrated in Figure 1.8. For systems already in use, it is critical that we establish a

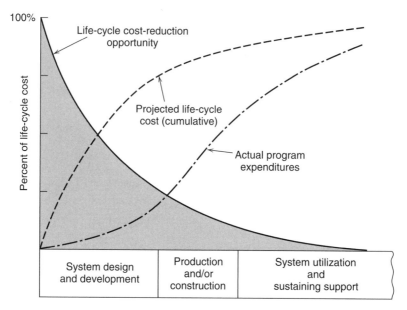

FIGURE 1.8 Commitment of life-cycle cost.

systematic approach to reviewing their requirements and subsequently implementing an effective *evaluation and continuous product/process improvement methodology.* In any event, the current environment, as highlighted herein, is certainly conducive to the implementation of the principles and concepts discussed throughout this text.

1.3 THE NEED FOR SYSTEM ENGINEERING

The trends and concerns conveyed in Section 1.2 are only a sample of the major issues that need to be addressed. The challenge is to be more effective and efficient in the development and acquisition of new systems (i.e., any time that there is a newly identified need and a new system requirement has been established), as well as in the operation and support of those systems already in use. This can be accomplished through the proper implementation of system engineering concepts, principles, and methods.

In exploring topics such as *systems, system engineering, system analysis*, and the like, one will find a variety of approaches in existence. The specific terms may be defined somewhat differently, depending on individual backgrounds, experiences, and on the organizational interests of practitioners in the field. Thus, with the objective of providing some clarification relative to the material throughout this text, it seems appropriate to consider a few concepts and definitions at this point.[9]

---

[9] The bibliography presented in Appendix F includes a variety of publications covering system engineering and related areas.

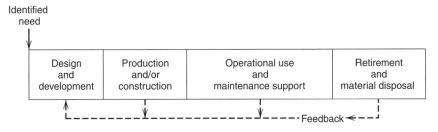

FIGURE 1.9   The system life cycle.

1.3.1 The System Life Cycle

As shown in Figure 1.9, the *life cycle* includes the entire spectrum of activity for a given system, commencing with the identification of need and extending through system design and development, production and/or construction, operational use and sustaining maintenance and support, and system retirement and material disposal. As the activities in each phase interact with the activities in other phases, it is essential to consider the overall life cycle in addressing system-level issues, particularly if one is to properly assess the risks associated with the decision-making process throughout.

Although the life-cycle phases conveyed in Figure 1.9 reflect a more generic sequential approach, the specific activities (and the duration of each) may vary somewhat, depending on the nature, complexity, and purpose of the system. Needs may change, obsolescence may occur, and the levels of activity may be different, depending on the type of system and where it fits in the overall hierarchical structure of activities and events. In addition, the various phases of activity may overlap somewhat, as illustrated in the two examples presented in Figure 1.10.

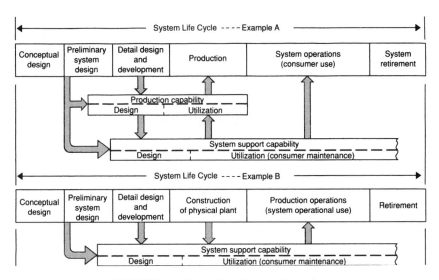

FIGURE 1.10   Examples of system life cycles.

Figure 1.10 shows how an airplane, a ground transportation vehicle, or an electronic device may progress through conceptual design, preliminary design, detail design, production, and so on, as reflected through the series of activities for example A. When this example is evaluated further, the top row of activities is applicable to those elements of the system that relate directly to the accomplishment of the mission (e.g., an automobile). At the same time, there are two closely related life cycles of activity that must also be considered. The design, construction, and operation of the production capability, which can have a significant impact on the operations of the prime elements of the system, should be addressed concurrently along with the system maintenance and support activity. Further, these activities must be addressed early during the conceptual and preliminary design of those prime elements represented by the top row. Although all of these activities may be presented through an illustrated single flow, as conveyed in Figure 1.9, the breakout in Figure 1.10 is intended to emphasize the importance of addressing *all* aspects of the total system process and the various interactions that may occur.

Example B in Figure 1.10 is presented to cover the major phases associated with a manufacturing plant, a chemical processing plant, or a satellite ground tracking facility, where the construction of a "one-of-a-kind" system configuration is required. Again, the maintenance and support capability is identified separately in order to indicate degree of importance and to suggest that there are many interaction effects that must be considered.

Although there may be variations in approaches, the nomenclature used, the duration of the different phases, and so on, it is still appropriate that systems be viewed in the context of their respective life cycles. This is further complicated in the *system-of-systems (SOS)* situation where each of the identified systems within a given SOS structure will likely have a different life cycle. Nevertheless, a total *life-cycle approach* must be assumed in the decision-making process. The past is replete with examples in which major decisions have been made in the early stages of system acquisition based on the "short-term" only. In other words, in the design and development of a new system, the consideration for *production/construction, maintenance and support*, and/or *retirement and disposal* for that system was inadequate. These activities were considered later and, in many instances, the consequences of this 'after-the-fact' approach were costly, as discussed in Section 1.2.[10]

1.3.2   Definition of System Engineering

System engineering may be defined in a number of ways, depending on one's background and personal experience. The inaugural issue of *Systems Engineering*, published by the International Council on Systems Engineering (INCOSE), describes

---

[10]Referring to Figure 1.10, the emphasis as presented addresses the three life cycles, including (1) the life cycle pertaining to the mission-related elements of the system, (2) the production capability, and (3) the maintenance and support capability. There is a fourth life cycle that is equally important but not highlighted in the figure, and this pertains to the design and implementation of the *retirement and material recycling/disposal* capability. One needs to design for producibility, design for supportability/serviceability, and design for recyclability and disposability.

a variety of approaches.[11] However, there is a basic theme throughout that deals with a top-down process, which is life-cycle oriented, involving the integration of functions, activities, and organizations.

The International Council on Systems Engineering (INCOSE) defines it as follows:[12]

> Systems engineering is an interdisciplinary approach and means to enable the realization of successful systems. It focuses on defining customer needs and required functionality early in the development cycle, documenting requirements, and then proceeding with design synthesis and system validation while considering the complete problem. Systems engineering considers both the business and technical needs of all customers with the goal of providing a quality product that meets the user needs.

The Department of Defense (DOD) defines system engineering as follows:

> An approach to translate approved operational needs and requirements into operationally suitable blocks of systems. The approach shall consist of a top-down, iterative process of requirements analysis, functional analysis and allocation, design synthesis and verification, and system analysis and control. Systems engineering shall permeate design, manufacturing, test and evaluation, and support of the product. Systems engineering principles shall influence the balance between performance, risk, cost, and schedule.

More specifically:

> The systems engineering process shall:[13]

1. Transform approved operational needs and requirements into an integrated system design solution through concurrent consideration of all life-cycle needs (i.e., development, manufacturing, test and evaluation, deployment, operations, support, training, and disposal; and
2. Ensure the interoperability and integration of all operational, functional, and physical interfaces. Ensure that system definition and design reflect the requirements for all system elements: hardware, software, facilities, people, and data; and
3. Characterize and manage technical risks.

> The key systems engineering activities that shall be performed are requirements analysis, functional analysis/allocation, design synthesis and verification, and system analysis and control.

---

[11] Inaugural Issue "System Engineering" *Journal of the International Council on Systems Engineering*. vol. 1, no. 1 (July/September 1994).

[12] INCOSE-TP-2003-002-03.1, *Systems Engineering Handbook: A Guide for System Life Cycle Processes and Activities*, Version 3.1, (Seattle, WN: INCOSE, August 2007).

[13] Department of Defense Regulation 5000.2R. "Mandatory Procedures for Major Defense Acquisition Programs (MDAPS) and Major Automated Information System (MAIS) Acquisition Programs." Chapter 5. Paragraph C5.2 (April 5, 2002).

A slightly different definition (preferred by the author) states that system engineering is:

> The application of scientific and engineering efforts to: (1) transform an operational need into a description of system performance parameters and a system configuration through the use of an iterative process of definition, synthesis, analysis, design, test and evaluation, and validation; (2) integrate related technical parameters and ensure the compatibility of all physical, functional, and program interfaces in a manner that optimizes the total definition and design; and (3) integrate reliability, maintainability, usability (human factors), safety, producibility, supportability (serviceability), disposability, and other such factors into a total engineering effort to meet cost, schedule, and technical performance objectives.[14]

Basically, system engineering is *good* engineering with certain designated areas of emphasis, a few of which are noted as follows:

1. A top-down approach is required, viewing the system as a *whole*. Although engineering activities in the past have very adequately covered the design of various system components, the necessary *overview* and an understanding of how these components effectively fit together has not always been present.

2. A *life-cycle* orientation is required, addressing all phases to include system design and development, production and/or construction, distribution, operation, sustaining maintenance and support, and retirement and material phaseout. Emphasis in the past has been placed primarily on system design activities, with little (if any) consideration given to their impact on production, operations, support, and disposal.

3. A better and more complete effort is required relative to the initial *identification of system requirements*, relating these requirements to specific design goals, the development of appropriate design criteria, and the follow-on analysis effort to ensure the effectiveness of early decision making in the design process. In the past, the early "front-end" analysis effort, as applied to many new systems, has been minimal. This, in turn, has required greater individual design efforts downstream in the life cycle, many of which were not well integrated with other design activities and required modification later on.

4. An *interdisciplinary* effort (or team approach) is required throughout the system design and development process to ensure that all design objectives are met in an effective manner. This necessitates a complete understanding of the many different design disciplines and their interrelationships, particularly for large projects.

Inherent within the system engineering process is a "top-down/bottom-up" development approach, as illustrated in Figure 1.11. The emphasis throughout this text is

---

[14]This is a slightly modified version of the definition of systems engineering that was included in the original version of MIL-STD-499. *Systems Engineering* (Washington, DC: Department of Defense, July 1969).

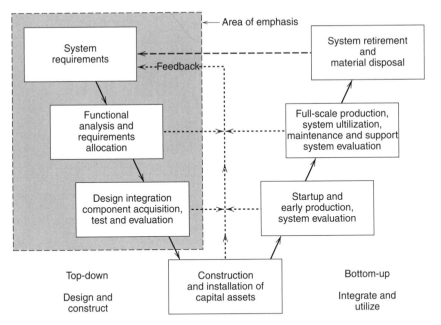

FIGURE 1.11   Top-down/bottom-up system development process.

the shaded area; that is, the front-end requirements analysis activity. Traditionally, the requirements have not been well defined from the beginning, resulting in some rather extensive and costly efforts during the final integration and test activity.

Figure 1.12 presents an extension of the basic life-cycle phases shown in Figure 1.9, describing typical activities that occur in each phase, identifying various configuration *baselines* that should be established as one progresses from the initial identification of need to the development of a fully operational system, and including the iterative steps inherent within the system engineering process. Although the presentation of information in the figure may lead the reader to believe that the system acquisition process is very complex, the objective is to show this as a *process* in itself. Every time there is a newly identified *need*, there are certain steps through which a design engineer should evolve—that is, conceptual design, preliminary design, and so on. Even if the effort (in terms of the resources expended) is minimal, there is still the requirement for design activities at the *system level* and on down. The objective is to view these phase-related activities as a process within itself and to identify the baselines where the design evolves from one level of definition to the next. Tailoring the activities in Figure 1.12 to the system in question is essential for the successful implementation of the system engineering process.

The system engineering process per se includes the basic steps of requirements analysis, functional analysis, requirements allocation, design optimization and tradeoffs, synthesis, evaluation, and so on (refer to blocks 0.1, 0.2, 0.3, etc., in Figure 1.12). These steps are *iterative* by nature, evolving from the system-level definition to the subsystem level, detailed level, and on down to the component. Further,

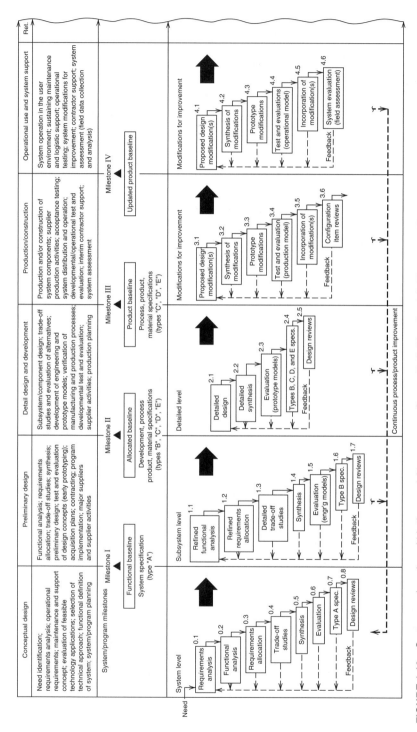

FIGURE 1.12 System engineering within the acquisition process. *Source:* B. S. Blanchard, D. Verma, and E. Peterson, *Maintainability: A Key to Effective Serviceability and Maintenance Management* (New York: John Wiley & Sons, Inc., 1995). This material is used by permission of John Wiley & Sons, Inc.

these steps are not necessarily accomplished in a serial sequence, but are interactive with the appropriate *feedback* provisions at each step in the process. Although the requirements may vary somewhat from program to program, the purpose of this figure is to provide a baseline for future reference as different topics are presented throughout this text.

In block 0.2 (Figure 1.12), the accomplishment of the *functional analysis* will lead to the identification of resources in terms of the need for hardware, software, people, facilities, data, and the like. The functional analysis identifies the "whats" from a requirements perspective, and this leads to the accomplishment of trade-offs and the description of the "hows" pertaining to the completion of functions. Figure 1.13 illustrates the identification of hardware, software, and human requirements (from the functional analysis), and the subsequent life cycles associated with the development of each of these resources. One of the goals of system engineering is to justify these resource requirements through a top-down approach and to ensure the proper development of each through a fully integrated system as one progresses through the design of its various elements.

Figure 1.14 presents the system engineering approach from a different perspective. As one progresses through the life cycle, there is a need to ensure the full traceability of requirements from the system level and on down to the component. As *technical performance measures* (TPMs), or the applicable *metrics*, are established for the system, these measures must be allocated or apportioned to the next level, appropriate *design criteria* are identified, and these criteria must be reflected and supportive from the top down. Further, the appropriate methods/tools must be applied in the design process to ensure that the overall objectives of the system are met. Inherent within the system engineering process is the need to ensure that this traceability is maintained and to cause the integration of the appropriate techniques/methods/tools to facilitate the development process in an effective and efficient manner.

In summary, the system engineering process is continuous, iterative, and incorporates the necessary feedback provisions to ensure convergence. Figure 1.15 illustrates the *feedback* capability that must be built into the process, applied at the system level, to the subsystem level, and so on, as illustrated in Figure 1.12.

System engineering per se is not considered an engineering discipline in the same sense as civil engineering, mechanical engineering, reliability engineering, or any other design specialty area. Actually, system engineering involves efforts pertaining to the overall design and development process employed in the evolution of a system from the point when a need is first identified, through production and/or construction and the ultimate installation of that system for consumer use. The objective is to meet the requirements of the consumer in an effective and efficient manner. The system engineering process is covered further in Chapter 2. Finally, the concepts and principles associated with system engineering are not necessarily new or novel. A review of the literature in Appendix F indicates that many of the principles identified herein were being promoted back in the 1950s and early 1960s. However, in many instances, the system engineering process has not been implemented very well (if at all). Yet, at this point in time, there is a need to emphasize these concepts more than ever.

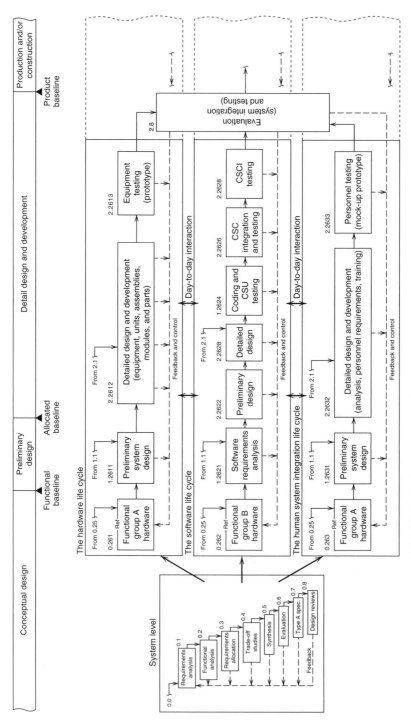

FIGURE 1.13 The integration of the hardware, software, and human life cycles.

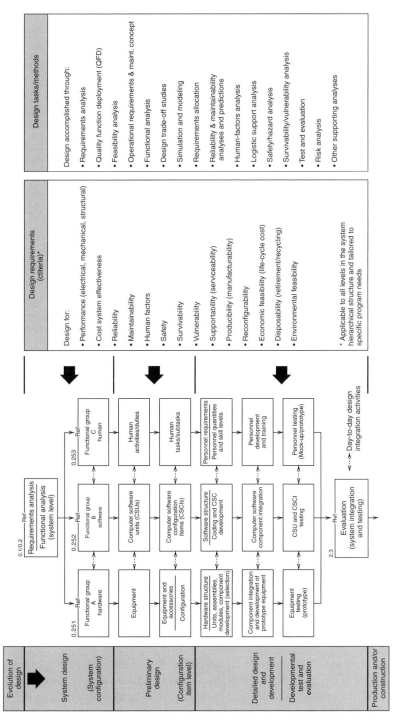

FIGURE 1.14 The top-down traceability of requirements.

23

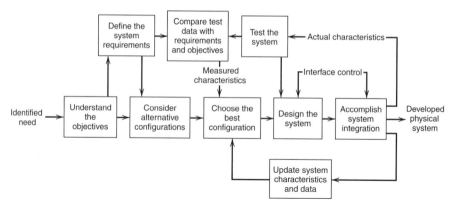

FIGURE 1.15 Feedback in the system engineering process.

1.3.3 Requirements for System Engineering

A primary objective in the implementation of system engineering is to evolve through the *process* illustrated in Figure 1.12, stemming from the initial identification of a *problem* and the subsequent definition of a *need* for a new system to accomplish certain specific functions that are not currently being performed. Every time that a new system requirement is identified, it is essential that we progress through a series of steps that will logically and effectively lead to an end solution. Hopefully, following the general steps described in Section 1.3.2 will facilitate this objective. This is not to indicate that the process in Figure 1.12 is complex by nature; it is the *thought process* (i.e., the "way of thinking") that is important. Whether one is dealing with a large complex system (in the context of SOS) or a relatively small system, the same basic process should be followed. In other words, there is a requirement for the implementation of system engineering (and its process) for any type of system, whether large or small. This, in turn, should enable the development of a system that is both timely and cost-effective in its operation and support.

Traditionally, a designer will often start with a bunch of existing system components, determine just how they might be improved to respond to a new system requirement, add some new items, modify these, and basically assume a trial-by-error approach in arriving at some configuration that will do something! In other words, only a bottom-up approach has been pursued which, in turn, has resulted in a rather costly outcome. The requirement here is to follow a more *top-down/bottom-up approach* as shown in Figure 1.11.

1.3.4 System Architecture

In general, systems are composed of many different interacting elements. Although each is unique in itself and has its own capabilities and characteristics, the composite of these must be arranged and integrated into some framework intended to accomplish the required function(s) or mission(s); i.e., the ultimate system configuration.

This framework of elements represents the system's *architecture*. More specifically, *architecture* can be defined as follows:[15]

> The fundamental organization of a system, embodied in its components, their relationships to each other and to the environment, and the principles guiding its design and evolution.

An architecture deals with a top-level description of system structure (configuration), its operational interfaces, anticipated utilization profiles (mission scenarios), and the environment within which it is to operate; then, it describes how these various requirements for the system interact. This, in turn, leads into a description of the *functional architecture*, which evolves from the functional analysis and its description of the system in "functional" terms. From this analysis, and through the requirements allocation process and the definition of the various resource requirements necessary for the system to accomplish its mission, the system's *physical architecture* is defined. Through application of this process, one is able to evolve from the *whats* to the *hows*.

Referring to Figure 1.12, system architecture is basically described at the "system" level through the requirements analysis (system operational requirements and the maintenance and support concept), functional analysis, requirements allocation, trade-off analysis, and design synthesis (the steps shown by blocks 0.1 through 0.5).[16]

1.3.5 System Science

Often, in addressing the subject of system engineering, one uses the terms *system science* and *system engineering* interchangeably. For the purposes of this text, *system science* deals primarily with the observation, identification, description, experimental investigation, and theoretical explanation of facts, physical laws, inter relationships, and so on, associated with natural phenomena. Science deals with basic concepts and principles that help to explain how the physical world behaves. In the sense that they are applied sciences, the disciplines of biology, chemistry, and physics cover many of these relationships. In any event, system engineering includes the application of scientific principles throughout the system design and development process.[17]

---

[15]IEEE 1471-2000, IEEE Recommended Practice for Architectural Description of Software-Intensive Systems (New York: Institute of Electrical and Electronic Engineers (IEEE), 2000).

[16]Two good references that include coverage of architecting, architecture, system architecture, etc., are (1) E. Rechtin and M. Maier, *Systems Architecting: Creating and Building Complex Systems*, 2nd ed., (Boca Raton, FL: CRC Press, 2000); and (2) C.S. Wasson, *System Analysis, Design, and Development: Concepts, Principles, and Practices* (Hoboken, NJ: John Wiley & Sons, 2006).

[17]Systems science is a major subject by itself, and adequate coverage is not included here. Three excellent references are R. L. Ackoff, S. K. Gupta, and J. S. Minas, *Scientific Method: Optimizing Applied Research Decisions* (New York: John Wiley & Sons, Inc., 1962); G. M. Sandquist, *Introduction to System Science* (Upper Saddle River, NJ: Prentice-Hall, 1985); and L. Von Bertalanffy, *General Systems Theory* (New York: George Braziller, 1968).

1.3.6 System Analysis

Inherent within the system engineering process is an ongoing analytical effort. In a somewhat puristic sense, analysis refers to a separation of the whole into its component parts, an examination of these parts and their interrelationships, and a follow-on decision relative to a future course of action.

More specifically, throughout system design and development there are many different alternatives (or trade-offs) requiring an evaluation effort in some form. For instance, there are alternative system operational scenarios, alternative maintenance and support concepts, alternative equipment packaging schemes, alternative diagnostic routines, alternative manual versus automation applications, and so on. The process of investigating these alternatives, and the evaluation of each in terms of certain criteria, constitute an ongoing analytical effort.

To accomplish this activity effectively, the engineer (or analyst) relies on the use of available analytical techniques/tools to include operations research methods such as simulation, linear and dynamic programming, integer programming, optimization (constrained and unconstrained), and queuing theory to help solve problems. Further, mathematical models are used to help facilitate the quantitative analysis process.

In essence, *system analysis* includes that ongoing analytical process of evaluating various system design alternatives, employing the application of mathematical models and associated analytical tools as appropriate. Analytical methods and models are discussed further in Chapter 4.[18]

1.3.7 Some Additional System Models

In the early 1980s, when the makeup of systems became more *software intensive* there were a number of models developed with the objective of portraying the system life cycle. The *waterfall model* is probably the oldest and most widely used of the system development models in this category at the time.[19] This model, shown in Figure 1.16, is based on a top-down approach for software development and includes the steps of initiation, requirements analysis, design, testing, and so on. Often, in its implementation, the steps were viewed as being relatively independent from one another and were to be executed in a strict sequence, and the feedback effects were not emphasized. In addition, the required interfaces with the other elements of the system (e.g., hardware, the human factor, facilities, data) were not usually considered.

---

[18]System analysis is covered further in an number of references listed in the Bibliography in Appendix F. Some of the operations research tools utilized in accomplishing system analyses are included in: B. S. Blanchard and W. J. Fabrycky, *Systems Engineering and Analysis*, 4th ed. (Upper Saddle River, NJ: Pearson Prentice Hall, 2006); F. S. Hillier and G. J. Lieberman, *Introduction to Operations Research*, 6th ed. (New York: McGraw-Hill, 1995); and H. A. Taha, *Operations Research: An Introduction*, 8th ed. (Upper Saddle River, NJ: Pearson Prentice Hall, 2006).

[19]B. W. Boehm. *Software Engineering Economics* (Upper Saddle River, NJ: Prentice Hall, 1981), p. 36.

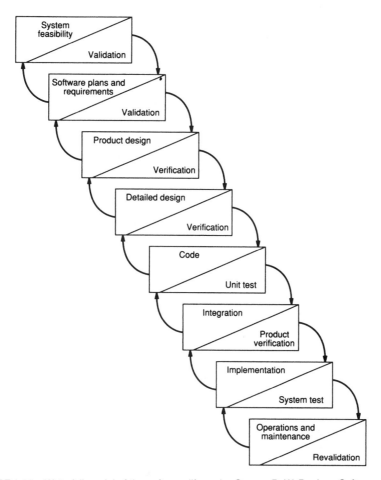

FIGURE 1.16 Waterfall model of the software life cycle. *Source:* B. W. Boehm, *Software Engineering Economics*, © 1981. Reprinted by permission of Pearson Education, Inc., Upper Saddle River, NJ.

In the mid-1980s a generic *spiral model* was developed for software-intensive systems.[20] In this method, the analyst continually examines objectives, strategies, design alternatives, and validation methods. System development results through several iterations of this model. Figure 1.17 illustrates a modified version of the original generic approach, evolving from a prototype model. Note that rapid prototyping is used in each cycle and that the model emphasizes risk analysis. This approach is

---

[20]The generic spiral model was presented by B. W. Boehm, "A Spiral Model of Software Development," in *Software Engineering Project Management*, R. H. Thayer and M. Dorfman, eds. (Washington, DC: IEEE Computer Society Press, 1988). This was modified in Figure 1.17 and is included in A. P. Sage, *Systems Engineering* (New York: John Wiley & Sons, Inc., 1992) pp. 53–54.

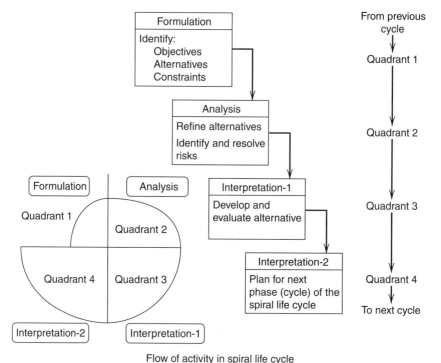

Flow of activity in spiral life cycle

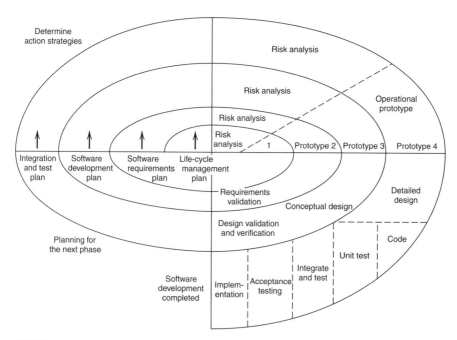

FIGURE 1.17 The spiral model for the software life cycle. *Source:* A.P. Sage, *Systems Engineering* (New York: John Wiley & Sons, Inc., 1992). This material is used with permission.

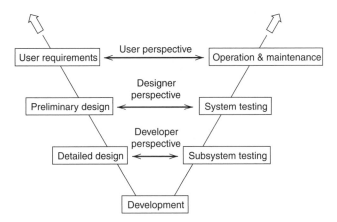

FIGURE 1.18   Generic "Vee" developmental model.

particularly useful in high-risk developments because design sometimes evolves as detailed requirements emerge.

The *Vee model*, introduced in the early 1990s, reflects a top-down and bottom-up approach to system development.[21] In Figure 1.18, the left side of the Vee represents the evolution of user requirements into preliminary and detail design, and the right side represents the integration and verification of system components through subsystem and system testing. This model most nearly reflects the approach conveyed in Figure 1.11 (Section 1.3.2).

Figure 1.19 represents an extension of the Vee model concept.[22] Of particular note is the interface between the system and the software subsystem. Quite often, individuals refer to *software systems*. Although software may be predominant within the structure of a system, it is *not* the "system" per se. It does not fulfill a functional requirement by itself. Software requirements are identified through functional analysis and are subsequently developed through the steps illustrated in Figure 1.13. Figure 1.19 emphasizes that there are system engineering activities that lead into the software development process.

Numerous models have been introduced through the past few decades, with the objective of providing a logical approach to the overall process of system design and development. The few identified here are only representative of the total population. Most of these models are directed primarily to the system acquisition process only and/or to some element of the system (such as software); hence, they lack a certain degree of completeness. If implemented properly, they are excellent in terms of accomplishing their intended objectives. However, it should be recognized that their

[21]K. Forsberg and H. Mooz, "The Relationship of System Engineering to the Project Life Cycle," *Proceedings of the 4th Annual Symposium* (Seattle, WA: International Council on Systems Engineering, INCOSE, 1991), p. 289.
[22]B. G. Downward, "A Brave New World: Melding Systems and Software Engineering," *Proceedings of the 4th Annual Symposium* (Seattle, WA: International Council on Systems Engineering, INCOSE, 1991), p. 157.

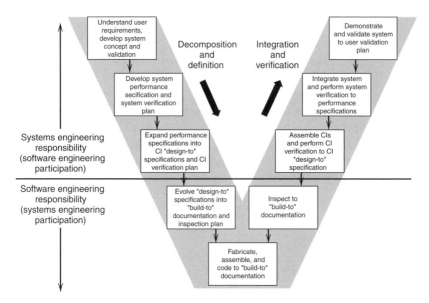

FIGURE 1.19   The systems versus software engineering boundary. *Source:* B. G. Downward, "A Brave New World: Molding Systems and Software Engineering," *Proceedings of the Symposium of the International Council on Systems Engineering* (Seattle, WA: INCOSE, 1991), 157.

application may be limited unless utilized within the broader spectrum of system engineering described in Section 1.3.2.[23]

1.3.8   System Engineering in the Life Cycle (Some Applications)

The system engineering process is applicable in all phases of the life cycle, as illustrated in Figure 1.12. In the early stages of conceptual design, the emphasis is on understanding the true needs of the consumer (user) and in developing the actual requirements for the system. These requirements, which constitute the baseline that needs to be established from the beginning, must be traceable from the top and on down to the component level as necessary. This top-down approach (with the appropriate feedback incorporated), reflected in the left side of the Vee diagram in Figure 1.11, is critical for the successful implementation of a system engineering

[23]There are numerous other models, including prototype models, the Sashimi Model, the Scrum Model, the Handcuff Model, the Hollywood Model, the Evolutionary Development Model, and so on. A good reference covering some of these models (in a summary manner) is R. S. Scotti and S. S. Gulu Gambhir, "A Conceptual Framework for a Customer-Centered System Development Life-Cycle Model." *Proceedings of the 6th Annual Symposium* (Seattle, WA: International Council on Systems Engineering, INCOSE, 1996), p. 547. The reader may also refer to the Vee model, Tufts Systems Engineering Process Model. Plowman's Model, and INCOSE's model, described in *INSIGHT Vol.* 5, no. 1, published by INCOSE (April 2002) pp. 7–16. Additionally, for more current information on models, in general, one should review the various issues of *Systems Engineering: The Journal of the International Council on Systems Engineering (INCOSE)*, published quarterly by John Wiley & Sons, Hoboken, NJ.

program. It is the establishment of these early requirements that has a great impact on the ultimate life-cycle cost for a given system (see Figure 1.8).

Given the basic requirements, the emphasis then shifts to an iterative process of synthesis, analysis, design optimization, and validation. Trade-off studies are conducted, with the objective of providing a well-balanced system design. There are many different design objectives that must be met, some of which may be somewhat conflicting, and the role of system engineering is to identify, prioritize, integrate, and to cause the development of a system configuration that will meet all customer requirements in a timely, effective, and efficient manner. Such an ultimate configuration must consider the system in total to include the development of the production, maintenance and support, and retirement/material recycling capabilities, as shown in Figure 1.10.

System engineering activities continue through the construction and/or production phase to ensure that the designed system configuration is compliant with the initially specified requirements. Next, there is an ongoing and iterative process of *assessment* (and validation) throughout the operational use and maintenance support phase, and during the system retirement and material recycling stage. This assessment, with the proper feedback, is important to ensure not only that the initially specified system requirements being met, but also that any changes in requirements that take place in the user environment are properly reflected back into the design process (through redesign, reengineering, etc.). In other words, there is a continuous product/process improvement feedback loop, included at the bottom in Figure 1.12, which is critical in the implementation of system engineering.

Experience related to the evaluation and assessment of a system, which is operational and being maintained in the user's environment, must be captured. A baseline configuration (with the appropriate metrics) must be established for the purposes of *benchmarking* and the initiation of possible changes for improvement. This, of course, requires that a good comprehensive data collection, analysis, and evaluation capability be implemented to provide the necessary feedback.

This knowledge of what really is happening to the system in the user's environment is critical but is often lacking because there is no good assessment capability in place. Hence, we often end up introducing the same mistakes again and again as we design new systems. Assessment is an inherent part of the system engineering process.

Finally, when changes are being initiated (whether for corrective action or for product/process improvement), the consequences of such changes must be evaluated from a top system-level perspective; that is, assessing the impact of a change on the overall system. The principles of configuration management and change control must be implemented to ensure that the end results are consistent with the basic requirements in terms of both effectiveness and life-cycle cost (i.e., both sides of the spectrum in Figure 1.6). Such changes may be applicable to the prime mission-related elements of the system, the construction/production capability, the maintenance and support infrastructure, and/or the retirement and material recycling capability. The interaction effects, both upward and downward, must be properly addressed in a *systems* context.

The system engineering process is applicable in all phases of the life cycle, and the successful implementation of such is dependent on many different organizational groups working in a cooperative and integrated manner. Although a single organizational entity may ultimately be responsible for overall leadership in fulfilling system engineering objectives, accomplishing many of the various individual required tasks may be the responsibility of the different participating organizations. In other words, an integrated *team* approach is required.

Additionally, this process is applicable to any type of system such as a communication network, commercial shopping plaza, defense system, information processing network, healthcare capability, transportation system, space system, power distribution network, supply chain network, environmental control capability, and so on. Further, this process can be applied at any level in the overall hierarchal structure presented in Figure 1.3.

1.4 RELATED TERMS AND DEFINITIONS

The preceding framework for the basics in describing the principles and concepts associated with system engineering is now extended to include a few additional but related concepts. Figure 1.20 presents the system development process and life-cycle activities in a slightly different context. System design activities, emphasized in Figure 1.11 and in the first three columns in Figure 1.12, are addressed in blocks 1. 2, and 3; construction and/or production activities are included in blocks 3 and 4; and then there are the system operational and support activities reflected in blocks 5, 6, 7, and 8.

In Figure 1.20, it should be noted that there is a *forward* flow of activities covering not only the design and development of the system, but also the construction and/or production of the system and its elements, the transportation and distribution of these elements to the various operational sites, and the subsequent installation of the system for sustaining operational use. Throughout this flow, there are production-and logistics-related activities that are essential if the ultimate system is to accomplish its objectives. For instance, the lack of an effective and efficient transportation capability, the absence of a supplier-produced component in a timely manner, or the lack of appropriate information may preclude the successful accomplishment of a mission requirement(s). Thus, having the appropriate "logistics" available where and when needed is critical.

At the same time, there is a *reverse* (or backward) flow of activities that deal with the maintenance and support of the system throughout its life cycle, which may be necessary in the event of a system failure. An unreliable system that is nonoperational when needed will obviously not be able to perform its designated mission, and there must be an effective and efficient infrastructure that is readily available and in place that can respond to problems, with the objective of repairing the system and returning it to full operational status in a timely manner. The lack of a needed spare part, an available transportation capability, a necessary item of test equipment, required maintenance software, an appropriate repair facility, or the right data, may prevent

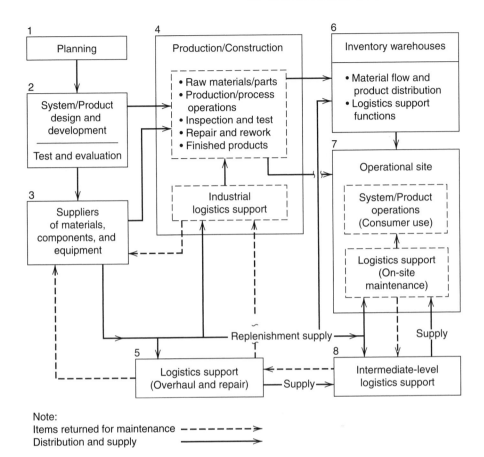

FIGURE 1.20 System operational and maintenance flow.

the system from performing its intended function(s). Thus, there are a variety of logistics and maintenance support functions needed, which are inherent within this reverse flow of activities (as illustrated in blocks 3, 4, 5, 7, 8, and the dotted lines in Figure 1.20).

These activities, associated with both the *forward* and *reverse* flows in Figure 1.20, are characteristic and applicable for any system. However, they are usually addressed downstream and "after the fact" in the life cycle (constituting one of the problems highlighted earlier). In the past, there has been little emphasis on *the design for reliability and maintainability, the design for producibility, the design for packaging, the design for transportation and handling, the design for supportability and serviceability, the design for disposability and recyclability, design for the environment,* and so on. Yet these factors should be considered as critical system-level parameters, along with *the design for performance*, and emphasized in the early phases of the system engineering process illustrated in Figure 1.11.

Given this background information, a number of terms and definitions are discussed further with the objective of strengthening the understanding of system engineering before addressing the process described in Chapter 2.

1.4.1   Concurrent/Simultaneous Engineering

In the mid-1980s the term *concurrent engineering* became popular, with the objective of placing additional emphasis on "concurrency" as it applies to the design and development of the prime mission-related elements of a system, the construction and/or production capability, the maintenance and support infrastructure, and the retirement and material recycling capability. In Figure 1.10, the various life cycles should be viewed on a concurrent basis, which is directly supportive of system engineering objectives.

One of the first formal definitions resulted from a Department of Defense study, in which *concurrent engineering* is defined as "a systematic approach to the integrated, concurrent design of products and their related processes, including manufacture and support. This approach is intended to cause the developers, from the outset, to consider all elements of the product life cycle from conception through disposal, including quality, cost, schedule, and user requirements."[24] As such, concurrent engineering should be included within the system engineering process.

1.4.2   Some Major Supporting Design Disciplines

Throughout the life cycle, and particularly in system design and development, there are many different individual disciplines that contribute in providing the ultimate system/product configuration. An objective in system engineering is to integrate all of these various design disciplines and specialty groups into a team effort by creating a structured process that proceeds from conception, through development and production, and into system operation. An abbreviated description of nine of these critical disciplines follows:

1. *Software engineering.* From its beginning, software engineering has influenced modern system engineering practices to a great degree and, in some instances, the development of software has proceeded down an independent path without considering the other elements of a system (refer to Figure 1.13 and the software life cycle). Although the element of *software* often constitutes a major part of a system's overall hierarchical structure, it must be properly integrated with the applicable system hardware, people, facilities, information/data, elements of support, etc., on a timely basis.

2. *Reliability engineering.* This design discipline has the objective of ensuring that a system will meet the customer's expectations for good performance

---

[24]R. I. Winner, J. P. Pennell, H. E. Bertrand, and M. M. G. Slusarczuk, *The Role of Concurrent Engineering in Weapons Systems Acquisition.* Report R-338 (Alexandria, VA: Institute of Defense Analysis, 1988). Additional references are included in Appendix F.

throughout its life cycle. Reliability is inherent within and is a major factor in the overall *availability* of a system, and can be measured in terms of probability of success, mean time between failure (MTBF), and failure rate (λ).

3. *Maintainability engineering.* This design discipline has the objective of ensuring that a system incorporates the necessary inherent built-in characteristics such that it can be maintained with ease, accuracy, and economy in the performance of both corrective and preventive maintenance actions as required. Maintainability is the *ability* of a system to be maintained, whereas maintenance constitutes those actions taken to restore a system to (or retain a system in) a specified operating condition. Some typical measures include mean time between maintenance ($\overline{\text{MTBM}}$), maintenance downtime (MDT), mean corrective maintenance time ($\overline{\text{Mct}}$), mean preventive maintenance time ($\overline{\text{Mpt}}$), maintenance labor hours per operating hour (MLH/OH), and cost per maintenance action ($/MA). As such, maintainability (along with reliability) is inherent within and a major factor in the overall *availability* of a system.

4. *Human factors and safety engineering.* This design discipline addresses the *human being* as a major element of a system (refer to Figure 1.13 and the human system integration life cycle). It deals with the anthropometric, human sensory, physiological, and psychological factors in system design. Safety engineering is a design discipline with the objective of ensuring that a system is *safe* to both operate and maintain throughout its life cycle. Some good measures include the number of successful operational task sequences (in the accomplishment of operator, maintenance, and support tasks) without error, human error rates, personnel safety/hazardous rates, number of training sequences and rates, and so on.

5. *Security engineering.* With the advent and increased emphasis on *terrorism*, there is a need to design a system such as to preclude the introduction of faults/failures (whether unintentional or through planned actions), which, in turn, will cause serious damage to equipment, facilities, loss of human life, and/or result in some other catastrophic system degradation. The objective is to incorporate good condition monitoring and diagnostic capabilities to ensure that the system is secure in its operation and supported in accomplishing its intended mission(s). Knowing the status of a system's condition at all times is essential. Several measures may include hours/days/years of successful operation without violating system security, number of system faults per unit of time resulting in system damage, loss of life, threat to the environment and society, and so on.

6. *Manufacturing and production engineering.* This discipline addresses both (a) the design of the prime mission-related elements of a system to ensure that they can be produced and/or constructed effectively and efficiently; and (b) design of the production and/or construction capability to ensure that these prime elements can be easily processed as planned (refer to Figure 1.10 and the applicable life cycles as shown). The objective is to incorporate good industrial engineering, reliability and maintainability, human factors, and safety

characteristics inherent and within the overall production/construction process. Some good measures include the overall availability of the production process, the number of items produced in a given time frame, the unit cost of a produced item, and the overall equipment effectiveness (OEE) rate.

7. *Logistics and supportability engineering.* Logistics and the system maintenance support infrastructure have generally been considered after-the-fact and as a separate entity, independent of the basic system that it is supporting. However, if a system is to successfully accomplish its intended mission, this support infrastructure must be in place and available, in a timely manner, in the event of need (refer to Figure 1.20, blocks 4–8). If a specific time frame (i.e., date) is established for system operation, then the appropriate logistics support must be available before-the-fact (and throughout) in order for this to happen; if a system failure occurs during the period of system utilization, then the right maintenance support infrastructure must be available and responsive in order for the system to complete its objective; and so on. To ensure that all system requirements are fulfilled, the basic mission-related elements of a given system must be *designed for supportability*, and the logistics and maintenance support infrastructure must also be designed such that it can both effectively and efficiently support these elements. In other words, this infrastructure must be considered as a major element of the system, and supportability requirements must be "designed-in" along with reliability, maintainability, human factors, safety, and other related factors (Figure 1.20, blocks 1–3).

8. *Disposability engineering.* This discipline primarily addresses the design of a system and its components such that, in the event of obsolescence, they can be either recycled for other applications or processed for disposal. In the event of disposal, the results should not have any degradation or negative impact on the environment (i.e., not result in any residual solid waste, toxicity, noise pollution, water pollution, or equivalent).[25]

9. *Environmental engineering.* The *environment*, as defined herein, refers primarily to ecological considerations such as air pollution, water pollution, noise pollution, radiation, and solid waste. The design objective concerns itself both with the impact of the new system design configuration on those factors in the environment (i.e., the creation of negative impacts on the external environment), and the impact of similar factors from other outside sources on the new system being introduced (the external impact of these negative conditions on the system itself). Such concerns must be addressed in the system design process from the beginning.

The nine disciplines mentioned represent only an example of a few critical areas that require some emphasis in system design and development, along with the more

---

[25] Additional material on the "design for producibility and disposability" may be found in B. S. Blanchard and W. J. Fabrycky, *Systems Engineering and Analysis*, 4th ed. (Upper Saddle River, NJ: Pearson Prentice-Hall, 2006), Chapter 16, pp. 554–577.

traditional engineering disciplines such as aerospace and aeronautical engineering, chemical engineering, civil engineering, electrical engineering, industrial engineering, mechanical engineering, and others, as applicable. A key objective of system engineering is to create a "team" approach through the proper integration of these and other needed disciplines in a timely manner.

1.4.3 Logistics and Supply Chain Management (SCM)

The subject of *logistics* is introduced from an engineering design perspective in Section 1.4.2, but it is essential that a review of the overall scope of logistics be included as well. The term *logistics* can be described somewhat differently, depending on application (e.g., *commercial* versus *defense*), the companies and organizations involved, and one's personal background and experience. In the commercial sector, *logistics* is often defined as that part of the supply chain process that *plans, implements*, and *controls* the efficient, effective forward and reverse flow and storage of goods, services, and related information between the point of origin and the point of consumption in order to meet customer requirements. The *supply chain (SC)* refers to that group of organizations and activities pertaining to the overall flow of materials and services from various supplier sources to the ultimate customer(s).[26]

For many years, emphasis was primarily directed to the physical aspects of supply, materials handling, and transportation and distribution, as shown in Figure 1.21. However, during the past several decades, the area of *commercial* or *business* logistics has been expanded significantly with the introduction of the latest electronic commerce (EC) methods, information technology (IT), electronic data interchange (EDI), development of radio frequency identification (RFID) tags and global positioning system (GPS) technology, and the application of good business processes to the activities shown in Figure 1.21. Further, the trends toward more outsourcing, greater international competition, and increased globalization have resulted in the need for the establishment of coalitions and industry/government partnerships worldwide. This, in turn, has resulted in the currently popular concepts associated with *supply chain management (SCM)*. SCM pertains to *management of the supply chain, or group of supply chains, with the objective of providing the required customer services, both effectively and efficiently*. It requires a highly integrated approach, employing the appropriate resources (e.g., transportation, warehousing, inventory control, and information) and implementing the necessary business processes to ensure complete satisfaction. The Council of Supply Chain Management Professionals (CSCMP) has adopted this definition (refer to footnote 26).

In any event, no matter how one may wish to define it, the associated concepts and magnitude of SCM are certainly growing in recognition and importance. SCM, as described herein, basically includes those *forward* flow of activities reflected by blocks 3, 4, 6, and 7 in Figure 1.20.[27]

---

[26]This definition was developed by the Council of Supply Chain Management Professionals (CSCMP), 2805 Butterfield Road, Oak Brook, IL.

[27]It should be noted that the material presented thus far emphasizes "logistics" in the commercial (business) sector and not the entire spectrum of logistics as practiced in the defense sector. Historically, logistics

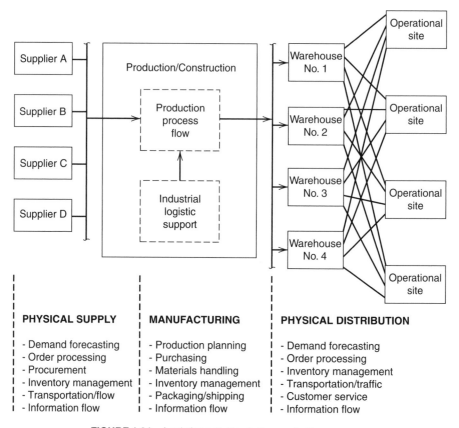

FIGURE 1.21   Logistics activities in the production process.

1.4.4   Integrated System Maintenance and Support

In Figure 1.20, there is also a *reverse* flow relating to the activities in blocks 3,4, 5,7, and 8, with the dotted lines indicating a path whereby faulty (or obsolete) items are sent for maintenance and repair (or disposal) as necessary. Figure 1.22 shows an expansion of this, presented in the form of a maintenance and support infrastructure.[28]

Figure 1.22 illustrates an example of a basic *three-levels-of-maintenance* approach (i.e., organizational maintenance, intermediate-level maintenance, and depot/producer/supplier maintenance). Depending on the type of system, the mission to be

---

as it applies for defense systems, has been covered within the context of *integrated logistic support (ILS)*, which, in turn, has included not only what is described in Section 1.4.3 but also the maintenance and support infrastructure that is discussed further in Section 1.4.4. Recently, the term *acquisition logistics* has become popular within the Department of Defense (DOD) and includes the principles and concepts of SC/SCM and system maintenance and support, as applicable in each and every phase of the system life cycle.

[28]The system maintenance and support infrastructure (i.e., the maintenance concept and showing major repair policies) is discussed further in Chapter 2, Section 2.5.

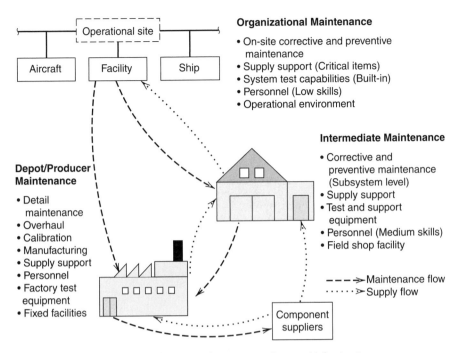

FIGURE 1.22 System maintenance and support infrastructure.

accomplished, the system's complexity and reliability, the geographical location and where utilized, customer desires, overall cost, and so on, there may be some variation in repair policies and the infrastructure may include only two levels of maintenance. Further, there may be one or more third-party maintenance contractors involved in the overall network, and this configuration may well change as the system ages and the demand for support changes. In any event, there must be some form of a maintenance and support infrastructure, in place and readily available, to ensure that the system will continue to be operational when required.

Referring to the network in Figure 1.23, it can be seen that the functional elements of support include maintenance and support planning, maintenance personnel, supply support (spares/repair parts and associated inventories), test and support equipment, packaging and transportation, maintenance facilities, computer resources, technical data, and related management requirements. These various elements must be properly integrated, with the support of an effective management information capability, as illustrated in Figure 1.23.

Traditionally, the subject of maintenance has been addressed after the fact and downstream in the system life cycle, and the maintenance and support infrastructure has not been considered as an element of a system but rather as a separate and somewhat unrelated entity. Through the years, the results of such a practice have been rather costly, with a large portion of the total life-cycle cost for many systems being attributed to maintenance activities accomplished downstream (refer to Figures 1.6 and 1.7). In response, there have been some efforts leading to the recognition of

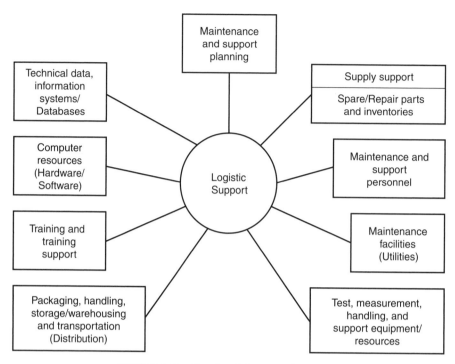

FIGURE 1.23   Functional elements of logistics.

the maintenance and support infrastructure early in the system life cycle and as an inherent element of a system (refer to Figure 1.8).

In the defense sector, the DOD initiated the concept of *integrated logistic support* (ILS) in the mid-1960s. ILS is a management function that provides the initial planning, funding, and controls that help to ensure that the ultimate consumer (or user) will receive a system that will not only meet performance requirements, but can be supported expeditiously and economically throughout its programmed life cycle. More specifically, ILS can be defined as "a disciplined, unified, and iterative approach to the management and technical activities necessary to (1) integrate support considerations into system and equipment design; (2) develop support requirements that are related consistently to readiness objectives, to design, and to each other; (3) acquire the required support; and (4) provide the required support during the operational phase at minimum cost." A major ILS objective is to ensure the proper and timely integration of the elements of support, as shown in Figure 1.23.[29]

[29]This definition initially evolved from Department of Defense Instruction 5000.2, "Defense Acquisition Management Policies and Procedures," Part 7, 1991. It should be noted that this definition is the result of an evolving set of definitions stemming from the original *ILS Planning Guide for DOD Systems and Equipment*, 4100.35, 1967.

More recently, in the interest of economy, the DOD has been emphasizing (1) increased reliance on the utilization of best commercial logistics and supply chain practices in meeting the support requirements for defense systems, and (2) an increased emphasis on the *design for supportability/sustainability* and consideration of the maintenance and support infrastructure within the context of the systems engineering process. Relative to the first area, there has been a great deal of growth in adopting many of the principles of supply chain management in the defense sector, particularly in view of the growth in information technology (IT) and electronic commerce (EC) methods. In regard to the second area of emphasis, the concept of *acquisition logistics* has become popular, and its focus is segmented into three interrelated parts: (1) designing the system for support, (2) designing the support system, and (3) acquiring the support elements.[30] In essence, logistics in the defense sector has become an integrated mix of the activities depicted in Figures 1.21 and 1.22.

When addressing the overall issue of maintenance, another consideration must be properly integrated within this overall spectrum of system support. In the commercial sector, and primarily oriented to the maintenance of equipment in a typical manufacturing plant, the concept of *total productive maintenance (TPM)* has become quite popular and is currently being implemented worldwide. TPM, a concept originally developed by the Japanese in the late 1960s and early 1970s, is a system-oriented, life-cycle approach to maintenance, with the objective of maximizing productivity in the commercial manufacturing plant. TPM:

1. Promotes the overall effectiveness and efficiency of equipment in a factory. It includes *maintenance prevention* (MP) and *maintainability improvement* (MI), which consider the appropriate incorporation of reliability and maintainability characteristics in design.
2. Establishes a complete preventive maintenance program for factory equipment based on life-cycle criteria (similar to the reliability-centered maintenance approach used in establishing preventive maintenance requirements).
3. Is implemented on a "team" basis, involving various departments to include engineering, production operations, and maintenance.
4. Involves every employee in the company, from the top management to the workers on the shop floor. Even equipment operators are responsible for the care and maintenance of the equipment they operate.
5. Is based on the promotion of preventive maintenance through "motivational management" (the establishment of autonomous small-group activities for the maintenance and support of equipment).

The introduction of TPM was motivated by the high costs of producing products in the factory, combined with the fact that a good percentage of these high costs was attributed to equipment maintenance on the production line. The objective is

---

[30]MIL-HDBK-502. *Acquisition Logistics* (Washington, DC: Department of Defense, May 1997). It should be noted that this document includes a major section on systems engineering.

to reduce the costs of producing products by minimizing maintenance costs in the factory. The implementation of TPM, which can be measured in terms of *overall equipment effectiveness (OEE)*, has become quite popular, particularly during the past several decades.[31]

As a final note, it should be reemphasized that the logistics and maintenance support infrastructure (i.e., the overall structure described in Sections 1.4.3 and 1.4.4) must be considered and included as a major "element" of a system, and that this element must be properly integrated, along with the other system elements, within the system engineering process from the beginning.

1.4.5 Data and Information Management

Characteristic of most programs is the large amount of data and information associated with the design and development, operation, and the sustaining maintenance and support of systems throughout their respective life cycles. Included in such data requirements are specifications and plans, engineering drawings and associated design data, system test and evaluation data, logistics and maintenance data, technical data (in the form of system operating instructions, maintenance and overhaul manuals, and component parts lists), engineering change data, subcontractor and supplier data, and so on. Additionally, there is a vast amount of information (e-mail, Internet communications, and various categories of messages) that is generated and distributed daily among the various individuals and organizations that are involved throughout the different stages of the system life cycle.

The challenge is to properly plan for, coordinate, and integrate such data/information into a unified data package, from the beginning. Such a data package not only includes a historical basis for what has been accomplished and why it was done, but also reflects the proper configuration of a system at any given point in time. The generation of too much data, or too little data, can be very costly. Thus, it is critical that the right amount of data/information be available when required and in a timely manner, not too early or too late. This, in turn, requires some coordination and integration covering all aspects of system-level activities and across organizational lines (including the customer, major contractor, subcontractor, and supplier organizations).

The initial determination of data requirements stems from the early development of system-level requirements in conceptual design. These data requirements are then expanded as the system is further defined through the functional analysis and allocation of requirements (refer to Figure 1.12). From this point on, there are likely to be many different requirements, in different forms, throughout a given program. Thus, it is important that the data/information integration and management function be included within the overall spectrum of system engineering.

---

[31]The concept of TPM was initiated in Japan, through the Japan Institute of Plant Maintenance (JIPM), in the late 1960s. A good initial reference is S. Nakajima (ed.), *Total Productive Maintenance (TPM) Development Program* (Portland, OR: Productivity Press, 1989, English translation). Refer to Appendix F for additional references.

1.4.6 Configuration Management (CM)

Configuration management (CM) is a management approach that includes identifying, documenting, and auditing the functional and physical characteristics of an item, recording the configuration of the item, and controlling the changes to the item and its documentation. The purpose is to provide a complete audit trail of design decisions and system modifications. CM is a concept of *baseline* management, which includes the definition of the *functional* baseline for a system, the *allocated* baseline, and the *product* baseline identified in Figure 1.12. Successful fulfillment of system engineering requirements is heavily dependent on a good disciplined approach to baseline management. This is particularly true in considering the current trends toward evolutionary design and the introduction of new technologies into a system configuration on a continuing basis.[32]

1.4.7 Total Quality Management (TQM)

Total quality management (TQM) can be described as a totally integrated management approach that addresses system/product quality during all phases of the life cycle and at each level in the overall system hierarchy. It provides a before-the-fact orientation to quality, and it focuses on system design and development activities as well as manufacturing and production, maintenance and support, and related functions. TQM is a unification mechanism linking human capabilities to engineering, production, and support processes. The emphasis is on total customer satisfaction, the iterative practice of "continuous improvement," and a total integrated organizational approach. As part of the initial system design and development effort, consideration must be given to (1) the design of the processes that will be used to manufacture and produce the components of the system and (2) the design of the support infrastructure that will provide the necessary ongoing maintenance of that system throughout its planned life cycle. In this regard, the principles of TQM must be inherent within the system engineering process.

1.4.8 Total System Value and Life-Cycle Cost (LCC)

A system should be measured in terms of its total value to the consumer. For the purposes of discussion, it is necessary to consider both sides of the balance, as illustrated in Figure 1.24; that is, the *technical factors* and the *economic factors*. Of particular interest within the domain of system engineering is the issue of *life-cycle*

---

[32]Department of Defense Regulation 5000.2R, "Mandatory Procedures for Major Defense Acquisition Programs (MDAPS) and Major Automated Information System (MAIS) Acquisition Programs," Chapter 5, 2002. Additional references include: (1) MIL-HDBK-61, "Military Handbook: Configuration Management Guidance," Washington, DC, 1997; and (2) ANSI/EIA 646, "National Consensus Standard for Configuration Management," Electronic Industries Alliance, Arlington, VA, 1998.

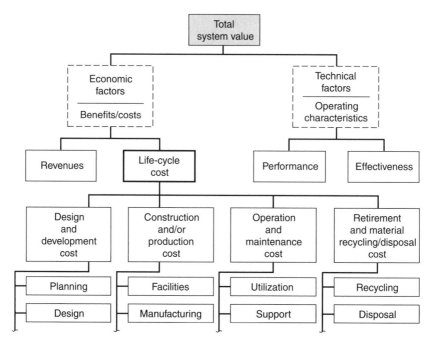

FIGURE 1.24 Total system value.

cost (LCC). LCC includes all costs associated with the system life cycle, which can be broken down as follows:

1. *Research and development (R&D) cost.* This includes the cost of feasibility studies; developing operational and maintenance requirements; system analyses; detail design and development; fabrication, assembly, and test of engineering models; initial system test and evaluation: and associated documentation.

2. *Production and construction cost.* Included here are the cost of fabrication, assembly, and test of operating systems (production models); operation and the sustaining maintenance and support of the manufacturing capability; facility construction; and the acquisition of an *initial* system support capability (e.g., test and support equipment, spare/repair parts, and technical documentation).

3. *Operation and maintenance cost.* This includes the cost of system operation and the sustaining maintenance and support of the system through its planned life cycle (e.g., manpower and personnel, spare/repair parts and related inventories, test and support equipment, transportation and handling, facilities, software, modifications, and technical data).

4. *System retirement and phase-out cost.* This final expense is the cost of phasing the system and its components out of the inventory because of obsolescence or wearing out; recycling of items for further use; condemnation; and the disposal of materials.

Life-cycle costs can be categorized many different ways, depending on the type of system and the sensitivities desired in cost-effectiveness measurement. The objective is to ensure *total cost visibility* (see Figure 1.7). This is necessary if one is to be able to properly assess the risks associated with each of the major design and management decisions made throughout the life cycle. Life-cycle cost (LCC) is a major theme throughout this text, and the process for conducting a life-cycle cost analysis is highlighted in Appendix B, and illustrated through a case study in Appendix C.

1.5 SYSTEM ENGINEERING MANAGEMENT

The successful realization of system engineering principles and concepts is dependent not only on the *technology* issues and the process for implementing such, but on the *management* issues as well. As illustrated in Figure 1.25, there are two sides of the spectrum, and each is highly dependent on the other. The.best tools/models may be available to implement the process shown in Figure 1.12. However, there is no guarantee for success unless the proper organizational environment has been created and an effective and efficient management structure is in place. Top management must first *believe in* and then *provide the necessary support* to enable the application of system engineering methods to all applicable projects, both in-house and external. Specific objectives must be defined, policies and procedures must be developed and properly implemented, and an effective review and reward structure must be supportive. This structure must prevail throughout the customer, prime

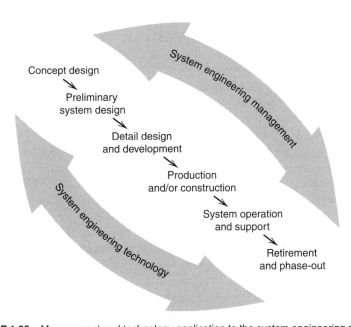

FIGURE 1.25   Management and technology application to the system engineering process.

contractor, and down through the various applicable supplier organizations as required. The challenge is that of *proper implementation*.

Although there are variations from one program to the next, Figure 1.26 presents a baseline for discussion. The major program phases and milestones are noted, along with a few selected activities and events that are considered to be significant from a system engineering perspective. It should be noted that the emphasis in the figure is primarily on the system acquisition (procurement) process, and not on the downstream operation and support and retirement phases. These phases of activity, which are discussed in detail in subsequent chapters, are briefly summarized as follows:

1. During the early stages of conceptual design, it is essential that good communications between the producer and the consumer(s) be established from the beginning. Defining the true need, conducting feasibility analyses, developing operational requirements and the maintenance concept, and identifying specific quantitative and qualitative requirements at the system level are critical. These requirements must be properly conveyed through a well-prepared system specification (type A). This top-level system specification constitutes the most important *technical* document, from which all lower-level specifications evolve. Without a good foundation from the beginning, all subsequent lower-level requirements may be questionable (refer to Chapter 3).

2. During the latter stages of conceptual design, a comprehensive System Engineering Management Plan (SEMP), or System Engineering Plan (SEP), must be developed to ensure the implementation of a program that will lead to a well-coordinated and integrated product output. The SEMP, which evolves from the top-level Program Management Plan (PMP), integrates all lower-level planning documents. It includes the design-related tasks necessary to enhance the day-to-day system development effort, the implementation of concurrent engineering methods, and the integration of the appropriate organizational entities into a "team" approach. The SEMP must directly support the requirements in the system specification (type A) from a *management* perspective, and the two documents must "talk to each other." The SEMP is addressed in detail in Chapter 6.

3. During the latter stages of conceptual design, a Test and Evaluation Master Plan (TEMP), or equivalent, must be developed for the purposes of assessment and ultimate validation. As requirements are initially specified in the system specification (type A) and planned through the tasks described in the SEMP, the methods/techniques to be used for measuring and evaluating the system to ensure compliance with these requirements must be described. This plan must address test and evaluation activities on a fully integrated basis, employing the appropriate combination of simulation and other analytical tools, mockups, laboratory models, and prototype models. Test and evaluation are covered further in Chapter 2.

4. As system design and development progresses, there is a need to schedule a series of formal design reviews at discrete points where the design configuration evolves from one level of definition to another; that is, conceptual, system,

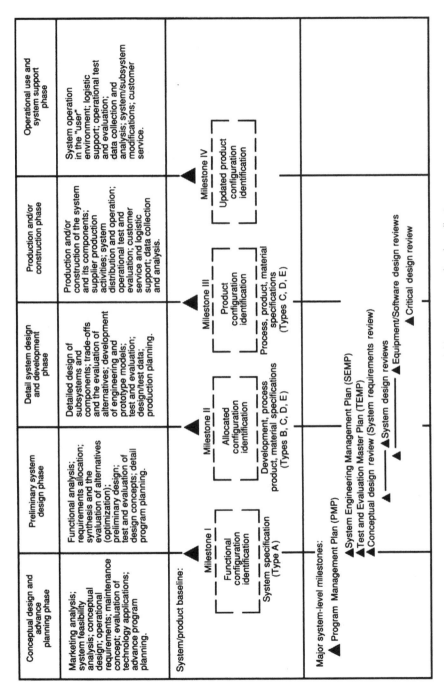

FIGURE 1.26　The system acquisition process and major milestones.

47

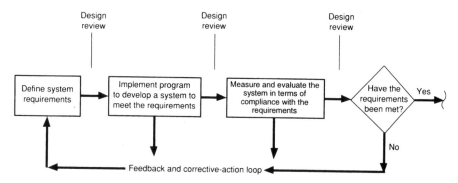

FIGURE 1.27   The basic system requirements, evaluation, and review process.

equipment/software, and critical design reviews. The purpose of these reviews is to ensure that the specified requirements are being met prior to entering into a subsequent phase of effort, and to ensure that the necessary communications exist across organizational lines. See Chapter 5 for further discussion of design reviews and evaluation requirements.

5. Toward the latter stages of detail design, throughout the construction/ production phase, and during the operational use and maintenance support phase, there is a need to provide ongoing assessment and validation of the system. The objective is to ensure that the consumer requirements are being met and to establish a "baseline" for the purposes of benchmarking and the initiation of *a continuous process improvement* activity. Design changes are initiated as required to correct any noted deficiencies.

The successful implementation of system engineering principles is highly dependent on proper management of the simplified process depicted in Figure 1.27. Inherent in this process is the application of different technologies employed to facilitate the steps of requirements analysis, functional analysis and allocation, synthesis, design optimization, and validation.

1.6   SUMMARY

This chapter provides an abbreviated introduction to some of the key terms and definitions, principles and concepts, and critical issues in the implementation of *system engineering* and associated requirements in the design and development, production/construction, operation and support, and retirement of systems. Such terms as a *system, system of systems (SOS), system architecture, system science, system analysis, logistics, integrated system maintenance and support, configuration management, total quality management,* and *system value and life-cycle cost* are introduced. Hopefully, this will stimulate the thought processes needed for the material ahead. The information presented herein, and particularly the concepts illustrated in Figures 1.12, 1.13, and 1.14, are a natural introduction to the system engineering process discussed in Chapter 2.

QUESTIONS AND PROBLEMS

1. Provide, in your own words, a definition of a *system*. Include some examples.
2. Select a system of you choice and describe the system life cycle. Construct a detailed flow diagram tailored to your situation.
3. Describe what is meant by a *system of systems (SOS)*. Provide an illustrated example.
4. When referring to the basic *elements* of a system, what is included?
5. Define *system engineering*. What is included? Why is it important? How does system engineering differ from system science and system analysis?
6. What are the differences (or similarities) between system engineering and some of the more traditional disciplines such as civil engineering, electrical engineering, mechanical engineering, and so on?
7. Refer to Figure 1.10 (Example A). Describe the interrelationships between the three illustrated life cycles.
8. Refer to Figure 1.13. What are some of the key system engineering objectives that can be applied?
9. Refer to Figure 1.14. What are some of the key system engineering objectives that can be applied?
10. What is the significance of the feedback process illustrated in Figure 1.15?
11. What are the major system engineering functions in conceptual design? Preliminary design? Detail design and development? System operational use and life-cycle support?
12. Describe the basic differences between the waterfall model, the spiral model, and the Vee model. How do they compare with the model proposed by the author?
13. Refer to Figure 1.20. Briefly highlight the activities that are critical for the successful implementation of the system engineering process. When in the life cycle must these activities be addressed?
14. Refer to Figure 1.21. Describe how these activities might affect/influence system engineering (if at all).
15. Refer to Figure 1.22. Describe how these activities might affect/influence system engineering (if at all).
16. Refer to Figure 1.23. Explain why these elements should be considered (or not considered) as inherent elements of a system.
17. Refer to Section 1.4.2. Although each of these supporting design disciplines is important in the implementation of system engineering requirements, from your perspective which one(s) should receive a greater degree of emphasis than

the others? Why? If some degree of prioritization is required, how would you accomplish such?

18. The successful implementation of the system engineering process is dependent on both technological and management issues. Explain why. Provide an example of how one can affect the other.

19. Why is the system specification (type A) important? Develop an outline for a system of your choice.

20. What is the purpose of design reviews?

21. What is *concurrent engineering?* How does it relate to system engineering?

22. What is *configuration management?* Why is it important in system engineering?

23. Why is *logistics* important? How does it relate to system engineering (if at all)?

24. What is *life-cycle cost?* What is included? Why is it important to consider such cost in the decision-making process?

25. Describe, in you own words, some of today's challenges relative to the implementation of system engineering, when considering the current environment (refer to Section 1.2).

2

THE SYSTEM
ENGINEERING PROCESS

The *system engineering process* is inherent within and throughout the overall system life cycle, as illustrated in Figure 1.12. The emphasis is on a top-down, integrated, life-cycle approach to system design and development, conveyed through the activities identified in blocks 0.1 through 4.6 of the figure. This approach includes an initial definition of the problem (to be solved) and the identification of a consumer need, conductance of a feasibility analysis, development of system operational requirements and the maintenance and support concept, accomplishment of a functional analysis, allocation of requirements, and development of the top-level architecture for a given system. Subsequently, there is the iterative process of assessment and validation, and the incorporation of changes for product/process improvement as required. Although the process is more directed to the early stages of system design and development, consideration of the activities in the latter phases of production/construction, operational utilization, and system maintenance and support is essential for understanding the consequences of earlier decisions and the establishment of guidelines and benchmarks for the future. In other words, the feedback loop (as identified in the figure) is critical and is an integral part of the system engineering process.

This chapter addresses the system engineering process and the basic activities reflected in Figure 2.1. These activities represent a *process* that should be applied each time that there is a newly identified requirement for a system. A new requirement may evolve when a new system performance factor has been identified (e.g., when the required production rate in a factory is doubled, the capacity of a transportation vehicle is increased, a radar range increased, the speed of data transmission is increased, the weight of a product is reduced, and so on). Also, there may be some current deficiency, and there is a need to respond through the development of a

51

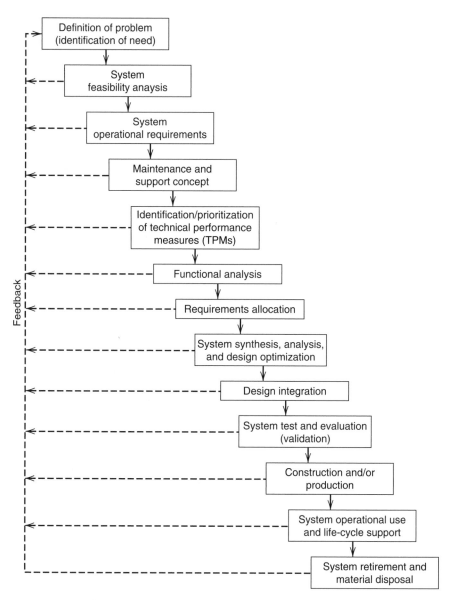

FIGURE 2.1   The system engineering process in the life cycle.

system to meet some completely new requirement. In any event, this is not to imply an additional amount of work or excessive costs, contrary to the perception of many that the implementation of system engineering requirements is time consuming and costly. However, it does require a change in thinking, a shift in emphasis in approaching a system design objective, and a new way of doing business.

The steps shown in Figure 2.1 must (of course) be tailored to both system and program requirements. There are often many iterations that occur, not only within the process overall but within each of the blocks. Analyses and trade-off studies are conducted at each stage, functions are identified in several of the blocks, and so on, and it is impossible to show graphically everything that occurs throughout. However, for the purposes of discussion and for better understanding, the steps illustrated in the figure are presented herein, in the general order indicated.[1]

2.1 DEFINITION OF THE PROBLEM (CURRENT DEFICIENCY)

The system engineering process generally commences with the identification of a "want" or "desire" for something and is based on a real (or perceived) deficiency. For instance, suppose that a current capability is not adequate in terms of meeting certain required performance goals, is not reliable or available when needed, cannot be properly supported, or is too costly to operate. As a result, a new system *requirement* is defined, along with a priority for introduction, the date when the new system capability is required for consumer use, and an estimate of the resources necessary for acquiring the new system capability. To ensure a good start, a complete *description of the problem* should be presented in specific qualitative and quantitative terms, in enough detail to justify progressing to the next step. More specifically, one should pose the following question: What is the nature and magnitude of the problem, and what are the associated risks if the problem is not addressed?

The requirement for identifying the need (as a starting point) may seem to be rather basic or self-evident. However, it often happens that a design effort is initiated as a result of a personal interest or a political whim, without the requirements being adequately defined. In the software area (in particular), there is a tendency to accomplish a lot of coding before identifying the functional need for such. In addition, there are instances in which the engineer sincerely believes that he or she knows what the customer needs, without first having involved the customer in the process. In essence, the attitude "design it now, fix it later" often prevails. As a result, it is not uncommon for someone to proceed with the design and ultimately produce a product that really was not wanted (or needed) in the first place. This approach, of course, can be rather expensive.

Defining the problem is sometimes the most difficult part of the process, particularly if one is in a rush to "get going." Yet, the number of false starts and the ultimate

---

[1] As an example of an "iteration," the *functional analysis* actually commences with the identification of the problem and a description of the functions that are not currently being accomplished; these functions are then established as a basis for the evaluation of possible alternative technology applications in a feasibility analysis; *operational* functions are identified in defining the operational requirements for the system; *maintenance* functions are described in the development of the maintenance concept; and an overall *functional baseline* for the system ultimately evolves from these earlier activities. In other words, the functional analysis is actually accomplished within the first six blocks shown in Figure 2.1, with the appropriate "feedback" as one progresses downward.

risks can be rather significant unless a good foundation is laid from the beginning. A complete description of the need, expressed in quantitatively stated *performance* parameters where possible, is essential. It is important that the results reflect a *true* customer requirement, especially in today's environment where resources are limited.

2.2 SYSTEM REQUIREMENTS (NEEDS ANALYSIS)

Given the problem definition, a *needs analysis* must be accomplished, with the objective of translating a broadly defined "want" into a more specific system-level requirement(s). At this point, the following questions should be addressed:

1. What is required of the system, stated in functional terms?
2. What specific functions must the system accomplish?
3. What are the primary functions to be accomplished?
4. What are the secondary functions to be accomplished?
5. What must be accomplished to completely alleviate the stated deficiency?
6. Why must these functions be accomplished?
7. When must these functions be accomplished?
8. Where is this to be accomplished, and for how long?
9. How many times must these functions be accomplished?

There are many basic questions of this nature that may be applicable, and it is important to describe the anticipated consumer (customer) requirements in a *functional* manner in order to avoid a premature commitment to a specific design concept or configuration and, thus, the unnecessary expenditure of valuable resources at this point in time. The basic objective is to define *what* is required before "locking in" on the *how* it is to be accomplished.[2]

Accomplishing a needs analysis in a satisfactory manner can best be realized through a team approach involving the customer, the ultimate consumer/user (if different from the customer), the prime contractor or producer, and major suppliers (as appropriate). The objective is to ensure the proper communications between the various parties involved. In particular, the *voice of the customer* must be heard, and the system developer must respond accordingly. Methods such as conducting surveys and interviews, the utilization of good checklists, the application of such tools as *quality function deployment* (*QFD*), and related techniques/models may be employed.[3]

---

[2] While the specific responses to the eight questions may be somewhat overlapping, or appear to be redundant, it is critical that a somewhat "exhaustive" approach be assumed at this point in the life cycle in order "force the issue" and to fully identify a good definition of requirements as early as possible in the sysstem life cycle; thus, reducing the risks associated with the subsequent decision-making processes.

[3] An excellent technique, often utilized as an aid in defining requirements and ensuring that the proper communications exist between the customer/consumer and producer, is the *Quality Function Deployment*

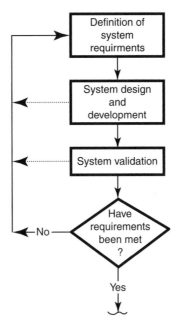

FIGURE 2.2   System requirements.

The "requirements definition process" commences from the beginning, as illustrated by the top block in Figure 2.2. On the one hand, while it may not be possible to go to the depth suggested earlier, it is essential that as much as possible be accomplish and as early as possible! On the other hand, there is always the temptation to delay such and to determine the requirements later during the system design and development process (or even during system test and verification). In such cases, program schedules are likely to slip, the resultant costs could be significant, and the ultimate needs of the customer (consumer) will not be met!

2.3   SYSTEM FEASIBILITY ANALYSIS

Through the needs analysis, the functions that a system must perform are identified. On the one hand, there may be a single function such as *transport product XYZ from A to point B*, or *communicate between points D, E, and F*, or *produce X quantity of Y products by time Z*, or *provide a maintenance and support capability for Systems G and H*. On the other hand, there may be any number of multiple functions to be

(QFD) method. The QFD method was developed initially at the Kobe shipyard of Mitsubishi Heavy Industries, Japan, and has evolved considerably since. It is used to facilitate the translation of a prioritized set of subjective customer requirements into a relevant set of system-level requirements during the conceptual design phase. The application of the QFD method is demonstrated further in Section 2.6.

performed, some primary and some secondary. To ensure that a good design concept is established from the beginning, all feasible functions should be identified, with the most rigorous functions being selected as the basis for defining system-level design requirements. It is important that *all* possibilities be addressed to ensure that the proper overall "technical" approach is selected for design consideration.

A *feasibility analysis* is accomplished with the objective of evaluating different technological approaches that may be considered in responding to the specified functional requirements. In considering different design approaches, alternative technology applications are investigated. For instance, in the design of a communications capability, should one use fiber-optics technology, cellular (wireless), or the conventional hardwired approach? In designing an aircraft structure, to what extent should one incorporate composite materials? When designing an automobile, should one apply very-high-speed integrated electronic circuitry in certain control applications, or should one select a more conventional electromechanical approach?

It is necessary to (1) identify the various possible design approaches that can be pursued to meet the requirements; (2) evaluate the most likely candidates in terms of performance, effectiveness, logistics and maintenance support requirements, and life-cycle economic criteria; and (3) recommend a preferred approach. The objective is to identify and select an overall *technical* approach to the future system design, and *not* to select specific components (e.g., hardware, software, facilities, etc.). There may be many alternatives; however, the number of possibilities must be narrowed down to a few feasible options, consistent with the availability of anticipated resources (i.e., human power, materials, and money).

It is at this early stage in the system life cycle (i.e., during the conceptual design phase) when major design decisions are made relative to adopting a specific design approach. When there is not enough critical information available, a research project activity may be initiated with the objective of developing new methods/techniques for specific applications. In some programs, the completion of applied research tasks and preliminary design activity is accomplished sequentially, whereas in other situations, there may be a number of mini-projects underway at the same time.

The results of the feasibility analysis will have a significant impact not only on the future operational characteristics of the system, but on the production and sustaining maintenance support requirements as well. The selection (and follow-on application) of a given technology has reliability and maintainability implications, may affect human factors and safety requirements, may impact producibility and manufacturing methods, may significantly affect the requirements for logistics and future maintenance support, and will certainly impact life-cycle cost.

With the early feasibility analysis being so critical and having such a potentially large impact on the follow-on system design and development activity, the role of the *system engineer* becomes extremely important. In most situations, the detailed investigations and evaluation efforts leading to specific design approaches are highly technical and are accomplished by specialists in a given engineering discipline. Often, these specialists are not oriented to the "system" as an overall entity, or its manufacturing process, or its maintenance and support capability, or the factors affecting

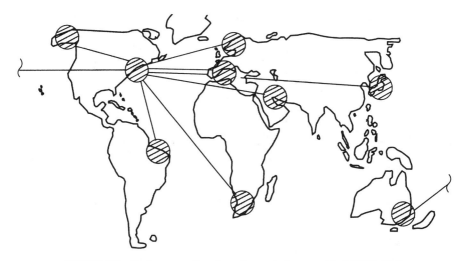

FIGURE 2.3 System operational requirements (geographical distribution).

life-cycle cost. Yet, major design decisions are made, the results of which end up in the system specification, and all subsequent design activity must comply accordingly. Thus, the need for a strong system engineering thrust at this early stage in the life cycle is critical.

2.4 SYSTEM OPERATIONAL REQUIREMENTS

Having defined the basic need and the selection of a feasible technical design approach, it is necessary to project this information further through a more comprehensive description of anticipated system *operational requirements*. The objective is to reflect the needs of the consumer (i.e., user) in terms of system deployment, utilization, effectiveness, and the accomplishment of its intended mission(s). The operational concept, as defined herein, includes the following general information:

1. *Operational distribution or deployment.* The number of user sites where the system will be operated, the geographical distribution and time frame, and the type and quantity of major system components at each location. These factors respond to this question: *Where is the system to accomplish its mission(s), and for how long?* Figure 2.3 presents a sample worldwide distribution scheme.

2. *Mission profile(s) or scenario(s).* Description of the prime mission of the system and its alternative or secondary mission(s). This refers to this question: *What specific function(s) must the system accomplish in order to respond to the identified need?* This information may be presented through the development of a series of operational profiles, illustrating the "dynamic" aspects

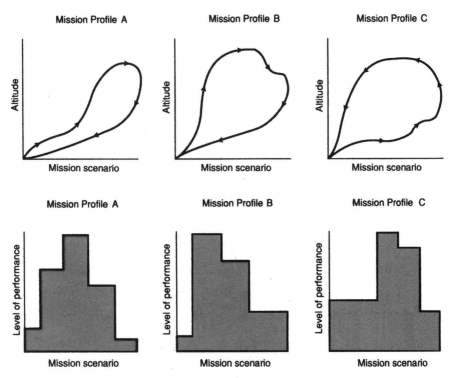

FIGURE 2.4   Sample system operational profiles. *Source*: Benjamin S. Blanchard, *Logistics Engineering and Management*, 6th edition, © 2004, p. 130. Reprinted by permission of Pearson Education, Inc., Upper Saddle River, NJ.

required in accomplishing a given mission. Examples may include an aircraft flight path between two cities, an automobile routing, a shipping route for the transportation of materials, a communication "on-off" profile over time, the distribution of electrical power over time, and so on. Figure 2.4 shows an example of three possible profiles.

3. *Performance and related operational parameters.* Definition of basic operating characteristics of the system expressed in quantitative terms. This refers to such parameters as range, accuracy, speed, rate, capacity, throughput, power output, message clarity, size, weight, operational availability, and so on. *What are the critical system performance parameters necessary to accomplish the mission at the various user sites?* These parameters should be related directly to the operational profiles shown in Figure 2.4, or equivalent (as applicable).[4]

4. *Utilization requirements.* Anticipated usage of the system and its elements in accomplishing its mission. This refers to hours of system operation per day,

---

[4]Included in this category are *key performance parameters (KPPs)* and supporting *key system attributes (KSAs)* as applied in the development of major requirements for defense systems.

duty cycle, on-off cycles per month, percentage of total capacity utilized, facility loading, and so on. *How are the system and its elements being utilized?* This investigation leads to determining some of the stresses that will be imposed on the system by its operator and its environment.

5. *Effectiveness requirements.* System requirements, specified quantitatively (as applicable), include such factors as cost/system effectiveness, operational availability (also included as a performance requirement mentioned earlier), dependability, system reliability (e.g., failure rate), mean time between maintenance (MTBM), readiness rate, maintenance downtime (MDT), facility utilization (in percent), logistics response time, life-cycle cost, personnel skill levels, safety rate, security level, and so on. *Given that the system will perform, how effective or efficient will it be?* (Refer to footnote 4 for defense systems.)

6. *Major system interface or interoperability requirements.* This refers to the primary interfaces with other systems in an overall SOS hierarchical structure, and the performance requirements that this system must provide as an input to, or receive as an output from, other systems in such a structure. *What are the specific input-output requirements in a system-of-systems (SOS) context?* (Refer to Chapter 1, Figure 1.3.)

7. *Environment.* Definition of the environment in which the system is expected to *operate* in an effective and efficient manner; for example, high-low temperature range, shock and vibration, noise, humidity, arctic or tropics, mountainous or flat terrain, airborne, ground, and/or shipboard. Following a set of mission profiles may result in specifying a range of values in each area. *To what will the system be subjected during its operational use, and for how long? What impact will the new system have on the current external environment?* In addition to system operation(s), environmental considerations should address transportation, handling, and storage modes. It is possible that the system (and/or some of its components) will be subjected to a more rigorous environment during transportation than during its operation.

The establishment of system operational requirements provides the foundation and constitutes a baseline for all subsequent system design and development effort. The objective is to respond to the questions presented in Section 2.2: *What* functions must the system perform, *when* must these be accomplished, *where* will the system be utilized and for how long, and *how* will the system accomplish its objective?

Although conditions may well change, some initial assumptions are required at this point. For example, the initial definition of need and the assumptions made in conducting the feasibility analysis may change, specific system utilization requirements at different operational sites may vary from one location to the next, the length of the life cycle may change as a result of obsolescence, and so on. Nevertheless, the development and documentation of this type of information, from the beginning, is essential.

In the past, the operational requirements for many new systems were either (1) developed later and "downstream" in the life cycle after the system design configuration became somewhat fixed, or (2) were developed early by a separate organizational entity (e.g., an outside consultant, marketing group, or equivalent), placed in a file awaiting a decision to proceed with preliminary design, and then forgotten when subsequent design activity resumed. At this point, with the need for this type of information readily apparent (but with such not being available), various individual design groups would generate their own assumptions, different design functions were not referencing the same baseline, and conflicting requirements would evolve. This, in turn, usually led to the development of a system configuration that did not meet the needs of the consumer, and that required the subsequent initiation of corrective action through costly system modifications. In other words, if the applicable system operational requirements are not well defined from the start and integrated as an input to the design process, the follow-on results can be quite costly.[5]

This is another critical area of activity where a strong system engineering thrust is necessary. The operational requirements for the system in question must be thoroughly defined and integrated, and the appropriate information must be documented and disseminated in a timely manner throughout all applicable design (and related) organizations. Every one involved in the design process must "track" the same baseline, and system engineering must provide the leadership in developing such requirements and the ongoing integration of such as the design and development process evolves.

2.5 THE LOGISTICS AND MAINTENANCE SUPPORT CONCEPT

In addressing system requirements, the normal tendency is to deal primarily with those elements of the system that relate directly to the steps involved in the actual *performance of the mission*; that is, prime equipment, operator personnel, operational software, operating facilities, and associated data. At the same time, very little attention is given to logistics and the maintenance support of the system until later and downstream in the life cycle. In general, emphasis in the past has been directed to only *part* of the system, and not the *entire* system. This, of course, has led to some of the problems discussed in Section 1.2 (refer to Figures 1.6, 1.7, and 1.8, in particular).

To meet the overall objectives of system engineering, it is essential that *all* aspects of the system be considered on an integrated basis from the beginning (see Figures 1.2 and 1.8). This includes not only the prime mission-oriented segments of the system, but the entire logistics and maintenance support capability as well. System support, which includes those activities illustrated in Figures 1.21 and 1.22, must be considered as part of the early needs analysis, during the feasibility analysis when new technologies are being evaluated for possible application, and a before-the-fact

---

[5]It is agreed that it is often easier (and considered to be safer) to delay such actions until a time later on when more information becomes available. However, by doing such, the ultimate risk(s), and associated costs often become much greater. Thus, it is important to pursue these objectives in somewhat of an "exhaustive" manner from the beginning, even if this initial baseline evolves somewhat over time.

logistics and maintenance support concept must be developed as to how the system is to be supported on a life-cycle basis.

The *logistics and maintenance support concept,* developed during early conceptual design, evolves from the definition of system operational requirements, as illustrated in Figure 2.5. Initially, one must deal with the flow of activities and materials from design through production, and to the consumer's operational site(s) where the system is utilized (see Figure 1.21). In addition, there is a flow of activities and materials involving the sustaining system maintenance support capability (see Figure 1.22). In the case of the latter, such activities include distribution and transportation of spare/repair parts, personnel, test equipment, and data from the various suppliers to the producer (depot) and intermediate levels of maintenance, and to the operational sites as required. The flow chart in Figure 2.5 reflects the activities that are related to the overall system sustaining support capability, which must be in place and available throughout the system life cycle.[6]

Included within the framework illustrated in Figure 2.5 is a description of the various levels of maintenance, major repair policies, projected organizational responsibilities, effectiveness requirements, and overall environment within which system maintenance and support will be accomplished. Although there may be some variations from one application to the next, the maintenance concept generally includes the following information:

1. *Levels of maintenance.* Corrective and preventive maintenance may be performed on the system itself (or an element thereof) at the site where the system is used by the consumer, in an intermediate shop near the consumer, and/or at a depot or manufacturer's facility. Maintenance level pertains to the division of functions and tasks for each area where maintenance is performed. Anticipated frequency of maintenance, task complexity, personnel skill-level requirements, special facility needs, and so on, dictate to a great extent the specific functions to be accomplished at each level. Depending on the nature and mission of the system, there may be two, three, or four levels of maintenance. However, for the purposes of further discussion, maintenance is classified as organizational, intermediate, and supplier/depot.

   (a) *Organizational maintenance.* Organizational maintenance is performed at the operational site (e.g., airplane, vehicle, manufacturing production line, or communication facility). Generally, it includes tasks performed by the using

---

[6]The logistics and maintenance support *concept,* as defined in this text, is a before-the-fact series of illustrations and statements pertaining to how the system is to be supported throughout the life cycle, from inception and through system retirement and material recycling and/or disposal. It forms the basis for the early definition of system design requirements and the *design for supportability* (e.g., two versus three levels of maintenance support, system/component packaging, degree of diagnostics and condition monitoring to be incorporated, quantitative effectiveness requirements for various elements of the support infrastructure, etc. Refer to Section 1.4.2, items 2, 3, and 7, in particular. A logistics and maintenance support *plan,* often developed later, defines the specific requirements for support based on a known configuration and on the results of supportability and related analyses. The concept is an *input* to design, and the plan is the *result* of design. Refer to B.S. Blanchard, *Logistics Engineering and Management,* 6th ed. (Upper Saddle River, NJ: Pearson Prentice Hall, 2004).

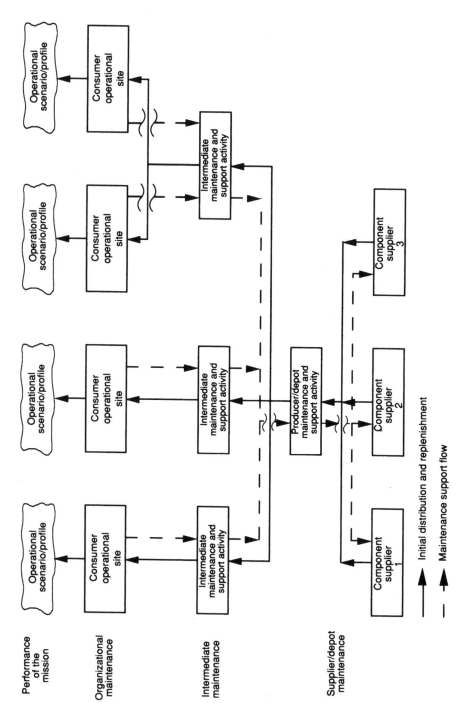

FIGURE 2.5 System operational and maintenance flow.

organization on its own equipment. Organizational-level personnel are usually involved with the operation and use of equipment and have minimum time available for detail system maintenance. Maintenance at this level normally is limited to periodic checks of equipment performance, visual inspections, cleaning of system elements, verification of software, some servicing, external adjustments, and the removal and replacement of some components. Personnel assigned to this level generally do not repair the removed components, but forward them to the intermediate level. From the standpoint of maintenance, the least skilled personnel are assigned to this function. The design of equipment must take this fact into consideration (i.e., design for simplicity).

(b) *Intermediate maintenance.* Intermediate maintenance tasks are performed by mobile, semimobile, and/or fixed specialized organizations and installations. At this level, end items may be repaired by the removal and replacement of major modules, assemblies, or piece parts. Scheduled maintenance requiring equipment disassembly may also be accomplished. Available maintenance personnel are usually more skilled and better equipped than those at the organizational level and are responsible for performing more detail maintenance.

Mobile or semimobile units are often assigned to provide close support to deployed operational systems. These units may be vans, trucks, or portable shelters containing some test and support equipment and spares. The mission is to provide on-site maintenance (beyond that accomplished by organizational-level personnel) to facilitate the return of the system to its full operational status on an expedited basis. A mobile unit may be used to support more than one operational site. A good example is the maintenance vehicle that is deployed from the airport hangar to an airplane parked at a commercial airline terminal gate and needing extended maintenance.

Fixed installations (permanent shops) are generally established to support both the organizational-level tasks and the mobile or semimobile units. Maintenance tasks that cannot be performed by the lower levels, due to limited personnel skills and test equipment, are performed here. High personnel skills, additional test and support equipment, more spares, and better facilities often enable equipment repair to the module and component part level. Fixed shops are usually located within specified geographical areas.

Rapid maintenance turnaround times are not as imperative here as at the lower levels of maintenance.

(c) *Depot or supplier maintenance.* The depot level constitutes the highest type of maintenance and supports the accomplishment of tasks above and beyond the capabilities available at the intermediate level. Physically, the depot may be a specialized repair facility supporting a number of systems/equipment in the inventory or may be the equipment manufacturer's plant. Depot facilities are fixed, and mobility is not a problem. Complex and bulky equipment, large quantities of spares, environmental control provisions, and so on, can be provided if required. The high-volume potential in depot facilities fosters the use of assembly-line techniques, which, in turn, permits the use of relatively unskilled labor for a large portion of the workload, with a concentration of highly skilled specialists in such certain key areas as fault diagnosis and quality control.

The depot level of maintenance includes the complete overhauling, rebuilding, and calibration of equipment, as well as the performance of highly complex maintenance actions. In addition, the depot provides an inventory supply capability. The depot facilities are generally remotely located to support the needs of a specific geographical area or designated product lines.

The three levels of maintenance are presented in Figure 2.6.[7]

2. *Repair policies.* Within the constraints illustrated in Figures 2.5 and 2.6, there may be a number of possible policies specifying the extent to which repair of a system component will be accomplished (if at all). A repair policy may dictate that an item should be designed to be nonrepairable, partially repairable, or fully repairable. Repair policies are established initially, criteria are then developed, and system design progresses within the bounds of the repair policy that is selected. An example of a repair policy, for system XYZ, developed as part of the maintenance concept during conceptual design, is illustrated in Figure 2.7.[8]

3. *Organizational responsibilities.* The accomplishment of maintenance may be the responsibility of the consumer, the producer (or supplier), a third party, or a combination thereof. In addition, the responsibilities may vary, not only with different components of the system, but over time, through operational use of the system and the sustaining support phase. Decisions pertaining to organizational responsibilities may impact system design from a diagnostic and packaging standpoint, as well as dictate repair policies, contract warranty provisions, and the like. Although conditions may change, some initial assumptions are required at this point.

4. *Maintenance support elements.* As part of the initial maintenance concept, criteria must be established relating to the various elements of maintenance support. These elements include supply support (spare and repair parts, associated inventories, provisioning data), test and support equipment, personnel and training, transportation and handling equipment, facilities, data, and computer resources. Such criteria, as an input to design, may cover self-test provisions, built-in versus external test require ments, packaging and standardization factors, personnel quantities and skill levels, transportation and handling factors and constraints, and so on. The maintenance concept provides some initial system design criteria pertaining to the activities

---

[7]The criteria presented in Figure 2.6 represent just an example of the guidance information that might be established as a start in attempting to develop various maintenance policies (or the extent of maintenance to be accomplished at each level). Actually, one will need to consider a number of factors (e.g., economics, technology, social and security factors, warranty provisions, cost, and component criticality) in determining what will be repaired and where, and these need to be tailored to the particular system being addressed.

[8]Repair policies are usually verified through a *level-of-repair analysis* (LORA), initially accomplished in conjunction with the maintenance concept development, later accomplished as part of a maintainability analysis and/or supportability analysis, and ultimately leading to the development of the mainlenance plan. Refer to Chapter 3 and the case study presented in Appendix C for additional discussion.

Criteria	Organizational Maintenance	Intermediate Maintenance		Depot Maintenance
		Mobile or semimobile units	Fixed units	
Done where?	At the operational site or wherever the prime equipment is located	Truck, van, portable shelter, or equivalent	Fixed field shop	Depot facility — — — — — Specialized repair activity, or manufacturer's plant
Done by whom?	System/equipment operating personnel (low maint. skills)	Personnel assigned to mobile, semimobile, or fixed units (intermediate maintenance skills)		Depot facility personnel or manufacturer's production personnel (mix of intermediate production personnel skills and high maintenance skills)
On whose equipment?	Using organization's equipment	Equipment owned by using organization		
Type of work accomplished?	Visual inspection Operational checkout Minor servicing External adjustments Removal and replacement of some components	Detailed inspection and system checkout Major servicing Major equipment repair and modifications Complicated adjustments Simple software maintenance Limited calibration Overload from organizational level of maintenance		Complicated factory adjustments Complex equipments repairs and modifications Overhaul and rebuild Detailed calibration Software maintenance (detailed modifications Supply support Overload from intermediate level of maintenance

FIGURE 2.6 Major levels of maintenance.

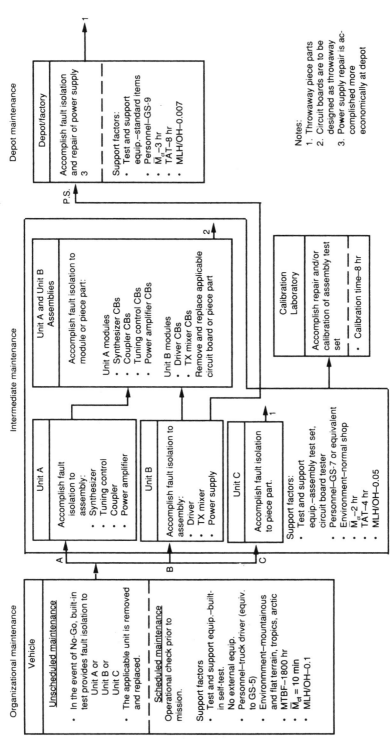

FIGURE 2.7 System maintenance concept flow (repair policy.) *Source:* Benjamin S. Blanchard, *Logistics Engineering and Management*, 6th edition, © 2004, p. 143. Reprinted by permission of Pearson Education, Inc., Upper Saddle River, NJ.

illustrated in Figure 2.5, and the final determination of specific logistic and maintenance support requirements will occur through the completion of a maintenance engineering analysis (or equivalent) as design progresses.

5. *Effectiveness requirements.* Effectiveness requirements pertaining to the logistics and maintenance support infrastructure must be properly integrated with the effectiveness factors associated with system operational requirements (described in Section 2.4, item 5), and may include such factors as the availability of the overall support infrastructure, logistics response time, maintenance downtime (e.g., downtime due to the unavailability of a required element of support), life-cycle cost associated with the logistics and maintenance support capability, and so on. Additionally, there may be any number of requirements applicable and important at a lower level. For example, in the supply support area, they may include a spare part demand rate; the probability of a spare part being available when required; the probability of mission success given a designated quantity of spares in the inventory; and the economic order quantity as related to inventory procurement. For test equipment, the length of the queue while waiting for test, the test station process time, and the test equipment reliability are key factors. In transportation, transportation rates, transportation times, the reliability of the transportation, and transportation costs are of significance. For personnel and training, one should be interested in personnel quantities and skill levels, human error rates, training rates, training times, and training equipment reliability. In software, the number of errors per mission segment or per line of code may be important measures. These factors, as related to a specific system-level requirement, must be addressed. It is meaningless to specify a tight quantitative requirement applicable to the repair of a prime element of the system when it takes six months to acquire a needed spare part. The effectiveness requirements applicable to the support capability must complement the requirements of the system overall.

6. *Environment.* Definition of the environment as it pertains to maintenance and support. This includes temperature, shock and vibration, humidity, noise, arctic versus tropical environment, mountainous versus flat terrain, shipboard versus ground conditions, and so on, as applicable to maintenance activities and related transportation, handling, and storage functions.

In summary, the *logistics and maintenance support concept* provides the basis for establishment of supportability requirements as an *input* to the system design process. Not only do these requirements impact the prime mission-oriented elements of the system, but they should also provide guidance in the design and/or procurement of the necessary elements of logistic support (identified through the *forward* and *reverse* flow requirements shown in Figure 1.20). In addition, this concept forms the baseline for the detailed logistics and maintenance support plan, prepared during the detail design and development phase shown in Figure 1.12.[9]

---

[9]It should be noted that these requirements must be properly integrated with those for other systems in a common SOS configuration. They may be either more, or less, stringent, but must be properly tailored to the particular system in question.

2.6 IDENTIFICATION AND PRIORITIZATION OF TECHNICAL PERFORMANCE MEASURES (TPMs)

Given the development of system operational requirements and the logistics and maintenance support concept, it is necessary for the designer to review these requirements in terms of specific quantitative values, their relative degrees of importance, criticality from the standpoint of accomplishing the desired mission(s), and priorities in design in the event that trade-offs are necessary. In the design of a vehicle, is *speed* more important than *size*? For a manufacturing plant, is *production quantity* more important than *product quality*? In a communication system, is *range* more important than *clarity of message*? For a computer capability, is *capacity* more important than *speed*? Or, for any type of system, is *operational availability* more important than *logistics response time*, or is *system effectiveness* more important than *life-cycle cost*?[10]

The number of objectives may be numerous, and the designer needs to understand which are more important than others and the relationships between them. In addition, it is desirable to express these objectives in quantitative terms where feasible. It is difficult (if not impossible) to proceed with the design in a satisfactory manner unless there are some measurable goals specified from the beginning. These goals, in turn, must reflect the customer's (consumer's) requirements.

In developing system requirements, there may be any number of goals, often initially expressed in very general qualitative terms such as *the system must be designed to meet customer requirements effectively and efficiently.* Or, *the system must he designed for maximum availability.* Or, *the system must be designed to be reliable.* The question is, *how does one respond to such requirements and how does one measure the results for the purpose of system validation?* Additionally, and given such, *which becomes more important and of a higher priority than the others?* In helping to respond, and in the interest of further clarification, the use of an *objectives tree* (or something of an equivalent nature) may aid in facilitating this prioritization process.

Referring to Figure 2.8, and in the absence of better guidance, the designer will need to interpret the specified requirements and make some assumptions as to what is meant by "effectively" and "efficiently." Although the objective is to design a system in response to consumer requirements, it may not always happen unless there is a good communications link between the designer and the customer. Through a team effort, the approach conveyed in Figure 2.8 can help clarify the requirements. Initially, it may be necessary to express design objectives in qualitative terms,

---

[10]The objective is to derive, from the basic system requirements (described in Sections 2.4 and 2.5), the overall *performance* goals to which the system must be designed. "Performance," used in this context, refers to *all* of the major parameters against which the system design must comply, and *all* elements of the system must be addressed in responding to the total mission(s) requirements. Such goals may include operational factors, logistics factors, maintenance support factors, KPPs, KSAs, and/or equivalent.

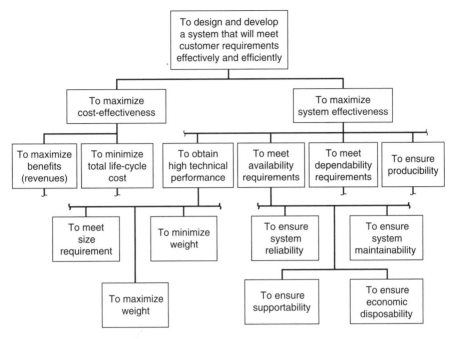

FIGURE 2.8   Objectives tree (partial).

showing their relationships in a top-down hierarchical manner. Subsequently, an attempt should be made to establish *quantitative* measures for each block in the figure and ensure that the appropriate "traceability" exists both downward and upward. Applying this approach to the system breakdown in Figure 1.14, *what measures should be. applied and to what level in the overall hierarchical structure for the system? Further, what design criteria should be established for each level? Is reliability more important than maintainability? Are human factors more important than cost?* Establishing these relationships will, in turn, help the designer to identify areas where emphasis must be applied in the design process and the areas that can be traded off in the event that something has to give.

An excellent tool that can be applied to aid in establishing the necessary communications between designers and the consumer (i.e., the customer) is the *quality function deployment* (QFD) method. QFD constitutes a *team approach* to help ensure that the voice of the customer is reflected in the ultimate design. The purpose is to establish the necessary requirements and to translate those requirements into technical solutions. Consumer requirements and preferences are defined and categorized as *attributes*, which are then weighted according to the degree of importance. The QFD method gives the design team an understanding of customer desires, forces the customer to prioritize those desires, and enables a comparison of one design

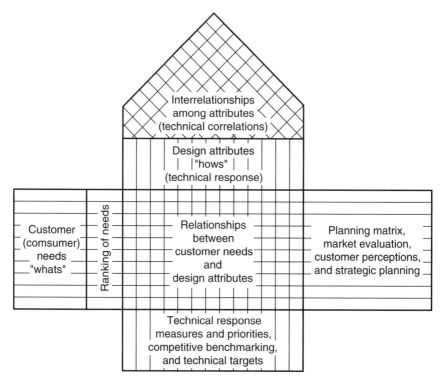

FIGURE 2.9 House of Quality (modified).

approach against another. Each customer attribute is then satisfied by a technical solution.[11]

The QFD process involves constructing one or more matrices, the first of which is often referred to as the house of quality (HOQ). A modified version of the HOQ is presented in Figure 2.9. Starting on the left side of the structure is the identification of customer needs and the ranking of those needs in terms of priority, the levels of importance being specified quantitatively. This reflects the "whats" that must be addressed. A team, with representation from both consumer and design organizations, determines the priorities through an iterative process of review, evaluation, revision, reevaluation, and so on. The top part of the HOQ identifies the designer's *technical* response relative to the attributes that must be incorporated into the design in order to respond to the needs (i.e., the voice of the customer). This constitutes the "hows,"

---

[11]Three good references pertaining to the QFD process are (1) Y. Akao, (ed.), *Quality Function Deployment: Integrating Customer Requirements into Product Design* (New York: Productivity Press, Inc, 2004); (2) L. Cohen, *Quality Function Deployment: How to Make QFD work for You* (Upper Saddle River, NJ: Prentice Hall, 1995); and (3) J. Revelle, J.W. Moran, and C. Cox, *The QFD Handbook* (Hoboken, NJ: John Wiley & Sons, 1997).

and there should be at least one technical solution for each identified customer need. The interrelationships among attributes (or technical correlations) are identified, as well as possible areas of conflict. The center part of the HOQ conveys the strength of the proposed technical response or its impact on the identified requirement. The bottom part allows for a comparison between possible alternatives, and the right side of the HOQ is used for planning purposes.[12]

The QFD method is used to facilitate the translation of a prioritized set of subjective customer requirements into a set of *system-level* requirements during conceptual design. A similar approach may be used to subsequently translate system-level requirements into a more detailed set of requirements at each stage in the design and development process. In Figure 2.10, the "hows" from one house become the "whats" for a succeeding house. Requirements may be developed for the system, subsystem, component, the manufacturing process, the support infrastructure, and so on. The objective is to ensure the required justification and traceability of requirements from the top down. Further, requirements should be stated in *functional* terms.

Although the QFD method may not be the only approach used in defining the requirements for system design, it does constitute an excellent tool for creating the necessary visibility from the beginning. One of the largest contributors to risk is the lack of a good set of requirements and an adequate system specification. Inherent within the system specification should be the identification and prioritization of technical performance measures (TPMs), as illustrated in Figure 2.11. The TPM, its associated measure (i.e., metric), its relative importance, and benchmark objective in terms of what is currently available will provide designers with the necessary guidance for accomplishing their task. This is essential for establishing the appropriate levels of design emphasis, for defining the criteria as an input to the design, and for identifying the levels of possible risk should the requirements not be met.

2.7 FUNCTIONAL ANALYSIS

An essential element of early conceptual and preliminary design is the development of a *functional* description of the system to serve as a basis for the identification of the resources necessary for the system to accomplish its objective(s). A function is a specific or discrete action (or series of actions) necessary to achieve a given objective; that is, an operation that the system must perform to accomplish its mission, or a maintenance action that is necessary to restore the system to operational use. Such actions may ultimately be accomplished through the use of equipment, people, software, facilities, data, or combinations thereof. However, at this point, the objective is to specify the "whats" and *not* the "hows"; that is, *what* needs to be accomplished versus *how* it is to be done. The functional analysis is an iterative process of breaking down requirements from the system level, to the subsystem, and as far down the

---

[12]J. Hauser and D. Clausing, "The House of Quality," *Harvard Business Review* (May–June 1988).

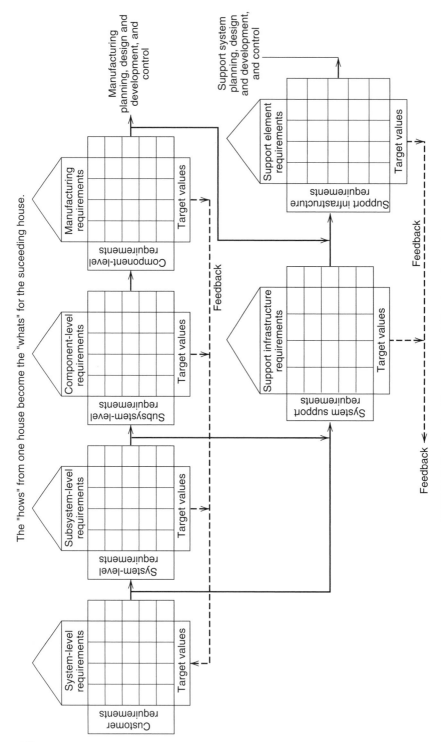

The "hows" from one house become the "whats" for the suceeding house.

FIGURE 2.10 Family of houses (traceability of requirements).

72

Technical performance measure (TPM)	Quantative requirement ("metric")	Current "benchmark" (competing systems)	Relative importance (customer desires)
Process time (days)	30 days (maximum)	45 days (system "M")	10
Velocity (mph)	100 mph (minimum)	115 mph (system "B")	32
Availability (operational)	98.5% (minimum)	98.9% (system "H")	21
Size (feet)	10 feet long 6 feet wide 4 feet high (maximum)	9 feet long 8 feet wide 4 feet high (system "M")	17
Human factors	Less than 1% error rate per year	2% per year (system "B")	5
Weight (pounds)	600 pounds (maximum)	650 pounds (system "H")	6
Maintainability (MTBM)	300 miles (minimum)	275 miles (system "H")	9
			100%

FIGURE 2.11   Prioritization of technical performance measures (TPMs).

hierarchical structure as necessary to identify input design criteria and/or constraints for the various elements of the system.[13]

In Figure 2.1, the functional analysis may be initiated in the early stages of conceptual design as part of the problem definition and needs analysis task, and functions that the system must perform in order to fulfill the needs of the consumer are identified. These *operating* functions are then expanded and formalized through the development of system operational requirements. Primary *maintenance and support* functions for the system, which evolve from the operational requirements, are identified as part of the maintenance concept development process. Subsequently, these functions must be expanded to include *all* of the activity, from the initial identification of need to the retirement of the system.

A functional analysis can be facilitated through the use of functional flow block diagrams, as illustrated in Figure 2.12. Block diagrams are developed primarily for

[13] In applying the principles of system engineering, not one piece of equipment, or element of software, or data item, or element of support should be identified and purchased without the need for such having first been justified through a functional analysis. On many projects, items are often purchased on the basis of what was initially perceived as a requirement, but which later turned out not to be needed. This practice, of course, can turn out to be quite costly.

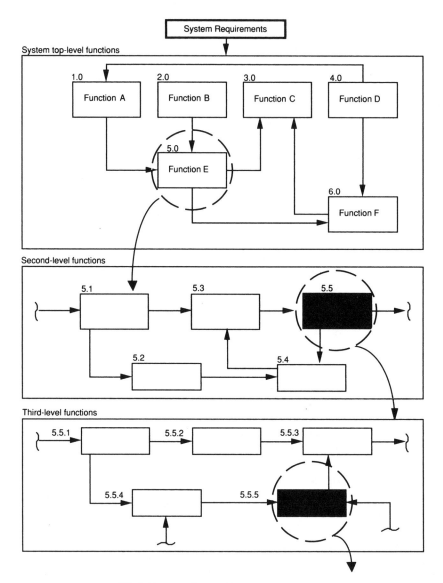

FIGURE 2.12 System functional breakdown.

the purpose of structuring system requirements into functional terms. They are developed to illustrate basic system organization and to identify functional interfaces. The functional analysis (and the generation of functional flow diagrams) is intended to enable the completion of the design, development, and system definition process in a comprehensive and logical manner. Top-level requirements are identified, partitioned to a second level, and on down to the depth required for the purposes of definition. More specifically, the functional approach helps to ensure the following:[14]

- All facets of system design and development, production, operation, support, and retirement are covered; that is, all significant activities within the system life cycle.
- All elements of the system are fully recognized and defined; that is, prime equipment, spare/repair parts, test and support equipment, facilities, personnel, data, and software.
- A means is provided for relating system packaging concepts and support requirements to specific system functions; that is, satisfying the requirements of good *functional* design.
- The proper sequences of activity and design relationships are established, along with critical design interfaces.

One of the objectives of functional analysis is to ensure traceability from the top system-level requirements down to the requirements for detail design. In Figure 2.13, it is assumed that there is a need for transportation between City A and City B. Through a feasibility analysis, trade-off studies were accomplished, and the results indicate that transportation by air is the preferred mode. Subsequently, through the definition of operational requirements, it was concluded that there is a requirement for a new aircraft system, demonstrating good performance and effectiveness characteristics, with quantitative goals specified for size, weight, thrust, range, fuel capacity, reliability, maintainability, supportability, cost, and so on. An aircraft must be designed and produced that will accomplish its mission in a satisfactory manner, flying through a number of operational profiles such as the one illustrated in Figure 2.13. Further, the maintenance concept indicates that the aircraft will be designed for support at three levels of maintenance by the user, will incorporate built-in test provisions, and will be in operational use for a life cycle of 10 years.

With this basic information, following the general steps in Figure 2.1. one can commence with the structuring of the system in functional terms. A top-level functional flow diagram can be developed to cover the primary activities identified within the specified life cycle. Each of these designated activities can be expanded through a second-level functional flow diagram, a second-level activity into a third-level functional flow, and so on.

Through this progressive expansion of functional activities, directed to defining the "whats" (versus the "hows"), one can evolve from the mission profile in Figure 2.13, down to a specific aircraft capability such as communications. A

---

[14]The preparation of functional flow block diagrams (FFBDs) may be accomplished through the use of any one of a number of graphical methods, including the Integrated DEFinition (IDEF) modeling method, the Behavioral Diagram method, and the N-Squared Charting method. Although the graphical descriptions are different, the ultimate objectives are similar.

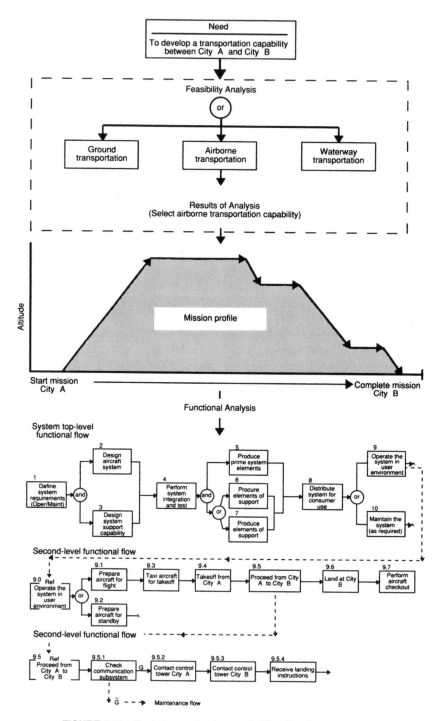

FIGURE 2.13 Evolutionary development of functional requirements.

communications subsystem is identified, trade-offs are accomplished, and a detail design approach is selected. Specific resources that are necessary to respond to the stated functional requirement can be identified. In other words, one can drive downward from the system level to identify the resources needed to perform certain functions (e.g., equipment, people, facilities, and data). Also, given a specific equipment requirement, one can progress "upward" for *justification* of that requirement. The functional analysis provides the mechanism for "down-up" traceability.

2.7.1 Functional Flow Block Diagrams (FFBDs)

In the development of functional flow diagrams, some degree of standardization is necessary, for the purpose of communication, in defining the system. Thus, certain basic practices and symbols should be used, whenever possible, in the physical layout of functional diagrams. The following eight guidelines should help:

1. *Function block.* Each separate function in a functional diagram should be presented in a single box enclosed by a solid line. Blocks used for reference to other flows should be indicated as partially enclosed boxes labeled "REF." Each function may be as gross or detailed as required by the level of the functional diagram on which it appears, but it should stand for a definite, finite, discrete action to be accomplished by equipment, personnel, facilities, software, or any combination thereof. Questionable or tentative functions should be enclosed in dotted blocks.

2. *Function numbering.* Functions identified on the functional flow diagrams at each level should be numbered in a manner that preserves the continuity of functions and provides information with respect to function origin throughout the system. Functions in the top-level functional diagram should be numbered 1.0, 2.0, 3.0, and so on. Functions that further indenture these top functions should contain the same parent identifier and should be coded at the next decimal level for each indenture. For example, the first indenture of function 3.0 would be 3.1, the second 3.1.1, the third 3.1.1.1, and so on. For expansion of a higher-level function within a particular level of indenture, a numerical sequence should be used to preserve the continuity of function. For example, if more than one function is required to amplify function 3.0 at the first level of indenture, the sequence should be 3.1, 3.2, 3.3, . . . , 3.*n*. For expansion of function 3.3 at the second level, the numbering will be 3.3.1, 3.3.2, . . . , 3.3.*n*. Where several levels of indentures appear on a single functional diagram, the same pattern should be maintained. Whereas the basic ground rule should be to maintain a minimum level of indentures on any one particular flow, it may become necessary to include several levels to preserve the continuity of functions and to minimize the number of flows required to functionally depict the system.

3. *Functional reference.* Each functional diagram should contain a reference to its next higher functional diagram through the use of a reference block. For example, function 4.3 should be shown as a reference block in the case where functions 4.3.1, 4.3.2, . . . , 4.3.*n,* are being used to expand function 4.3.

Reference blocks should also be used to indicate interfacing functions as appropriate.

4. *Flow connection*. Lines connecting functions should indicate only the functional flow and should not represent either a lapse in time or any intermediate activity. Vertical and horizontal lines between blocks should indicate that all functions so interrelated must be performed in either a parallel or a series sequence. Diagonal lines may be used to indicate alternative sequences (cases where alternative paths lead to the next function in the sequence).

5. *Flow direction*. Functional diagrams should be laid out so that the functional flow is generally from left to right, and the reverse flow, in the case of a feedback functional loop, from right to left. Primary input lines should enter the function block from the left side; the primary output, or *GO* line, should exit from the right; and the *NO-GO* line should exit from the bottom of the box.

6. *Summing gates*. A circle should be used to depict a summing gate. As in the case of functional blocks, lines should enter and/or exit the summing gate as appropriate. The summing gate is used to indicate convergence or divergence, or parallel or alternative functional paths, and is annotated with the term AND or OR. The term AND is used to indicate that parallel functions leading into the gate must be accomplished before proceeding to the next function, or that paths emerging from the AND gate must be accomplished after the preceding functions. The term OR is used to indicate that any of several alternative paths (alternative functions) converge to, or diverge from, the OR gate. The OR gate thus indicates that alternative paths may lead or follow a particular function.

7. *Go and no-go paths*. The symbols G and $\overline{G}$ are used to indicate go and no-go paths, respectively. The symbols are entered adjacent to the lines leaving a particular function to indicate alternative functional paths.

8. *Numbering procedure for changes to functional diagrams*. Additions of functions to existing data should be accomplished by locating a new function in its correct position without regard to sequence of numbering. The new function should be numbered using the first unused number at the level of indenture appropriate for the new function.

The functions identified should not be limited strictly to those necessary for the operation of the system, but must consider the possible effects of maintenance on system design. In most instances, maintenance functional flows will evolve directly from operational flows. An example of several functional flow diagrams is included in Appendix A.

2.7.2 Operational Functions

Operational functions, in this instance, constitute those that describe the activities that must be accomplished in order to fulfill the mission requirements. These may

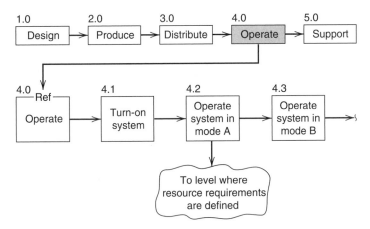

FIGURE 2.14   Functional block diagram (partial).

include both (1) those activities that involve the design, development, production, and distribution of a system for use and (2) those activities that are related directly to the completion of an operational mission scenario. In the second category, these may include a description of the various modes of system operation and utilization. For instance, typical gross operating functions may entail (1) "prepare aircraft for flight," (2) "transport material from the factory to the warehouse," (3) "initiate communications between the producer and the user," (4) "produce x quantity of units in a seven-day time frame," and (5) "process a data to eight company distribution outlets, in b time, with c accuracy, and in d format." System functions necessary to successfully complete the identified modes of operation are then described.

Figure 2.14 illustrates a simplified operational flow diagram. Note that the words in each block are action oriented and the block numbering allows for the downward–upward traceability of resource requirements. The functions are broken down to the depth necessary to describe the resources that will be required to accomplish the function—that is, equipment, software, people, facilities, and so on.

2.7.3 Maintenance and Support Functions

Once operational functions are described, the system development process leads to the identification of *maintenance and support* functions. For instance, there are specific performance expectations or measures associated with each block in an operational functional flow diagram. A check of the applicable functional requirement will indicate either a "go" or a "no-go" decision. A go decision leads to a check of the next operational function. A no-go indication (constituting a symptom of failure) provides a starting point for the development of a detailed maintenance functional flow diagram. The transition from an operational function to a maintenance function

Operational functions

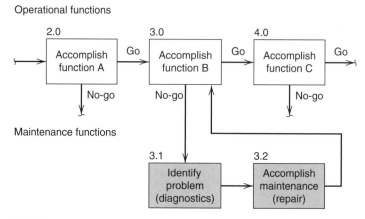

FIGURE 2.15 Transition from operational functions to maintenance functions.

is illustrated in Figure 2.15. Figure 2.16 presents a more in-depth functional flow diagram.[15]

2.7.4 Application of Functional Analysis

The functional analysis provides an initial description of the system and, as such, its applications are extensive. Figure 2.17 illustrates a top-level operational functional flow diagram for a manufacturing system, commencing with the identification of need (block 1.0) and extending through system retirement (block 7.0). In areas where a greater degree of definition is desirable, the applicable block(s) may be broken down to a second level, third level, and so on, in order to gain the appropriate level of visibility necessary for the determination of resource requirements. In this instance, the ultimate manufacturing operating functions have been identified in the break-out of block 5.1.

For each of the blocks in Figure 2.17, the analyst should be able to specify *input* requirements, expected *outputs,* external *controls* and/or *constraints,* and the *mechanisms* (or resources) necessary to accomplish the specific function in question. In the process of identifying the appropriate resource requirements, there may be a number of alternative approaches that should be considered. Trade-off studies are conducted, alternatives are evaluated against criteria developed from the established technical performance measures (i.e., the TPMs derived Section 2.6), and a preferred approach is recommended. It is at this point that one begins to identify the requirements for hardware, software, people, facilities, data, or combinations thereof. Figure 2.18 reflects the process that should be applied to each of the blocks in Figure 2.17.

---

[15]It should be noted that all of the *forward* and *reverse* flow activities shown in Figure 1.20 should be covered through either the "operational" or the "maintenance and support" functional flow block diagrams (FFBDs).

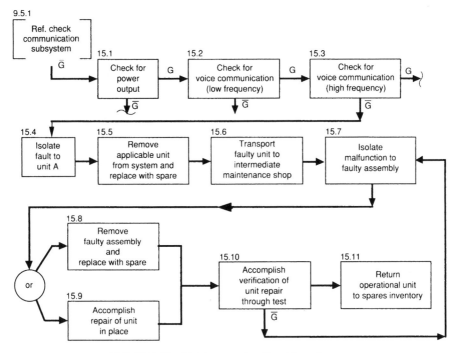

FIGURE 2.16   Maintenance functional flow diagram.

In the evaluation of each functional requirement, the alternatives may include the selection of "commercial off-the-shelf" (COTS) items readily available from a number of different sources of supply, COTS items that may require some degree of modification, and/or developmental items that are unique to a particular application or where some new design is required, Past experience has indicated that extensive time and cost savings can be realized through the selection of readily available COTS equipment or reusable software, the utilization of existing facilities, and so on. Figure 2.19 illustrates the various options in this area.

Figure 2.19 shows that it is essential that a *good* definition of the inputs and outputs (and the applicable metrics) be established if one is to fully understand not only the *interfaces* between the different functions identified in Figure 2.17, but the precise requirements in the process of resource identification. If these input-output requirements are not well defined, the decision-making process as to a preferred approach becomes difficult; thus leading to the possibility of initiating a new costly design and development effort when, in actuality, an existing off-the-shelf item could fulfill the need.

The functional analysis can facilitate an *open-architecture approach* to system design. A good comprehensive functional description of the system, with the interfaces well defined (both qualitatively and quantitatively), can lead to a structure that will not only allow for the rapid identification of resource requirements, but permit

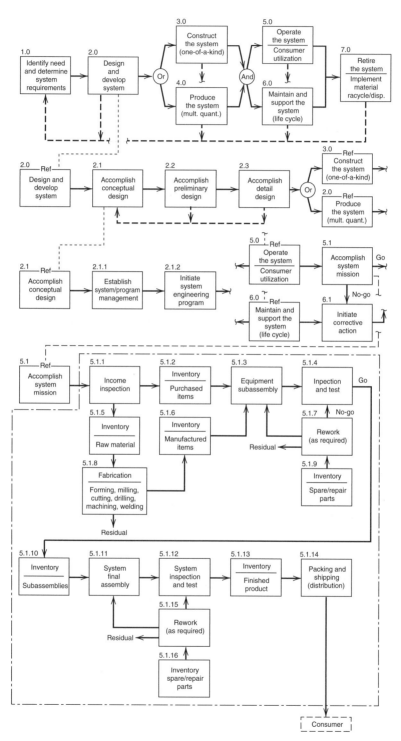

FIGURE 2.17 Functional flow diagram for a manufacturing system.

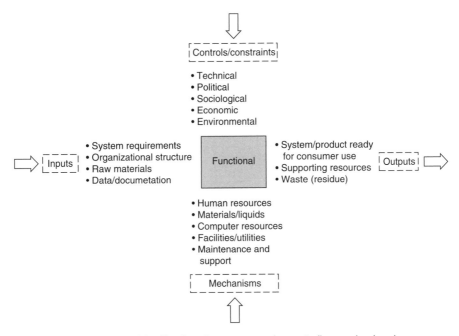

FIGURE 2.18   Identification of resource requirements (i.e. mechanisms).

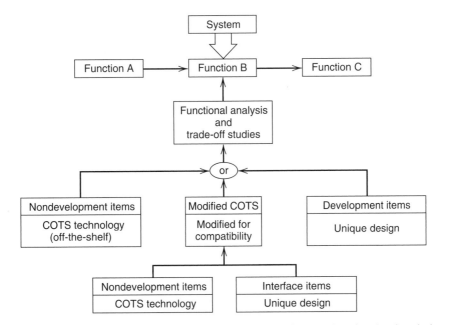

FIGURE 2.19   Identification of commercial off-the-shelf (COTS) items from functional analysis.

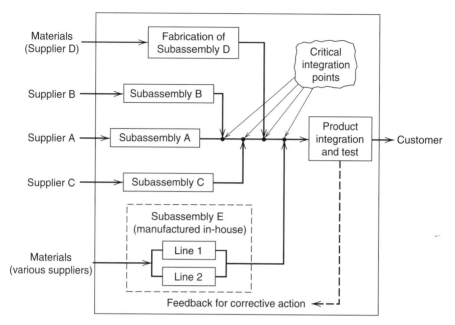

FIGURE 2.20 Manufacturing system (critical integration points).

the possible incorporation of new technologies later. The objective is to design and develop a system that can be easily modified, through the insertion of new technologies, without causing a "costly" redesign of all of the elements of the system in the process.

In many current situations, the requirements in design are changing from a detailed "design to the component level" to the design of systems using a *black-box integration approach*. Given the need to reduce acquisition times, while responding to an ever-changing set of requirements on a continuing basis and with many more suppliers involved, the system *architecture* must allow for the ease of upgrade and/or modification. In other words, the system *structure* must be such as to facilitate design on an *evolutionary* basis, and with minimum cost. This can be enhanced through a good and comprehensive functional definition of the system in the early conceptual design phase of the life cycle.

Figure 2.20 illustrates a manufacturing system in which there are many suppliers (on various locations throughout the world) who produce components for a consumer product that must be effectively integrated and tested. There are fabrication functions, subassembly functions, assembly functions, and test functions. Where, in many instances in the past, the manufacturing activity involved a bottom-up "build" approach, the challenges today relate to the *integration* of the various components into the end product. Without a good early definition and specification of the functional interfaces, the final integration and test activity may result in a costly trial-and-error process. In Figure 2.20, the example reflects a factory where the subprocesses were being accomplished effectively and efficiently; however, there were considerable problems associated with the "integration" activities—that is, the four critical integration points. The functional interfaces were not well defined from the beginning, causing a great deal of modification and rework downstream.

In completing a functional analysis, care should be taken to ensure that the required resources are properly identified for each function. A time-line analysis may be performed to determine whether the functions are to be accomplished in series or in parallel. It may be possible to share resources in some instances; that is, the same resources may be utilized to accomplish more than one function. The identified resources may be combined and integrated to the extent possible. Every effort should be made to avoid the specification of resources that are not necessary. Figure 2.21 illustrates a documentation format that can be applied to formalize the identification of such resources.

The functional analysis is a critical step in the early system design and development effort, and it forms a baseline for many activities that are conducted subsequently. For instance, it serves as a basis in the development of the following:

1. Electrical and mechanical design for functional packaging, condition monitoring, and diagnostic provisions
2. Reliability models and block diagrams
3. Failure mode, effect, and criticality analysis (FMECA)
4. Fault-tree analysis (FTA)
5. Reliability-centered maintenance (RCM) analysis
6. Maintainability analysis
7. Human-factors analysis
8. Operator task analysis (OTA)
9. Operational sequence diagrams (OSDs)
10. System safety/hazard analysis
11. Security analysis
12. Level of repair analysis (LORA)
13. Maintenance task analysis (MTA)
14. Logistics analysis (supply chain analysis)
15. Supportability/serviceability analysis
16. Operating and maintenance procedures
17. Producibility analysis
18. Disposability and material recycling analysis

In the past, the functional analysis has not always been completed in a timely manner, if completed at all. As a result, the various design disciplines assigned to a given program have had to generate their own analyses in order to comply with program requirements. In many instances, these efforts were accomplished independently, and many design decisions were made without the benefit of a *common* baseline to

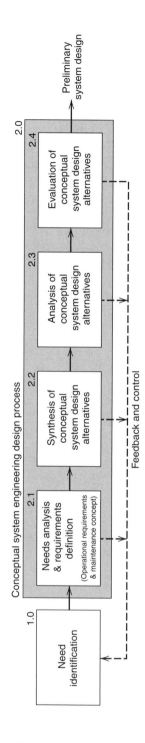

Conceptual system engineering design process

- 1.0 Need identification
- 2.1 Needs analysis & requirements definition (Operational requirements & maintenance concept)
- 2.2 Synthesis of conceptual system design alternatives
- 2.3 Analysis of conceptual system design alternatives
- 2.4 Evaluation of conceptual system design alternatives
- 2.0 Preliminary system design

Feedback and control

Activity number	Activity description	Required inputs	Expected outputs	Resource requirements (activities/techniques)
1.0	Need identification	Customer surveys; marketing inputs; shipping and servicing department logs; market niche studies; competitive product research.	A specific qualitative and quantitative needs statement responding to a current deficiency. Care must be taken to state this need in functional terms.	Benchmarking; statistical analyses of data (i.e., data collected as a result of surveys and consolidated from shipping and servicing logs, etc.)
2.1	Needs analysis and requirements definition	A specific qualitative and quantitative needs statement expressed in functional terms.	Qualitative and quantitative factors pertaining to system performance levels, geographical distribution of products, expected utilization profiles, user/consumer environment; operational life cycle, effectiveness requirements, the levels of maintenance and support, consideration of the applicable elements of logistic support, the support environment, and so on.	Quality Function Deployment (QFD); input-output matrix; checklists; value engineering; statistical data analysis; trend analysis; matrix analysis; parametric analysis; various categories of analytical models and tools for simulation studies, trade-offs, etc.
2.2	Synthesis of conceptual system design alternatives	Results from needs analysis and requirements definition process; technology research studies; supplier information.	Identification and description of candidate conceptual system design alternatives and technology applications.	Pugh's concept generation approach; brainstorming; analogy; checklists.
2.3	Analysis of conceptual system design alternatives	Candidate conceptual solutions and technologies; results from the needs analysis and requirements definition process.	Approximation of the "goodness" of each feasible conceptual solution relative to the pertinent parameters, both direct and indirect. This goodness may be expressed as a numeric rating, probabilistic measure, or fuzzy measure.	Indirect system experimentation (e.g., mathematical modeling and simulation); parametric analyses; risk analyses.
2.4	Evaluation of conceptual system design alternatives	Results from the analysis task in the form of a set of feasible conceptual system design alternatives.	A specific qualitative and quantitative needs statement responding to a current deficiency. Care must be taken to state this need in functional terms.	Design-dependent parameter approach; generation of hybrid numbers to represent candidate solution "goodness"; conceptual system design evaluation display.

FIGURE 2.21 Document format for resource requirements.

follow. This, of course, resulted in design discrepancies and costly modifications later in the system life cycle.

The functional analysis provides an excellent and very necessary baseline, and all applicable design activities must "track" the same data source in order to meet the objectives for system engineering, as stated in Chapter 1. For this reason, the functional analysis is considered a key activity in the system engineering process.

2.7.5 Interfaces with Other Systems in a SOS Configuration

Referring to Section 1.1.3 (and Figure 1.3), there may be any number of systems contained within some overall hierarchical structure. Each individual system must, of course, respond to some function need and, when combined, there may be some commonality among the functions being accomplished. In other words, in addressing an overall SOS configuration requirement, which includes the integration of two or more systems within, there may be a sharing individual functional entities such as shown in Figure 2.22. In this situation, there a common function that is both an integral part of system ABC and system DEF, and another such common function that supports both system DEF and system GHI.

In addressing the functional interfaces among systems, a good objective is to combine and share "common functions" wherever possible. Referring to Figure 2.18, for each possible application where a common function might be included to support two or more systems, the *inputs, outputs, controls/constraints,* and *mechanisms* must be identified and evaluated for each system requirement. Although the inputs and outputs must be consistent for all feasible applications, trade-offs can be accomplished with the objective of possibly reducing the overall supporting resource requirements (i.e., the mechanisms).

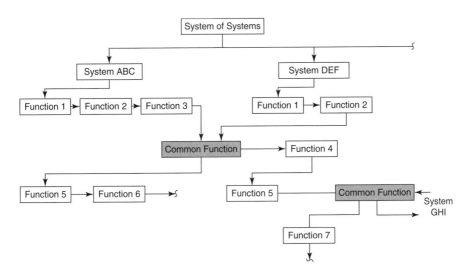

FIGURE 2.22 Functional interfaces in a system-of-systems (SOS) configuration.

However, extreme care must be exercised to ensure that none of the systems involved are in any way compromised. For example, as part of the design process, one should address the following questions:

1. Have all of the inputs, outputs, controls/constraints, and mechanisms for each "functional" system application been fully identified (refer to Figure 2.18)?
2. For each "common function" application where two or more systems are involved, have the requirements for *all* of the attached systems, when combined in a SOS configuration, been fully met?
3. For each of the attached systems in the SOS infrastructure, have all of the applicable requirements pertaining to reliability, maintainability, human factors and safety, life-cycle cost, etc. been fully met (refer to Section 1.4.2)? There should be no resulting degradation to any of these basic design requirements for any of the systems in a given SOS configuration.

The response to each question should, of course, be *yes!* Although addressing these issues may appear to be rather basic and obvious, it is often relatively easy to just go ahead and combine system functions, for the sake of promoting interoperability, without first really evaluating all of the interactions and feedback effects that may occur in the process.

2.8   REQUIREMENTS ALLOCATION

Having described the basic architecture for the system overall, the discussion continues with the requirements analysis process by defining the specific *input* design criteria for the various subsystems and major lower-level components of the system; i.e., specific qualitative and quantitative requirements to which the various elements of the system must be designed. With the top-level requirements already defined (refer to Figure 1.12, blocks 0.1/0.2), it is now necessary to identify and develop the specific "design-to" requirements for critical items of equipment, major software modules, applicable facilities, personnel, significant elements of support, and so on, that have been identified through the functional analysis. The requirements for the system must be *allocated* (or *apportioned)* down to its various major components, as appropriate. Conversely, the composite of the requirements for these components, when combined, must support the higher-level initially specified requirements for the overall system.

Basically, this is a top-down distribution process, which is somewhat iterative initially and often evolving from the results from trade-offs conducted horizontally across the spectrum of system components. The ultimate objective is, of course, to be able to define specific qualitative and quantitative *design requirements* for each significant element of the system, and to include such requirements in the appropriate specification for use in the procurement and acquisition process. In defining such

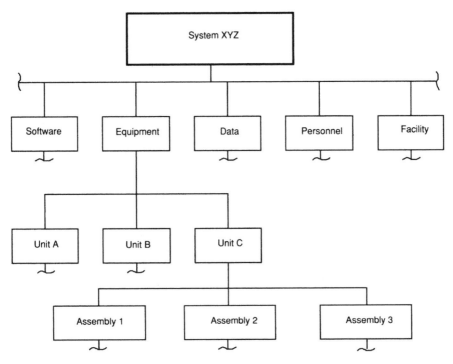

FIGURE 2.23   Hierarchy of system components.

requirements, one needs to address the anticipated *performance* and *effectiveness* goals for the item in question.[16]

2.8.1   Functional Packaging and Partitioning

Given a top-level description of the system, the next step is to break the system down into its components by *partitioning*.[17] This involves a breakdown of the system into subsystems and lower-level elements such as illustrated in Figure 2.23. Such elements are initially identified through the analysis and evaluation of each function on an individual-by-individual basis (see Figure 2.18). The challenge subsequently is

---

[16] As conveyed in Chapter 1 (Figure 1.4), there is a great deal of outsourcing taking place, and there appears to be considerable growth in the number of suppliers involved in a typical large-scale project. Whenever a new supplier is selected, a "specification" is prepared, which constitutes a critical part of the data package for the purposes of procurement and subcontracting. Because many of the major elements of a system are now being subcontracted, it is essential that a complete and well-defined set of *requirements* be included in each applicable specification. Further, there must be a top-down/bottom-up *traceability* of requirements throughout the hierarchy of specifications for a specific syssem. Specifications are discussed further in subsequent chapters.

[17] The concepts associated with system *architecture* and *partitioning* are discussed further in M. Maier and E. Rechtin, *The Art of System Architecting,* 2nd ed., (Boca Raton, FL: CRC Press, 2000).

to identify and group closely-related functions into packages, employing a common set of resources (e.g., equipment, software, facilities) to accomplish multiple purposes to the extent possible. Although it may be relatively easy to identify individual functional requirements and associated resources on an independent basis, the results may be quite costly when it comes to system packaging, weight, size, cost, and so on. The basic questions at this point are as follows:

- What hardware or software can be selected that will perform multiple functions?
- How can new functional capabilities be added in the future without adding any new physical elements to the system structure (i.e., growth potential)?
- Can any physical resources (e.g., equipment, software, facilities, people) be deleted without losing any of the required functional capabilities previously defined?

The partitioning of a system into its elements is evolutionary in nature. Common functions may be grouped or combined in such a way as to provide a system packaging scheme, to meet three objectives:

1. System elements may be grouped by geographical location, by nationality, by a common environment, or by similar types of equipment and/or software.
2. Individual system packages should be as independent as possible with a minimum of "interaction effects" in relation to other packages. A design objective is to be able to remove and replace a given package without having to remove and replace other packages in the process, or requiring an extensive amount of alignment and adjustment as a result.
3. In breaking down the system into subsystems, a configuration should be selected in which the communications between the subsystems is minimized. In other words, whereas the *internal* complexity in design may be high, the *external* complexity should be low. Breaking the system down into packages in which there are high rates of information exchange between these packages should be avoided.

An overall objective is to break the system down into elements so that only a very few critical events can influence or change the inner workings of the various packages that make up the overall system architecture. Accomplishing this objective should also facilitate the process of introducing new technology changes into the system for upgrading purposes and for the accomplishment of any system maintenance that may be required throughout the life cycle.[18]

Although the results of "partitioning" may constitute what is presented in Figure 2.23, the process for accomplishing it is better illustrated in Figure 2.24. System functions are identified, broken down into subfunctions, and grouped into

---

[18]The open-architecture approach to design is highly dependent on the functional packaging of system components and in meeting these objectives as stated.

three equipment units: Unit A, Unit B, and Unit C. The design should be such that any one of the three units can be removed and replaced without impacting the other units. In other words, there should be a minimum of interaction effects between the three units.[19]

2.8.2 Allocation of System-Level Requirements to the Subsystem Level and Below

With the identification of system elements, the next step is to *allocate* or *apportion* the requirements specified for the system down to the level desired to provide a meaningful *input* to design. This involves a top-down distribution of the quantitative and qualitative criteria developed through the QFD analysis described in Section 2.6. From the prioritized technical performance measures (TPMs), such as those identified in Figure 2.11, the designer needs to select and specify specific "design-to" requirements for each of the major elements of the system. For example, referring to Figure 2.24, what should be specified for each Unit A, Unit B, and Unit C in order to meet the system-level requirements in Figure 2.11?

The challenge is to first assign the appropriate factors at the unit level, considering complexity and utilizing historical experience and field data where available, prorating from the top down. Then, synthesize these factors at the unit level and determine whether they are realistic and that, when combined, they will support the requirements for the system. There may be times when a given requirement for one of the units will be too stringent, considering the available technology and possible sources of supply. In such cases, the specific design-to criteria for the unit may be changed (less restrictive), which, in turn, will require a tightening of a requirement for one or more of the other units. In other words, there may be both a top-down and a horizontal process in which trade-off studies are accomplished in arriving at a final recommended solution. There may be several iterations of this process before the specific requirements for the applicable major system elements are defined.

Figure 2.25 shows the results of an allocation (in this instance to four units). Utilizing an "objective-tree" approach (illustrated in Figure 2.8), the designer established the appropriate metrics for the system, then the metrics at the next lower level, and so on. There should be a *traceability* of requirements from the top down and, although the measures vary somewhat at each level, those identified at the lower levels must directly support the requirements for the overall system. Further, the depth to which requirements are specified is somewhat dependent on the priorities (i.e., the importance factors) identified in Figure 2.11. On the one hand, if there is a highly critical requirement from the perspective of the consumer, the allocation may be accomplished to the assembly level in Figure 2.25. On the other hand, if the allocation is accomplished unnecessarily to a very detailed level, the designer may be unduly

---

[19]Attaining this objective is critical, particularly in view of today's trends pertaining to the increasing utilization of commercial off-the-shelf (COTS) items and extensive amount of outsourcing in the purchase and acquisition of major subsystems and large-scale components. The question is, Can we acquire and integrate a variety of COTS items, with a minimum of interaction effects among these items, and without destroying the overall system configuration (architecture) in the process?

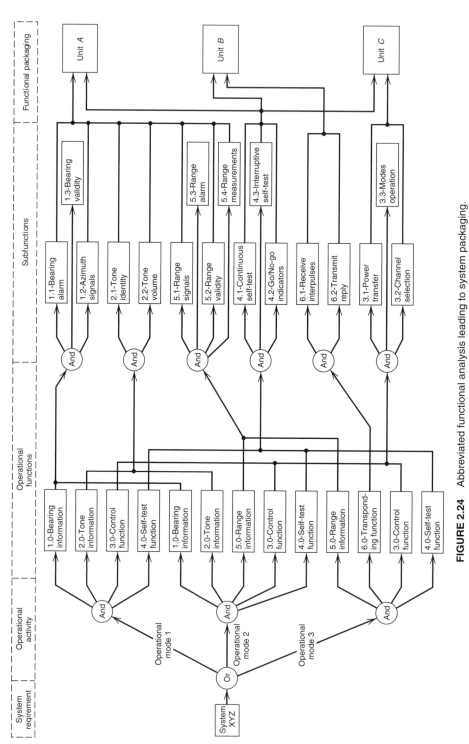

FIGURE 2.24 Abbreviated functional analysis leading to system packaging.

92

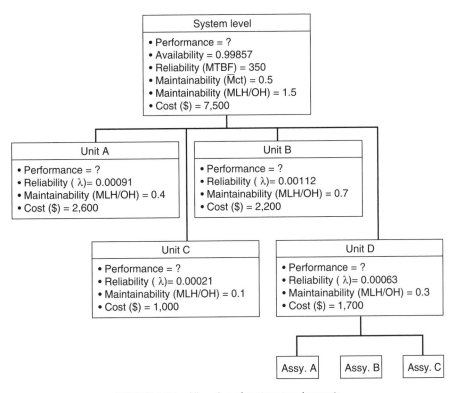

FIGURE 2.25   Allocation of system requirements.

constrained relative to what can be accomplished through the trade-off analysis and evaluation process.

The allocation process constitutes a top-down specification of design requirements to the depth necessary to provide *input criteria* for the appropriate system elements. Highly complex new designs will require a greater degree of coverage than would be necessary in utilizing commercial off-the-shelf (COTS) items. The results of the allocation process should be incorporated in the appropriate "Specification" identified in Figure 1.12. If the requirements are not properly specified from the top down, the results can be costly in terms of possible *overdesign, underdesign,* or both. The risks may be high if the requirements are not addressed from the beginning.[20]

2.8.3 Traceability of Requirements (Top-Down/Bottom-Up)

In the system engineering process and evolution of requirements illustrated in Figure 1.12, there are a series of specifications developed to cover various design-to

---

[20]The allocation of reliability, maintainability, human, economic, and related factors is discussed further in Chapter 3.

requirements, starting with the system level and including its various components. In Figure 1.12, a generic classification has been identified, commencing with the system specification—type A, the top-level specification, and including various lower-level specifications (types B, C, D, and E) covering new developments, the procurement of off-the-shelf products, processes, and materials. These generic categories of specifications are described in detail in Chapter 3 (Section 3.2); however, the important issue in this section is to ensure that (1) the proper *requirements* that have been defined for the system are included in the *system specification*, (2) the requirements for the various elements or components of the system that have been developed through the allocation process are included in the appropriate lower-level specification (i.e., B, C, D, or E specification as applicable), and (3) that there is a complete traceability of requirements from the system specification and on down. In Figure 2.26, which illustrates a partial "specification tree" for a typical project, such requirements must evolve from the top down and, at the same time, the combined requirements included in the lower-level specifications must support the requirements for the system as stated in the system specification. In other words, the *requirements* for the system and its components must be properly reflected through a good set of specifications. This is particularly important in view of the "outsourcing" and the large number of suppliers (from all over the world) that are likely to be responsible for the development and production of various system components.

2.8.4 Allocation of Requirements in a SOS Configuration

Referring to Section 2.7.5 (and Figure 2.22), the sharing of *common functions* between systems in a SOS configuration can provide some excellent benefits from the perspective of reducing (or minimizing) some of the resource requirements across the board. However, there are two definite imposed constraints in accomplishing the initial allocation of design requirements for new systems:

1. In the allocation of qualitative and quantitative "design-to" requirements (and in developing associated design criteria) for two or more systems within a totally new SOS configuration, the allocation process, along with the required trade-offs, must be accomplished across the entire spectrum of new systems. If, in Figure 2.22, both systems ABC and DEF are newly developed, then the allocation process must be accomplished for the applicable subsystems and major components in both systems on an integrated basis. In accomplishing such, care must be taken to assure that the basic functional requirements for each system are not compromised in the process.

2. In the allocation of qualitative and quantitative "design-to" requirements for a new system that is, in the interest of fulfilling an interoperability objective, being included within a given SOS configuration, then those already-existing "common functions" that are being shared with the new system will have a definite impact on the allocation process for the new system. If, in Figure 2.22, system ABC already exists and a newly designed system DEF is being added, the allocation process for the new system will be somewhat constrained

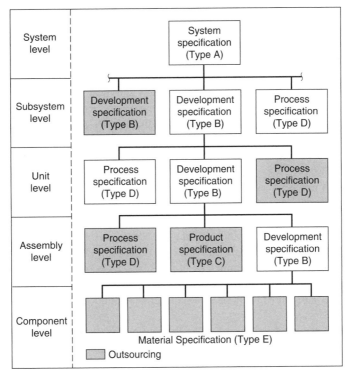

FIGURE 2.26   Specification tree (partial).

because the design requirements for the common function will be already be "fixed" with no variation allowed. Again, care must be taken to ensure that the functional requirements for the new system are not being compromised in any way.[21]

In any event, in accomplishing the system allocation process for any new system requirements, all of the various interfaces must be properly addressed, and there may be some new challenges when operating in a SOS configuration environment.

2.9 SYSTEM SYNTHESIS, ANALYSIS, AND DESIGN OPTIMIZATION

Synthesis refers to the combining and structuring of components in such a way as to represent a feasible system configuration. The requirements for a system have been

---

[21]There have been instances where the addition of a new system in a higher-level SOS configuration (to meet interoperability objectives) has had highly negative results, with the reliability of the new system being significantly reduced because of the poor reliability of a "common function" in an already existing system. Again, care must be taken to ensure that the requirements of a new system being added to an existing SOS structure will not be compromised due to the already existing capability.

established, some preliminary trade-off studies have been completed, and a baseline configuration must be developed to demonstrate the concepts discussed earlier. Synthesis is *design*. Initially, synthesis is employed to develop preliminary concepts and to establish basic relationships among the various components of the system. Later, when sufficient functional definition and decomposition have occurred, synthesis is used to further define the "hows" in response to the "what" requirements. Synthesis involves the selection of a configuration that can be representative of the form the system will ultimately take, although a final configuration is certainly not to be assumed at this point.[22]

The synthesis process usually leads to the definition of several possible alternative design approaches, which will be the subject of further analysis, evaluation, refinement, and optimization. As these alternatives are initially structured, it is essential that the appropriate technical performance parameters and associated measures be properly aligned to the applicable components of the system. For instance, technical performance parameters may include factors such as weight, size, speed, capacity, accuracy, security, volume, range, processing time, reliability, maintainability, and others as applicable. These parameters, or measures, must be prioritized and aligned to the appropriate elements of the system (e.g., equipment, unit or assembly, item of software, etc., as conveyed through the requirements allocation process described in Section 2.8).

In defining the initial requirements for the system, the technical performance measures (TPMs) are established based on their relationship and criticality to the accomplishment of the planned system mission; that is, the impact that a given factor has on cost-effectiveness, system effectiveness, and/or performance. These applicable TPMs are prioritized, and their relationships are presented in the form of design considerations, which, in turn, may be shown in the form of a hierarchical tree, as illustrated in Figure 2.27. The ranking of TPMs (and supporting design considerations), which will be built into the program management and review structure, will likely vary from one system to the next. A top-level measure for one system may be "reliability," whereas "availability" may be of greater importance in another example. In any event, the appropriate measures must be established, prioritized, and included in the specifications accordingly. As the design progresses, these measures will be used for the purposes of analysis and evaluation.[23]

---

[22] According to Sage and Armstrong, "synthesis" is the "step which involves searching tor, or hypothesizing, a set of alternative courses of action or options. Each alternative must be described in sufficient detail to permit analysis of the impacts of implementation and subsequent evaluation and interpretation with respect to the objectives. As part of this step, we identify a number of potential alternatives and associated alternatives measures." A. P. Sage and J. E. Armstrong, *Introduction to Systems Engineering* (New York: John Wiley & Sons. Inc., 2000), p. 55.

[23] It is important to emphasize the *process* whereby requirements are defined and prioritized as a result of a QFD analysis, allocated to the appropriate elements of the system, and subsequently addressed throughout the design in proportion to their respective level(s) of importance. In addition, these high-priority design factors must be inherent within and receive the appropriate attention through the applicable program management and review structure.

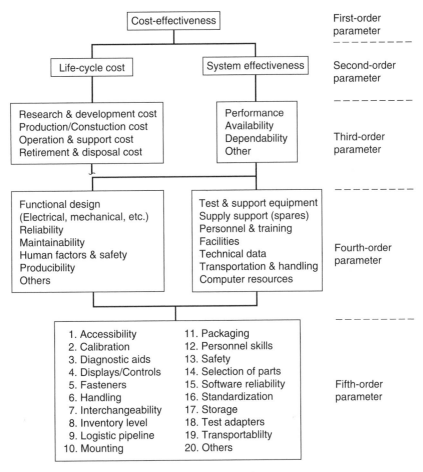

FIGURE 2.27   Order of evaluation parameters.

Given a number of alternatives, the evaluation procedure progresses through the general steps illustrated in Figure 2.28 and described as follows:

1. *Define analysis goals.* An initial step requires the clarification of objectives, the identification of possible alternative solutions to the problem at hand, and a description of the analysis approach to be employed. Relative to alternatives, all possible candidates must be initially considered; however, the more alternatives considered, the more complex the analysis process becomes. Thus, it is desirable to first list *all* possible candidates to ensure against inadvertent omissions, and then eliminate those candidates that are clearly unattractive, leaving only a few for evaluation. Those few candidates are then evaluated with the intent of selecting a preferred approach.

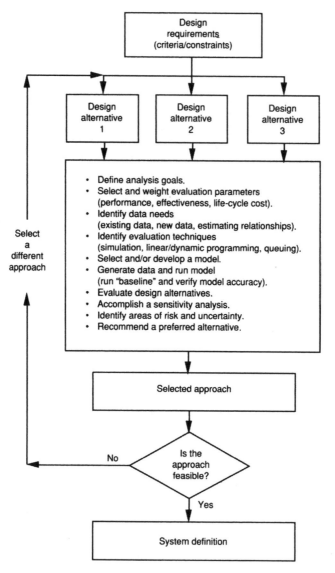

FIGURE 2.28   Evaluation of alternatives.

2. *Select and weight evaluation parameters.* The criteria used in the evaluation process may vary considerably, depending on the stated problem, the system being evaluated, and the depth and complexity of the analysis. In Figure 2.27, the parameters of primary significance include cost, effectiveness, performance, availability, and so on. At the detail level, the order of parameters

will be different. In any event, parameters are selected, weighted in terms of priority of importance, and tailored to the system being addressed.

3. *Identify data needs.* In evaluating a particular system configuration, it is necessary to consider operational requirements, the maintenance concept, major design features, production and/or construction plans, and anticipated system utilization and product support requirements. Fulfilling this need requires a variety of data, the scope of which depends on the type of evaluation being performed and the program phase during which the evaluation is accomplished. In the early stages of system development, available data are limited; thus, the analyst must depend on the use of various estimating relationships, projections based on past experience covering similar system configurations, and intuition. As the system development progresses, improved data are available (through analyses and predictions) and are used as an input to the evaluation effort. At this point it is important to initially determine the specific needs for data (i.e., type, quantity, and the time of need) and to identify possible data sources. The nature and validity of the data input for a given analysis can have a significant impact on the risks associated with the decisions made based on the analysis results. Thus, one needs to accurately assess the situation as early as practicable.

4. *Identify evaluation techniques.* Given a specific problem, it is necessary to determine the analytical approach to be used and the techniques that can be applied to facilitate the problem-solving process. Techniques may include the use of Monte Carlo simulation in the prediction of random events downstream in the life cycle, the use of linear programming in determining transportation resource requirements, the use of queuing theory in determining production and/or maintenance shop requirements, the use of networking in establishing distribution needs, the use of accounting methods for life-cycle costing purposes, and so on. Assessing the problem itself and identifying the available tools that can possibly be used in attacking the problem are necessary prerequisites to the selection of a model.

5. *Select and/or develop a model.* The next step requires the combining of various analytical techniques into the form of a model, or a series of models, as illustrated in Figure 2.29. A model, as a tool used in problem solving, aids in the development of a simplified representation of the real world as it applies to the problem being solved. The model should (a) represent the dynamics of the system configuration being evaluated, (b) highlight those factors that are most relevant to the problem at hand, (c) be comprehensive by including *all* relevant factors and be reliable in terms of repeatability of results, (d) be simple enough in structure to enable its timely implementation in problem solving, (e) be designed so that the analyst can evaluate the applicable system configuration as an entity, analyze different components of the system on an individual basis, and then integrate the results into the whole, and (f) be designed to incorporate provisions for easy modification and/or expansion to permit the evaluation of

additional factors as required. An important objective is to select and/or develop a tool that will help to evaluate the *overall* system configuration, as well as the *interrelations* of its various components. Models (and their applications) are discussed further in Chapter 4.[24]

6. *Generate data and model application.* With the identification of analytical techniques and the model selection task accomplished, the next step is to verify or test the model to ensure that it is responsive to the analysis requirement. Does the model meet the stated objectives? Is it sensitive to the major parameters of the system configuration(s) being evaluated? Evaluation of the model can be accomplished through the selection of a *known* system entity and the subsequent comparison of analysis results with historical experience. Input parameters may be varied to ensure that the model design characteristics are sensitive to these variations and will ultimately reflect an accurate output as a result.

7. *Evaluate design alternatives.* Each of the alternatives being considered is then evaluated using the techniques and the model selected. The required data are collected from various sources, such as existing data banks, predictions based on current design data, and/or gross projections using analogous and parametric estimating relationships. The required data, which may be taken from a wide variety of sources, must be applied in a consistent manner. The results are then evaluated in terms of the initially specified requirements for the system. Feasible alternatives are considered further. Figure 2.30 illustrates some considerations where possible feasible solutions fall within the desired shaded areas.

8. *Accomplish a sensitivity analysis.* In the performance of an analysis, there may be a few key system parameters about which the analyst is uncertain because of inadequate data input, poor prediction procedures, "pushing" the state of the art, and so on. There are several questions that must be addressed: How sensitive are the results of the analysis to possible variations of these uncertain input parameters? To what extent can certain input parameters be varied before the choice of alternatives shifts away from the initially selected approach? Experience shows that there are certain key input parameters in a life-cycle cost analysis, such as the reliability MTBF and the maintainability (mean corrective maintenance time, $\overline{M}ct$) that are considered to be critical in determining system maintenance and support costs. With good historical field data being very limited, there is a great deal of dependence on current prediction and estimating methods. Thus, with the objective of minimizing the risks associated with making an incorrect decision, the analyst may wish to vary the input MTBF

---

[24]There are many types, of models, including physical models, symbolic models, abstract models, mathematical models, and so on. *Model* as defined herein, refers primarily to a mathematical (or analytical) model. The development and application of various analytical methods are covered further in most texts on operations research. Two good references are (I) F. S. Hillier and G. J. Lieberman, *Introduction to Operations Research*, 6th ed. (New York: McGraw-Hill, 1995); and (2) H. A. Taha, *Operations Research: An Introduction*, 8th ed. (Upper Saddle River, NJ: Prentice-Hall, 2006).

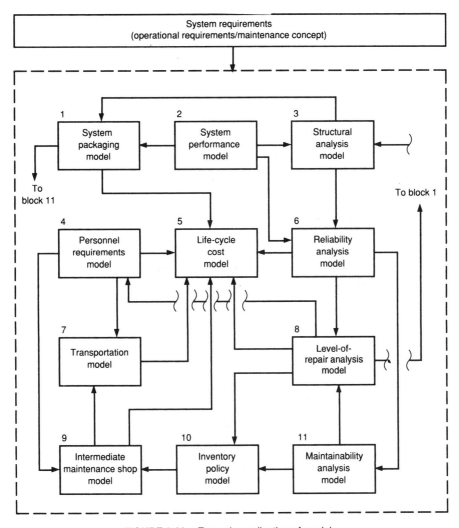

FIGURE 2.29 Example application of models.

and $\overline{M}ct$ factors over a designated range of values (or a distribution) to see what impact this variation has on the output results. Does a relatively *small* variation of an input factor have a *large* impact on the results of the analysis? If so, then these parameters may be classified as being critical TPMs in the overall design review and evaluation process, monitored closely as design progresses, and an additional effort may be generated to modify the design for improvement and to improve the reliability and maintainability prediction methods. In essence, a sensitivity analysis is directed toward determining the relationships between design decisions and output results.

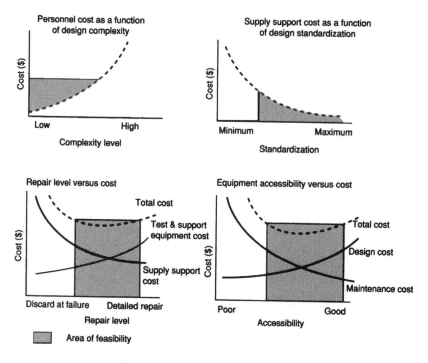

FIGURE 2.30 Example of evaluation results.

9. *Identify risk and uncertainty. The* process of design evaluation leads to decisions having a significant impact on the future. The selection of evaluation criteria, the weighting of factors, the selection of the life cycle, the use of certain data sources and prediction methods, and the assumptions made in interpreting analysis results will obviously influence these decisions. Inherent within this process are the aspects of "risk" and "uncertainty," because the future is, of course, unknown. Although these terms are often used jointly, *risk* actually implies the availability of discrete data in the form of a probability distribution around a certain parameter. *Uncertainty* implies a situation that may be probabilistic in nature, but one that is not supported by discrete data. Certain factors may be measurable in terms of risk or may be stated under conditions of uncertainty. The aspects of risk and uncertainty, as they apply to the system design and development process, must be integrated into the program risk management plan described in Chapter 6.

10. *Recommend preferred approach.* The final step in the evaluation process is the recommendation of a preferred alternative. The results of the analysis should be fully documented and made available to all applicable project design personnel. A statement of assumptions, a description of the evaluation procedure that was followed, a description of the various alternatives that were considered, and an identification of potential areas of risk and uncertainty should be included in this analysis report.

In Figure 1.12, the requirements for the system are established in conceptual design, functional analysis and the allocation of requirements are accomplished either late in conceptual design or at the start of preliminary design, and detail design is accomplished on a progressive basis from thereon. Throughout this overall series of steps, there is an ongoing effort involving synthesis, analysis, and design optimization. In the early stages of design, tradeoff studies may entail the evaluation of alternative operational profiles, technology applications, distribution schemes, or maintenance concepts. During early preliminary design, alternative methods for accomplishing a given function or alternative equipment packaging schemes may be the focus of the analysis. In detail design, the problems will be at a lower level in the overall hierarchical structure of the system.

In any event, the process illustrated in Figure 2.28 (and described herein) is applicable throughout the system design and development effort?[25] The only difference lies in the depth of analysis, the nature and type of data required, and the model used in accomplishing the analysis. For instance, one can perform a life-cycle cost analysis early in conceptual design, later in detail design, and as part of a system evaluation effort during the operational use phase. The same is true in accomplishing an FMECA, level-of-repair analysis, and so on. The process is the same in any case; however, the depth of analysis and the data requirements are different. The synthesis, analysis, and design optimization process must be tailored to the problem at hand. Too little effort will result in greater risks associated with decision making in design, and too much analysis effort may be expensive.[26]

2.10 DESIGN INTEGRATION

Accomplishing the overall design of a system, or a group of systems in a SOS configuration, requires an integrated *team* approach, commencing during the early stages of conceptual design and extending through detail system design and development, production and/or construction, system operation and sustaining support, and ultimate system retirement and the recycling/disposal materials. As the requirements for a new system are established, the *design team* is formed, performing system-level design functions, as illustrated in Figure 1.12. This includes accomplishing a needs analysis, feasibility analysis, operational requirements and the maintenance concept, and a preliminary definition of system architecture. At this stage, the design team may include only a small number of selected qualified indlviduals, with the objective of preparing a comprehensive system specification (type A) as an output. It is

---

[25] Although the emphasis here is primarily on the design and development of *new* systems, this process is equally applicable later in the life cycle in accomplishing system *validation* and/or *assessment* and in evaluating alternative ways in which a system can be improved/upgraded through modification.

[26] See Figure 1.14; one of the objectives in system engineering is to initiate and provide continuity in the application of various analytical methods/lools/models (i.e.. an integrated tool set) as one progresses from the system level of definition and on down to the development of the various system elements. There needs to be some type of "flow" process in this area as well.

important that the personnel selected have the appropriate backgrounds, can address the "big picture" as an entity, can recognize the many high-level interfaces that may exist, and can effectively work together and communicate on a day-to-day basis. The assignment of a large number of individual domain specialists, whose expertise lies in given technical fields, is not appropriate at this early stage. The organization of design teams is discussed further in Chapter 7.

As system development progresses, the appropriate design specialists are added to the team as required. The objective, from a system engineering perspective, is to ensure that the right specialists are available at the time required, and that their individual contributions are properly integrated into the whole. The selection of domain specialists is highly dependent on the specific requirements developed through the functional analysis and allocation process (described in Sections 2.7 and 2.8). As the input criteria for design will vary with the system and its mission, the emphasis in assigning those with the proper level of expertise to the team will be different from one project to the next. Additionally, the overall team make-up may be influenced by the activities (and assigned personnel) associated with the development of other systems within the same SOS configuration, In any event, Figure 2.31 identifies some of the considerations that must be addressed through the overall design integration effort.

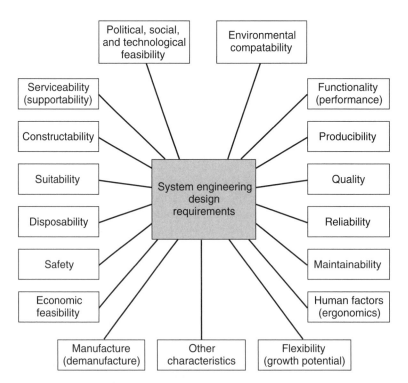

FIGURE 2.31   The integration of design requirements.

During the latter phases of the system life cycle (i.e., production/construction, system operation and support), the role of system engineering continues, but in the form of evaluation/validation and the introduction and processing of design changes as necessary. The requirement(s) for *change* may stem from some identified deficiency (i.e., failure to meet an initially specified requirement), or may be for the purposes of *continuous product/process improvement*. Further, changes in any one system may be required as a result of changes being introduced in other systems in a given SOS configuration, In any case, each proposed engineering change (i.e., "Engineering Change Proposal/ECP") must be evaluated, not just in terms of performance issues alone, but in terms of reliability, maintainability, supportability or serviceability, producibility, disposability, and life-cycle cost as well. The design change and modification process is described in Chapter 5.

Inherent in the established design team activity is, of course, the requirement for good communications on a day-to-day basis. Although the co-location of personnel in one geographical area (and the "eyeball-to-eyeball" contact) is preferred, the trends toward more *outsourcing* and *decentralization* usually result in the introduction of many different suppliers throughout the world.[27] Further, quite often there are design activities for the same system being conducted at remote locations and accomplished concurrently. Thus, the design team efforts are heavily dependent on the implementation and utilizations of various computer-aided tools, operating in a network such as illustrated in Figure 2.32.[28]

Successful implementation of the integrated computer-based network shown in Figure 2.32 is highly dependent on the structure of the design database and the language being utilized in the development and transfer of the design data. Such a database may include design drawings and layouts, the presentation of three-dimensional visual models, parts and materials lists, predictions and analysis reports, supplier data, interface data with other SOS systems, and whatever else that is necessary to describe the system configuration as designed. The designer must be able to gain access to the database and provide input easily, and the results must be transmitted to other members of the design team accurately and in a timely manner. The data, usually presented in a digital format, must be *standardized* across the applicable project and available to all members of the design team concurrently. Instead of many different data items flowing back and forth between different members (or organizations) of the design team, between the developer and various suppliers, between the producer and the customer, and so on, an integrated shared database structure is necessary, as presented conceptually in Figure 2.33. This, of course, should facilitate the

---

[27]The term *outsourcing* refers to the practice of soliciting the support of product suppliers to accomplish selected packages of work externally from the producer or prime contractor. Experience indicates that there is a greater use of external suppliers today than in the past. This, in turn, provides some additional challenges relative to maintaining the proper level of communications across and throughout the project organization.

[28]Included in this network is the proper mix of computer-aided design (CAD), computer-aided manufacturing (CAM), computer-integrated manufacturing (CIM), computer-aided support (CAS), and electronic commerce (EC) related tools utilized to varying degrees tn accomplish design, production, and system support activities.

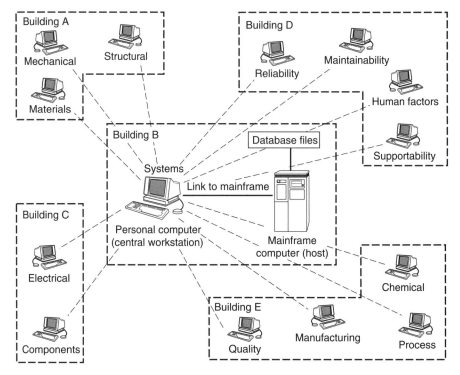

FIGURE 2.32 Design communication network.

process of communications, with every member of the design team having access to the same system configuration description.[29]

2.11 SYSTEM TEST AND EVALUATION

As the system design and development activity progresses, there needs to be an on-going measurement and evaluation (or validation) effort, as indicated in Figure 1.27. Realistically, a complete evaluation of the system, in terms of meeting the initially specified consumer requirements, cannot be accomplished until the system is produced and functioning in an operational environment. However, if problems occur and system modifications are necessary, the accomplishment of such an evaluation

---

[29]With the advent of new electronic commerce (EC) methods, electronic data integration (EDI) processes, and related integrated technology (IT) approaches on an almost continuing basis, it is anticipated that the nature of the *data environment* will be changing almost constantly. The objective here is to emphasize the need for good communications through the application, integration, and effective transfer of design data among the various members of the design team, supporting organizations, and management. The illustrations presented in Figures 2.32 and 2.33 are to be viewed conceptually and are not intended to reflect any particular technological approach.

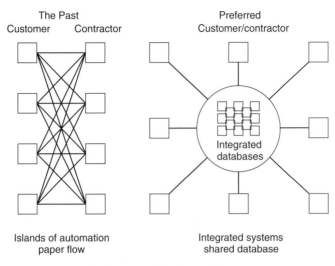

FIGURE 2.33   The data environment.

so far downstream in the life cycle may turn out to be quite costly. In essence, the earlier problems are detected and corrected, the better off the designer is in terms of both incorporating the required changes and the associated costs of modification.

In addressing the subject of evaluation, the objective is to acquire a high degree of confidence, as early in the life cycle as possible, that the system will ultimately perform as intended. Acquiring this confidence, through the accomplishment of laboratory and field testing involving a physical replica of the system (and/or its components), can be quite expensive. The resources required for testing are often quite extensive, and the necessary facilities, test equipment, personnel, and so on, may be difficult to schedule. Yet we know that a certain amount of formal testing is required in order to properly verify that system requirements have been met.

However, with a more comprehensive analysis effort and the use of prototyping, it may be possible to verify certain design concepts during the early stages of preliminary and detail design. With the advent of three-dimensional databases and the application of simulation techniques, the designer can now accomplish a great deal relative to the evaluation of system layouts, component relationships and interferences, human–machine interfaces, and so on. There are many functions that can now be accomplished with computerized simulation that formerly required a physical mock-up of the system, a preproduction prototype model, or both. The availability of computer-aided design (CAD), computer-aided manufacturing (CAM), computer-aided support (CAS) methods, and related technologies has made it possible to accomplish much in the area of system evaluation relatively early in the system life cycle, when the incorporation of changes can be accomplished with minimum cost.

In determining the need for test and evaluation, one commences with the initial specification of system requirements in conceptual design. As specific technical performance measures (TPMs) are established, it is necessary to determine the methods by which compliance with these factors will be verified.

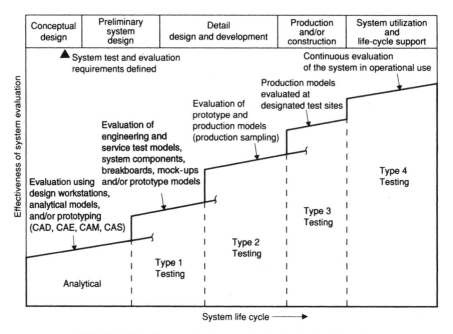

FIGURE 2.34 Stages of system evaluation during the life cycle.

How will these TPMs be measured, and what resources will be necessary to accomplish this? Responses to this question may entail using simulation and related analytical methods, using an engineering model for test and evaluation purposes, testing a production model, evaluating an operational configuration in the consumer's environment, or a combination of these. In essence, one needs to review the requirements for the system, determine the methods that can be used in the evaluation effort and the anticipated effectiveness of these methods, and develop a comprehensive plan for an overall integrated test and evaluation effort (i.e., Test and Evaluation Master Plan; refer to Figure 1.26). Figure 2.34 illustrates suggested categories of testing as they may apply in system evaluation.[30]

2.11.1 Categories of Test and Evaluation

In Figure 2.34, the first category is "analytical," which pertains to certain design evaluations that can be conducted early in the system life cycle using computerized techniques such as CAD, CAM, CAS, simulation, rapid prototyping, and related approaches. With the availability of a wide variety of models, three-dimensional databases, and so on, the design engineer is now able to simulate human–equipment interfaces, equipment packaging schemes, the hierarchical structures of systems, and

---

[30]The categories of test and evaluation may vary by type of system and/or by functional organization. These categories have been selected as a point of reference for discussion throughout this text.

activity/task sequences. In addition, through the utilization of these technologies, the design engineer is able to do a better job of predicting, forecasting, and accomplishing sensitivity/contingency analyses with the objective of reducing future risks. In other words, a great deal can be now accomplished in system evaluation that, in the past, could not be realized until equipment became available in the latter phases of detail design and development.

Type 1 testing refers primarily to the evaluation of system components in the laboratory using engineering breadboards, bench test models, service test models, rapid prototyping, and the like. These tests are designed primarily with the intent of verifying certain performance and physical characteristics and are developmental by nature. The test models used operate functionally, but do not by any means represent production equipment or software. Such testing is usually performed in the producer/supplier's laboratory facility by engineering technicians using "jury-rigged" test fixtures and engineering notes for procedures. It is during this initial phase of testing that design concepts and technology applications are validated and changes can be initiated on a minimum-cost basis.

Type 2 testing includes formal tests and demonstrations accomplished during the latter stages of the detail design and development phase when preproduction prototype equipment and software are available. Prototype equipment is similar to production equipment (that which will be delivered for operational use), but is not necessarily fully qualified at this point. A test program in this area may constitute a series of individual tests, tailored to the need, including the following:[31]

- *Environmental qualification.* Temperature cycling, shock and vibration, humidity, sand and dust, salt spray, acoustic noise, explosion-proofing, and electromagnetic interference.
- *Reliability qualification.* Sequential testing, life testing, environmental stress screening (ESS), and test, analyze, and fix (TAAF).
- *Maintainability demonstration.* Verification of maintenance tasks, task times and sequences, maintenance personnel quantities and skill levels, degree of testability and diagnostic provisions, prime equipment–test equipment interfaces, maintenance procedures, and maintenance facilities.
- *Support equipment compatibility.* Verification of the compatibility among the prime equipment, test and support equipment, and ground handling equipment.
- *Technical data verification.* The verification (and validation) of operating procedures, maintenance procedures, and supporting data.
- *Personnel test and evaluation.* Verification to ensure compatibility between the human and equipment, the personnel quantities and skill levels required, and training needs.

---

[31]"Qualified" equipment refers to the production configuration that has been verified through the *successful completion* of environmental qualification tests (e.g.. temperature cycling, shock and vibration), reliability qualification, maintainability demonstration, and supportability compatibility tests. Type 2 testing primarily refers to that activity associated with the qualification of a system.

- *Software compatibility.* Verification that software meets the system requirements, that there is compatibility between software and hardware, and that the
appropriate quality provisions have been incorporated. This includes computer
software unit (CSU) and computer software configuration item (CSC1) testing,
as reflected in Figure 1.13.

Another facet of testing in this category is production sampling tests, used when
multiple quantities of an item are being produced. Although the system (and its components) may have successfully passed the initial qualification tests, there must be
some assurance that the *same* level of quality has been maintained throughout the
production process. The process is usually dynamic by nature, conditions change,
and there is no guarantee that the characteristics that have been built into the design
will be retained throughout production. Thus, sample systems/components may be
selected (based on a percentage of the total produced), and qualification tests may be
conducted on a recurring basis. The results are measured and evaluated in terms of
whether improvement or degradation has occurred.

Type 3 testing includes the completion of formal tests at designated field test
sites by user personnel over an extended period of time. These tests are usually
conducted after initial system qualification and prior to the completion of the pro-
duction/construction phase. Operating personnel, operational test and support equipment, operational spares, applicable computer software, and validated operating and
maintenance procedures are used. This is the first time that *all* elements of the system (i.e., prime equipment, software, and the elements of support) are operated and
evaluated on an integrated basis. Such testing also includes a formal evaluation of
all of the major interfaces (i.e., the common functions and their interfaces) among
the different systems and within a given SOS configuration. A series of simulated
operational exercises are usually conducted, and the system is evaluated in terms of
performance, effectiveness, compatibility between the prime mission-oriented segments of the system and the elements of support, and so on. Although type 3 testing
does not completely represent a fully operational situation, the tests can be designed
to provide a close approximation.

Type 4 testing, conducted during the system operational use and life-cycle support phase, includes formal tests that are sometimes conducted to acquire specific
information relative to some area of operation or support. The purpose is to gain further knowledge of the system and its interfaces in the user environment, or of user
operations in the field. It may be desirable to vary the mission profile or the system
utilization rate to determine the impact on total system effectiveness, or it may be
feasible to evaluate several alternative maintenance support policies to see whether
system operational availability can be improved. Type 4 testing is accomplished at
one or more user operational sites, in a realistic environment, by operator and maintenance personnel, and is supported through the normal maintenance and logistics
capability. This is actually the first time that we will really know the true capability
of the system.

Of particular interest in type 4 testing is not only an assessment of the new system in operational use, but its impact(s) on other closely related operating systems

already in the inventory and, in turn, their impact(s) on the new system. In other words, if a newly developed system is introduced and included as part of a larger SOS configuration, there may be some impact(s), both outward and inward. The following questions should be addressed: *Does the operation of the new system affect (in any way) the operation of other systems within a given SOS configuration? Conversely, does the operation of other systems in the SOS structure affect the operation of the newly introduced system?* Any degradation in performance, reliability, or safety must be noted.

2.11.2 Integrated Test Planning

Test planning starts in the conceptual design phase when system requirements are initially established. If a requirement is to be specified, there must be a way to evaluate and validate the system at a later point in time to ensure that the requirement has been met. Thus, considerations for test and evaluation are intuitive from the beginning.

In Figure 1.26, initial test planning is included in a Test and Evaluation Master Plan (TEMP), prepared in the conceptual design phase. The document includes the requirements for test and evaluation, the categories of test, the procedures for accomplishing testing, the resources required, and associated planning information (i.e., tasks, schedules, organizational responsibilities, and cost).[32]

One of the key objectives of this plan, and of particular significance for system engineering, is the *complete integration* of the various test requirements for the overall system. By referring to the content of type 2 testing (Section 2.11.1), individual requirements may be specified for environmental qualification, reliability qualification, maintainability demonstration, software functionality, and so on. These requirements, stemming from a series of 'stand-alone' specifications, may be overlapping in some instances, and conflicting in other cases. Further, not all system configurations should be subjected to the same test requirements. In situations where there are new design technology applications, more up-front evaluation may be desirable, and the requirements for type 1 testing may be different from those in a situation involving the use of well-known state-of-the-art design methods. In other words, in areas where the potential technical risks are high, a more extensive evaluation effort early in the system life cycle may be feasible.

In any event, the TEMP represents a significant input relative to meeting the objectives of system engineering. Not only must one understand the system requirements overall, but knowledge of the functional relationships among the various components of the system is necessary. In addition, those involved in test planning must be familiar with the objectives of each specific test requirement, such as reliability qualification, maintainability demonstration, and so on. A total integrated approach

---

[32]In the defense sector, a TEMP is required for most large programs and includes the planning and implementation of procedures for Development Test and Evaluation (DT&E) and Operational Test and Evaluation (OT&E). DT&E basically equates to the analytical, type 1, and type 2 testing described in Section 2.11.1, and OT&E is equivalent to type 3 and type 4 testing.

to test and evaluation is essential, particularly when considering the costs associated with testing activities.

2.11.3 Preparation for Test and Evaluation

Prior to the start of formal testing, an appropriate period of time is designated for the purposes of test preparation. During this time, the proper conditions must be established to ensure effective results. These conditions will vary, of course, depending on the category of testing being undertaken.

On the one hand, during the early phases of design and development, as analytical evaluations and Type 1 testing are accomplished, the extent of test preparation is minimal. On the other hand, the accomplishment of type 2 and type 3 testing, in which the conditions are designed to simulate realistic consumer operations to the maximum extent possible, will likely require a rather extensive preparation effort. To promote a realistic environment, eight factors must be addressed:

1. *Selection of test item.* The system (and its components) selected for test should represent the most up-to-date design or production configuration, incorporating all of the latest approved engineering changes.

2. *Selection of test site.* The system should be tested in an environment that will be characteristic for user operations; that is, arctic or tropics, flat or mountainous terrain, airborne or ground. The test site selected should simulate these conditions to the maximum extent possible.

3. *Testing procedures.* The fulfillment of test objectives usually involves the accomplishment of both operator and maintenance tasks, and the completion of these tasks should follow formal approved procedures (e.g., validated technical manuals). The recommended task sequences must be followed to ensure proper system operation.

4. *Test personnel.* This includes (a) the individuals who will actually operate and maintain the system throughout the test and (b) the supporting engineers, technicians, data recorders, analysts, and administrators who provide assistance in conducting the overall test program. Personnel selected for the first category should be representative of user (or consumer) requirements in terms of the recommended quantities and skill levels.

5. *Test and support equipment/software.* The accomplishment of system operational and maintenance tasks may require the use of ground-handling equipment, test equipment, software, and/or a combination thereof. Only those items that have been approved for operation should be used.

6. *Supply support.* This includes all spares, repair parts, consumables, and supporting inventories that are necessary for the completion of system test and evaluation. Again, a realistic configuration, projected in a real-world environment, is desired.

7. *Test facilities and resources.* The conductance of system testing may require the use of special facilities, test chambers, capital equipment, environmental

controls, special instrumentation, and associated resources (e.g., heat, water, air-conditioning, power, telephone). These facilities and resources must be properly identified and scheduled.

8. *Interface requirements.* The conductance of system testing includes not only the verification and validation of the system itself operating in a relatively isolated environment, but the system operating in conjunction with other applicable systems, should it be part of a larger SOS structure. The applicable interfaces and test requirements as they pertain to the other systems in the SOS configuration must be defined herein and coordinated across the board as required.

In summary, the nature of the test preparation function is highly dependent on the overall objectives of the test and evaluation effort. Whatever the requirements may dictate, these considerations are important to the successful completion of these objectives.

2.11.4 Test Performance, Data Collection, Analysis, and Validation

With the necessary preparations in place, the next step is to commence with the formal test and evaluation of the system. The system (or elements thereof) is operated and supported in a designated manner, as defined in the TEMP. Throughout this process, data are collected and analyzed, and the results are compared with the initially specified requirements. With the system in operational status (either "real" or "simulated"), five questions arise:

1. How well did the system actually perform, and did it accomplish its mission objective?
2. What is the *true* effectiveness of the system?
3. What is the *true* effectiveness of the system support capability?
4. Does the system meet all of the requirements as covered through the specified technical performance measures (TPMs)?
5. Does the system meet all consumer requirements?

A response to these questions requires a formalized data-information feedback capability with the appropriate output in a timely manner. A data subsystem must be developed and implemented with the goal of achieving certain objectives, and these objectives must relate to these questions.

The process associated with formal testing, data collection, analysis, and evaluation is presented in Figure 2.35. Testing is conducted, data are collected and evaluated, and decisions are made as to whether the system configuration (at this stage) meets the requirements. If not, problem areas are identified and recommendations are initiated for corrective action.

The final step in this overall evaluation effort is the preparation of a final test report. The report should reference the initial test planning document (i.e., the TEMP),

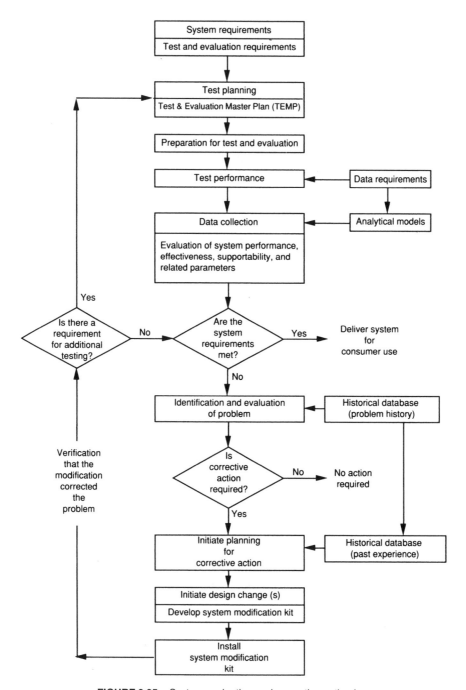

FIGURE 2.35 System evaluation and corrective-action loop.

describe all test conditions and the procedures followed in conducting the test, identify data sources and the results of the analysis, and include any recommendations for corrective action and/or improvement. Because this phase of activity is rather extensive and represents a critical milestone in the life cycle, the generation of a good comprehensive test report is essential to establish a good historical baseline.

2.11.5 System Modifications

The introduction of a change in an item of equipment, a software program, a procedure, or an element of support will likely affect many different components of the system. Equipment changes will likely affect software, spare parts, test equipment, technical data, and possibly certain production processes. Procedural changes will affect personnel and training requirements. Software changes may impact hardware and technical data. A change in any given component of a system will likely have an impact (of some kind) on most, if not all, of the other major components of that system.

Recommendations for changes, evolving from test and evaluation, must be dealt with on an individual basis. Each proposed change must be evaluated in terms of its impact on the other elements of the system, and on life-cycle cost, prior to a decision on whether to incorporate the change. The feasibility of incorporating the change will depend on the extensiveness of the change, its impact on the system in terms of its ability to perform the designated mission, and the cost of change implementation.

If a change is to be incorporated, the necessary change control procedures described in Chapter 5 must be implemented. This includes consideration of the time when the change is to be incorporated, the appropriate serial-numbered item(s) affected in a given production quantity, the requirements for retrofitting on earlier serial-numbered items, the development and "proofing" of the change modification kits, the geographic location where the modification kits are to be installed, and the requirements for system checkout and verification following the incorporation of the change. A plan should be developed for each approved change being implemented.

2.12 PRODUCTION AND/OR CONSTRUCTION

The earlier sections of this chapter have addressed primarily the design and development of systems and emphasized the importance of system engineering as an integrated and inherent activity therein. From this point, the system structure may assume several different forms, as illustrated in Figure 1.10. For a one-of-a-kind ground satellite tracking station (for example), the next phase in the system life cycle may involve *construction,* followed by the operational utilization of the tracking station in accomplishing its designated mission as required. For a system with many similar elements to be distributed throughout the world (as in Figure 2.3), the next phase will include *production,* followed by the utilization of these elements for the specified life cycle. Further, the production process includes numerous interfaces, as illustrated in Figure 1.21, where there are multiple suppliers, different transportation and warehousing requirements, and so on. In any case, however, it is assumed that the initial design and development of the system has been accomplished and the required performance and effectiveness characteristics have been verified (validated) through system test and evaluation (described in Section 2.11).

Given such verification, the challenge is to ensure that the system, its configuration, and its performance and effectiveness characteristics are *maintained* throughout the construction and/or production process. In the construction of a one-of-a-kind configuration, the introduction of poor quality (either through poor workmanship or through the use of substandard materials) in the building of a facility (for example) would certainly be degrading and have a negative impact on the system relative to the performance of its intended mission. In the production of multiple quantities of an item, even if the initial design has been verified through test and evaluation, there is no guarantee that subsequent models of the item being produced will exhibit the same characteristics. The production line is highly *dynamic,* and the lack of maintaining proper tolerances and the introduction of variances in manufacturing processes can significantly affect the output results. The question is, does the system that has been constructed and/or produced have the same inherent characteristics as the configuration that was designed and validated through system test and evaluation?[33]

Because an objective in system engineering is to facilitate the design and development of a system that will respond to customer requirements in a timely and effective manner, it is important not only that the final design configuration be ideal and well documented, but that the resultant output from the construction/production process be reflective and representative of what has been designed. Initially, this will involve an early emphasis on *design for constructability and producibility* (see Figure 2.31) and, later, the ongoing *evaluation* and *assessment* of construction and/or production activities. The system engineering process must encompass not only the initial design and development activities, but the follow-on assessment and feedback capabilities as well. Otherwise, one will never really know just how good the design/construction/production interfaces are and whether corrective action or improvement is needed.

2.13 SYSTEM OPERATIONAL USE AND SUSTAINING SUPPORT

As indicated in Figure 1.10, system engineering is *life-cycle* oriented. Given that a system has been properly designed and validated, constructed/produced, and installed at the user's operational site(s), the objective is to ensure that the resultant

---

[33] There has been a great deal of emphasis in recent years on the issue of *quality* (e.g.. total quality management, the implementation of six-sigma practices, the application of Taguchi methods, and so on). Much of this stems from experience indicating that although the design of a system may be good initially, a great deal of degradation can be introduced through the subsequent phases of ihe life cycle unless the practice of good "quality control" is maintained from the beginning. Refer to Appendix F for some excellent references on quality and quality control.

product will perform as intended and does indeed meet all of the customer requirements as initially defined. From this point, there are two key activities to include:

1. *Sustaining maintenance and support.* Throughout the system operational use phase, both scheduled and unscheduled maintenance will be required either to maintain the system in full operational status or to restore it to that status in the event of failure. It is essential that the proper maintenance actions be accomplished effectively and efficiently as required, and that the quality of the system not be degraded in the process[34]

2. *Incorporating new technologies aqd modifications for improvement.* With the increasing trends toward evolutionary system development and the incorporation of new technologies for the purposes of system "upgrading," on an almost continuous basis, care must be taken to ensure that the system is not degraded in the process. As stated in Section 2.11.5, proposed system modifications must be evaluated from a total life-cycle perspective, a plan for incorporation must be prepared, and the follow-on installation process must be of high quality (refer to the process shown in Figure 2.35).[35]

A major system engineering objective throughout the system operational use phase is that of *assessment,* to ensure that the system continues to perform as desired by the customer (user). The accomplishment of this objective is heavily dependent on the availability and implementation of a good *data collection, analysis,* and *feedback information capability.* The goal is twofold:

1. To collect and provide data on a continuing basis, covering the operations and support of the system as it performs its various mission scenarios throughout the planned life cycle. The purpose is to *assess* the actual performance and effectiveness of the system and its various elements (including the mission-related elements of the system and its maintenance and support infrastructure) and to ensure that all requirements are being met. Such an assessment may lead to the necessity for corrective action in the event of a problem.

2. To collect and provide data (covering an existing system in the field) for historical purposes and for feedback into the design process. Our engineering growth and potential for the future certainly depends on our ability to capture

---

[34]Poor maintenance practices (sloppy workmanship, the use of low-quality replacement parts, not utilizing the proper tools or following the proper procedures, the absence of follow-on quality inspections, and so on) can significantly degrade the system so that it will not be capable of performing its mission as intended.
[35]In the current environment (Section 1.2), one of the trends noted is the extension of the life cycle of many systems in use today, while at the same time the life cycles for many technologies are becoming shorter. This trend, combined with the emphasis on "evolutionary" design, leads to the conclusion that a system is likely to see many changes (modifications) as it evolves through its life cycle. Unless these changes are closely monitored for quality, and good configuration management and control practices are maintained in the process, there is a great possibility that an extensive amount of system degradation will occur with time.

experiences of the past and subsequently to apply the results in terms of *what to do* and *what not to do* for a new forthcoming design.[36]

The type and format of data may vary from one system to the next. It is important collect both success data and maintenance data. *Success data* refers to information covering system operations and utilization on a continuing basis. *How is the system doing on a day-to-day basis? Maintenance data* refers to information covering the various scheduled and unscheduled maintenance actions that occur throughout the life cycle. Maintenance event reports should reference the system and its operational status at the time a failure occurs (should this be the case). It is not uncommon to find that we do not pay much attention to what occurs in the field as long as things are going well. However, our reaction is often quite different when there are reported problems and "panic" occurs.[37]

In any event, the role of system engineering throughout the system use and sustaining support phase is continuous and very important. One may initially perceive that the application of system engineering principles and concepts during the design and development effort has been successful. However, the proof depends on what happens later.

2.14 SYSTEM RETIREMENT AND MATERIAL RECYCLING/DISPOSAL

With the concern for environmental impacts (and related costs) as they exist today, consideration must be given not only to the acquisition and utilization of a system and its elements throughout its intended life cycle, but also to the requirements associated with system retirement and the appropriate disposal of its components. There are many systems in use today that, upon becoming obsolete, will be costly to phase-out of the operational inventory. This may also be true for obsolete components that are replaced as a result of system modifications and the incorporation of new technologies for the purposes of system upgrading. Although some system components can be appropriately recycled, with the resulting materials available for other uses, there are a number of other components that cannot be consumed without causing a detrimental impact on the environment. Thus, an important activity within and throughout the system life cycle includes that of *retirement* and *material recycling and/or disposal*.

Although a system (as an entity) may continue to operate and be utilized until it can no longer serve a useful function, there may be many different components

---

[36]For the most part, and for many new system design efforts, we do a very poor job of capturing the experiences from earlier and similar systems that have been in operation in the past. This is due primarily to the fact that we do not have a good data collection and feedback capability in place. Hence, we tend to introduce the same problems over and over again as we evolve through new system developments, which, in turn, often results in costly modifications later on.

[37]One approach to a data collection, analysis, and system evaluation capability is described in B. S. Blanchard, *Logistics Engineering and Management,* 6th ed. (Upper Saddle River, NJ: Prentice-Hall, 2004). Section 8.2, pp. 353–358.

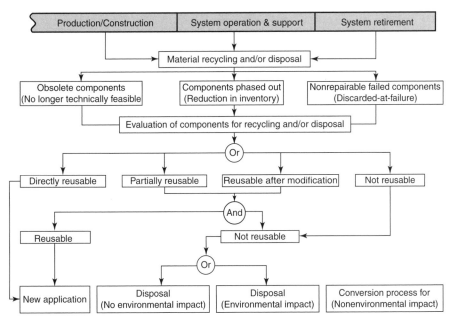

FIGURE 2.36   Component material retirement, recycling, and disopsal.

of that system that are removed and replaced for one reason or another. Referring to Figure 2.36, specific components may become obsolete because (1) they are no longer technically feasible as a result system modification and upgrading for improved performance, reliability, a reduction in life-cycle cost, and so on; (2) there has been a reduction in the operational inventory and there is no longer a need for them; and/or (3) component failures have occurred and they must be removed as part of an overall system repair action. Such components may then be evaluated further to determine whether they: (1) are needed and completely reusable in some other system application; (2) can be disassembled and reusable in part; (3) can be reusable after some modification; or (4) are not at all reusable. The residual must, of course, be processed through some form of *disposal.*

Inherent within the system engineering process are the early design considerations for system retirement and the recycling and/or disposal of its components as required—that is, *design for supportability, design for disposability, design for the environment.* Refer to Section 1.4.2 and Figure 2.31. Included within the basic definition of requirements for the system overall are the specific requirements associated with the various processes for material recycling and material decomposition and disposal. These requirements can be determined by extending and further developing the functional analysis (refer to Figure 2.17, block 7.0), and by determining the resources required for each applicable function (refer to Figure 2.18).

2.15   SUMMARY

Chapter 1 includes an introduction to system engineering and many of the terms and definitions associated with the subject area. The purpose of this chapter is to provide an overview of the *system engineering process* and just how these terms and definitions are directly related to the process. Further, the material in this chapter provides a *baseline* and a frame of reference for the subsequent discussion of individual design disciplines, design methods, and the activities associated with system engineering. The key area of emphasis here relates to the overall *process,* which is important and critical to understanding the principles and concepts associated with system engineering.

The material in this chapter is a necessary prerequisite to the information presented in Chapters 3 through 7. Chapters 3, *System Design Requirements,* builds on the process described in Chapter 2. Chapter 4, *Engineering Design Methods and Tools,* builds on the material in Chapter 2 and 3; and so on. It should also be noted that successful implementation of the system engineering process is highly dependent on effective planning and implementation of the system engineering management principles and concepts described in Chapter 6 and 7 (refer to Section 1.5 and Figure 1.25).

QUESTIONS AND PROBLEMS

1. Identify the basic steps in the system engineering process, and describe some of the *inputs* and *outputs,* associated with each step.

2. What is the purpose of *feasibility analysis?* What information is desired from such an analysis?

3. Why is the definition of system *operational requirements* important? What is included?

4. Why is the definition of the system *logistics and maintenance concept* important? What is included? How does the maintenance concept relate to the *maintenance plan?*

5. Identify a specific problem you wish to solve through the design and development of a new system. For your system:

   (a) Describe the current deficiency and identify the *need* for the new system.

   (b) Perform an abbreviated feasibility analysis and discuss the various alternative technical approaches you may wish to consider in designing the new system.

   (c) Define the basic operational requirements for the new system.

   (d) Define the maintenance concept for the new system.

   (e) Identify the critical technical performance measures (TPMs), based on the defined operational requirements and maintenance concept. Describe the process leading from the identification of TPMs to the identification of specific design characteristics.

6. What is meant by *quality function deployment* (QFD)? What are some of the benefits that can be derived from its application?

7. Identify a new system requirement and apply the QFD process (or something of an equivalent nature) in defining the specific characteristics that should be in cluded in the design (demonstrate the application by applying QFD to a real situation).

8. Describe how the QFD process can be beneficially applied in fulfilling the objectives of system engineering.

9. What is the purpose (i.e., possible benefits derived) of developing an *objectives tree* (or equivalent)?

10. What is meant by *functional analysis?* When should it be performed (if at all)? Why is it important in system engineering? What purpose(s) does it serve?

11. For the system selected in problem 5, perform a functional analysis. Construct a functional block diagram showing three levels of *operational* functions. From one of the blocks in the operational functional flow diagram, show two levels of *maintenance* functions. Show how the operational functions and the maintenance functions relate.

12. Select one block from the operational functional diagram and one block from the maintenance functional diagram in problem 11, and show inputs-outputs and how specific resource requirements are identified (e.g., hardware, software, people, facilities, data, etc.). Show an example by documenting the resource requirements using a format similar to that presented in Figure 2.21.

13. Refer to Figure 2.22. From your own experience, provide an example of two (2) different systems sharing two (or more) common functions. Describe some of the specific requirements in identifying and incorporating a *common function*.

14. Why are the identification and description of system-level *functional interfaces* important? What can happen if these interfaces are not well defined?

15. Identify some applications of functional analysis.

16. Describe what is meant by *allocation* or *partitioning*. What is its purpose? To what depth should it be applied? How can the process of allocation influence system design?

17. How are common functions addressed in the requirements *allocation* process? What cautions must be addressed in the allocation of quantitative requirements to functions?

18. For the system configuration described in problem 5, show a breakdown of the system into its subsystems and lower-level elements. Perform an allocation of requirements specified through the TPMs at the system level to the next level below.

19. What are the basic steps involved in *system analysis?* Construct a basic flow diagram illustrating the process, showing the steps, and including feedback provisions.

20. Describe what is meant by *synthesis*. How do the functions of *analysis, synthesis, and evaluation* relate to each other?

21. What is a *model?* Identify some of the basic characteristics of a model. List some of the benefits associated with the usage of mathematical models in system analysis. What are some of the problems/concerns?

22. What is meant by *sensitivity analysis?* What are some of the objectives of performing a sensitivity analysis? Benefits?

23. In your opinion, what are some of the major problems in implementing the process described in Figure 2.28? Identify at least three areas of concern.

24. What are some of the challenges associated with the day-to-day design process that must be addressed for successful implementation of the system engineering process?

25. How is a system *validated* in terms of compliance with the initially specified requirements?

26. How are test requirements determined?

27. Select a system of your choice and develop a comprehensive outline for a test and evaluation plan. Identify the categories of test, and describe the *inputs* and *outputs* for each category.

28. Describe some of the considerations associated with the initiation of design changes resulting from test and evaluation.

29. Describe the process associated with the initiation and implementation of design changes. What considerations must be incorporated to enhance the implementation of the system engineering process?

30. Why is system engineering important in the production/construction phase? Operational use and maintenance and support phase? Retirement and disposal phase?

31. Select a system of your choice and describe the process that you would implement in determining the specific resource requirements associated with the system *retirement and material recycling/disposal phase.*

32. Describe briefly how you would implement system engineering requirements for a proposed new SOS configuration.

3

SYSTEM DESIGN
REQUIREMENTS

This chapter addresses *total system design*, which can be defined as "the systematic activity necessary, beginning with the identification of the user need, to the selling and delivery of a product that will satisfy that need—an activity that encompasses product, process, people, and organization."[1] System design requirements evolve from the initial identification of a consumer/user need and are developed through the accomplishment of a feasibility analysis, the definition of system operational requirements and the maintenance concept, the development and prioritization of technical performance measures (TPMs), the accomplishment of a functional analysis, and the top-down allocation of requirements to the depth necessary. Given these basic requirements, the design process encompasses the activities of synthesis, analysis, and design optimization through the accomplishment of trade-off studies, and ultimately leads to the detail definition of the system configuration down to the component level. This is a continuous and iterative activity with the appropriate feedback, as illustrated in Figure 1.12. With the system configuration initially defined, the next step is *assessment* or *validation* through an ongoing test and evaluation effort, and any subsequent modification and refinement as necessary.

Figure 3.1 shows all of the basic functions in the system life cycle; *design* activities are included within each of the.blocks in the figure. The *design process* is continuous, commencing with the system-level configuration being developed (conceptual design), the subsystem-level next (preliminary system design), and, ultimately, the component level (detail design and development). At the same time, there are design requirements for the production/construction capability, the maintenance and

---

[1]This definition conveys the same thought, but is slightly modified from that included in S. Pugh, *Total Design: Integrated Methods for Successful Product Engineering* (Reading, MA: Addison-Wesley, 1990).

support infrastructure, and the system retirement and material recycling capability (refer to Figure 1.10). Subsequent to the detailed definition of the system and all of its elements, there is an ongoing data collection, analysis, and evaluation effort that is critical to ensure that the proposed design configuration will, indeed, meet the customer's requirements. If the initially specified requirements are not met, then there may be some changes required for corrective action and the system configuration will be modified accordingly.

The design process is "closed-loop" in a sense, as there are *development* activities, *evaluation and assessment* activities, and *redesign* activities as necessary. There is a general perception, adopted by many, that the design is *complete* when the system configuration is initially developed, independent of any subsequent evaluation and assessment activity. If this is the case, how does one really know whether the initially specified requirements have been met in a satisfactory manner? In any event, the design process must include all three of the aforementioned activity areas if it is to be complete.

The design process occurs as a result of a team effort, combining the necessary expertise from various organizations and technical specialties in the right place and in a timely manner. Depending on the system in question (and its mission objectives), there may be electrical requirements, mechanical requirements, structural requirements, material requirements, hydraulic requirements, reliability requirements, maintainability requirements, producibility and manufacturing requirements, environmental requirements, supportability requirements, quality requirements, and so on. These requirements will vary somewhat as one progresses through the steps in Figure 3.1. In the early stages, the design team may include only a few selected individuals with the appropriate system-level design experience. Later, additional personnel may be brought into the process as needs dictate—not too many or too few, but the proper mix of expertise at the time needed. Thus, the makeup of the design team will likely vary somewhat as the overall system development process evolves.

One of the objectives in system engineering is to assume a leadership role to ensure the proper and timely selection and integration of the required design disciplines, combined in such a manner as to enable the development of a system that will respond to the consumer/user need in a cost-effective manner. Not only will a typical program include designers representing different disciplines and with a wide variety of backgrounds and experiences, but the specific requirements for these areas of expertise will shift from one program phase to the next. Moreover, given the current emphasis relative to increased supplier participation on an international and global basis, there may a need to involve a number of suppliers from different nations as members of the design team.

Chapter 2 set the stage in defining the overall development process and the role of system engineering within its context. It is now appropriate to address some of the specifics relative to design requirements. The purpose of this chapter is to cover these requirements through the development of *specifications*, and then review some of the details as they pertain to individual design disciplines. An introduction to a select sample of disciplines is included, some commonalities are noted, and the

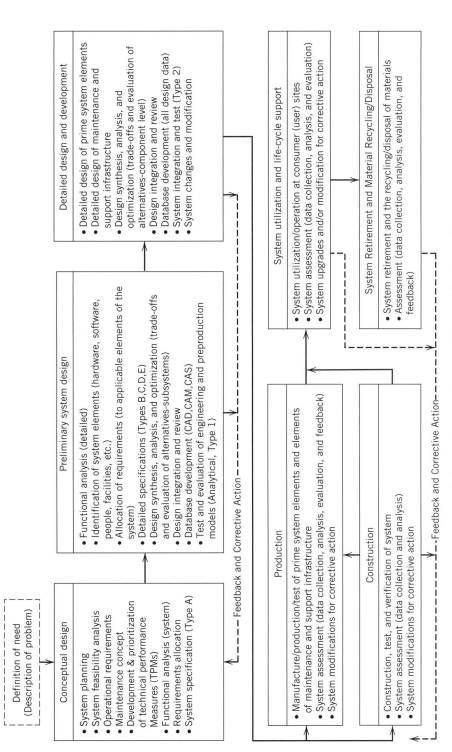

FIGURE 3.1 The major steps of system design and development.

Definition of need
(Description of problem)

Conceptual design
- System planning
- System feasibility analysis
- Operational requirements
- Maintenance concept
- Development & prioritization of technical performance Measures (TPMs)
- Functional analysis (system)
- Requirements allocation
- System specification (Type A)

Preliminary system design
- Functional analysis (detailed)
- Identification of system elements (hardware, software, people, facilities, etc.)
- Allocation of requirements (to applicable elements of the system)
- Detailed specifications (Types B,C,D,E)
- Design synthesis, analysis, and optimization (trade-offs and evaluation of alternatives–subsystems)
- Design integration and review
- Database development (CAD,CAM,CAS)
- Test and evaluation of engineering and preproduction models (Analytical, Type 1)

Detailed design and development
- Detailed design of prime system elements
- Detailed design of maintenance and support infrastructure
- Design synthesis, analysis, and optimization (trade-offs and evaluation of alternatives–component level)
- Design integration and review
- Database development (all design data)
- System integration and test (Type 2)
- System changes and modification

Production
- Manufacture/production/test of prime system elements and elements of maintenance and support infrastructure
- System assessment (data collection, analysis, evaluation, and feedback)
- System modifications for corrective action

Construction
- Construction, test, and verification of system
- System assessment (data collection and analysis)
- System modifications for corrective action

System utilization and life-cycle support
- System utilization/operation at consumer (user) sites
- System assessment (data collection, analysis, and evaluation)
- System upgrades and/or modification for corrective action

System Retirement and Material Recycling/Disposal
- System retirement and the recycling/disposal of materials
- Assessment (data collection, analysis, evaluation, and feedback)

─ ─ ─ Feedback and Corrective Action ─ ─ ─

125

importance of design integration through the application of system engineering methods is highlighted.[2]

3.1   DEVELOPMENT OF DESIGN REQUIREMENTS AND DESIGN-TO CRITERIA

As indicated in Figure 3.1, system *design requirements* evolve from the activities that occur throughout the conceptual design phase: definition of operational requirements, the maintenance concept, and the development and prioritization of the applicable technical performance measures (TPMs). As one determines these requirements for any given system, it will be necessary to define and allocate the appropriate *design-to criteria* for its different elements, as described in Chapter 2 (Section 2.8.2). As shown in Figure 2.23, the system may be partitioned and broken down into its elements, which may include various combinations of equipment, software, people, facilities, data, information, and so on. The question is, What specific quantitative and qualitative design-to criteria should be applied to the equipment, to the software, to the human element of the system, to the facilities, to the support infrastructure, to the information network, and to various combinations of these? An example of such requirements, as applied and allocated to equipment, is presented in Figure 2.25. However, the other elements of the system must be addressed as well.

Although these requirements (and design-to criteria) will vary from one system to the next, depending on the mission and the functions to be performed by the system, they may be defined through a combination of statements which, in turn, must be included in the applicable specification(s). A limited example follows:

1. The operational availability (A_o) for the system shall be greater than 98%.
2. The system software shall be designed such that a failure and its "cause" can be determined, with 99 percent accuracy, within 10 seconds from the time of its occurrence.
3. The system mean time between maintenance (MTBM) shall be 1000 hours or greater.
4. The mean maintenance downtime (MDT) for the system shall be 4 hours or less.
5. The equipment reliability mean time between failure (MTBF) shall be at least 3500 hours.
6. The mean corrective maintenance time ($\overline{\text{Mct}}$) for software shall be 30 minutes or less.
7. The response time for the logistics and maintenance support infrastructure shall not exceed 24 hours.

---

[2]It should be emphasized that no attempt has been made to cover all of the disciplines that may be required in the design of a system. Only a few are identified, with the intent of highlighting those areas that are not always properly addressed as part of a typical project activity.

8. The processing of a purchase order for the acquisition of a needed system component shall not exceed 1 hour.

9. The turnaround time for maintenance shall be 24 hours or less.

10. The data access time (i.e., time to search for and acquire a desired element of data), from the management information network, shall be 15 minutes or less.

11. The human operator error rate shall not exceed 1% per month, based on the operational scenarios completed by the system in accomplishing its mission.

12. The utilization of the facility, in direct support of system operations and maintenance, shall be 85%, or greater.

13. The unit life-cycle cost for the system shall not exceed x dollars, based on a 10-year planned life cycle.

14. The production facility shall be capable of producing x products, in y time, and at a z unit cost, with a defect rate not to exceed 1%.

15. The system shall be designed such that 95% of the material that makes up the system is recycleable.

16. The process time for removing an obsolete item from the operational inventory shall not exceed 12 hours, the cost per item processed shall be x dollars or less, and the resulting environmental degradation shall be zero.

The challenge, from a system engineering perspective, is to integrate and to view these requirements in the context of the whole. Although each specific requirement may be valid when addressed individually, there may be some conflicting issues that will require adjustment when they are considered in the context of the overall system. As an aid, it may be worthwhile to identify the different requirements and how they apply to the forward and reverse flow processes shown in Figure 1.20. This, in turn, may lead to the development of some lower-level and supporting requirements. In addition, a hierarchy of requirements should be developed and applied to the various elements of the system, broken down into its components as illustrated in Figure 2.23. This process is somewhat iterative and may be accomplished in several steps. The ultimate objective is to develop such requirements for the system and its elements for inclusion in the appropriate specification (see Figure 2.26).

3.2 DEVELOPMENT OF SPECIFICATIONS

The initial definition of system requirements is projected through a combination of formal specifications and planning documentation. Specifications basically cover the *technical* requirements for system design, and planning documentation includes all *management* requirements necessary to fulfill program objectives. The combination of specifications and plans is considered the basis for all future program engineering and management decisions.

The scope and depth of such documentation depend on the nature, size, and complexity of the system. In addition, the extent to which new design is feasible (where extra guidance and controls are desired), versus the selection of an off-the-shelf capabilily, will dictate the amount of documentation necessary. For small and relatively simple items, the technical specification and program planning requirements may be included in a single document. By contrast, for large-scale systems there may be a significant assemblage of documentation. In either case, the amount of documentation must be tailored to the need as dictated by the degree of technical and management controls necessary to accomplish program objectives.

In dealing with large systems, there are numerous elements that must be covered by specifications. Some components of the system may require an extensive amount of research-and-development effort, whereas other components are procured directly from existing supplier inventories. In regard to new items, some are developed by the major producer of the system and others are developed by suppliers remotely located in various parts of the world. In manufacturing, certain components may be produced in multiple quantities using conventional methods, whereas a special process may be required to produce other items. There may be a variety of specifications necessary to provide the guidance and controls associated with the development of the system and its components.

In preparing and applying specifications, there may be different classifications, depending on the type and nature of the item being designed or purchased off the shelf, as noted and illustrated in Figure 3.2.[3]

1. *System specification* (type A) Includes the technical, performance, operational and support characteristic for the system as an entity. It includes the allocation of requirements to functional areas, and it defines the various functional area interfaces. The major interfaces with other systems within the same SOS structure are also defined. The information derived from the feasibility analysis, operational requirements maintenance concept, and functional analysis is covered (refer to Sections 2.3 to 2.7).

2. *Development specification* (type B). Includes the technical requirements for any item below the system level where research, design, and development are accomplished. This may cover an equipment item, assembly, computer program, facility, critical item of support, and so on. Each specification must include the performance, effectiveness, and support characteristics that are required in the evolving of the design from the system level and down.

3. *Product specification* (type C). Includes the technical requirements for any item below the top system level that is currently in the inventory and can be procured off the shelf. This may cover standard system components (equipment, assemblies, units, cables), a specific computer program, a spare part, a tool, and so on.

---

[3]These specification classifications were originally taken from MIL-STD-490, "Specification Practices," Department of Defense, Washington, DC (latest revision).

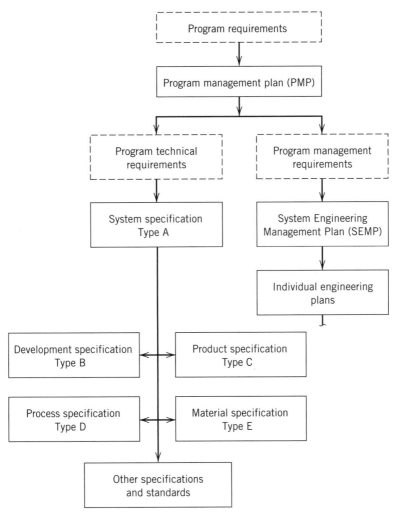

FIGURE 3.2   Hierarchy of technical specifications.

4. *Process specification* (type D). Includes the technical requirements that cover a service performed on any component of the system (e.g., machining, bending, welding, plating, heat treating, sanding, marking, packing, and processing).

5. *Material specification* (type E). Includes the technical requirements that pertain to raw materials, mixtures (e.g., paints, chemical compounds), and/or semi-fabricated materials (e.g., electrical cable, piping) that are used in the fabrication of a product.

The preparation of specifications is a key engineering activity. The system specification (type A) is prepared during the conceptual design phase. Development and product specifications are based on the results of "make-or-buy" decisions and are generally prepared during preliminary design.[4] Process and material specifications are more generally oriented to production and/or construction activities, and are usually prepared during the detail design and development phase. The relative timing of these specifications, in terms of program scheduling, is illustrated in Figures 1.12 and 1.26.

For large-scale systems, involving a wide mix of component suppliers, it is likely that many specifications will be generated and applied at varying stages in the system design and development process. In reviewing past experiences associated with different programs, it is clear that the generation and application of many different specifications on an independent basis has resulted in conflicts (i.e., contradictions relative to design criteria), as well as questions pertaining to which specification takes precedence in the event of conflict. In addition, there has been a tendency to specify not only the "whats" but the "hows" as well. This, in turn, can lead to a costly result. Specifications should be prepared to cover *performance* requirements, or *what is required* versus *how to do it.*

To help resolve the precedence problem, a *documentation tree* (or Specification tree) should be prepared, showing the hierarchy of specifications (and plans) from the top-level system specification and down. In the process of requirements allocation discussed in Section 2.8, it is necessary to establish requirements at the system level first, and then allocate these requirements down to the various components of the system. In developing specifications that dictate the design requirements for the various system components, it is essential that a good comprehensive system specification be developed first, and this specification then supplemented with the generation of good development, product, process, and/or material specifications as required.

In Figure 2.23 (Chapter 2), a preliminary hierarchy of system components is shown as a basis for the allocation of requirements. Figure 3.3 shows a variation of this hierarchy, converted into the form of a specification tree. Basically, the system specification is the top technical document for design. Other specifications supplement the system specification to varying degrees. Further, an order of precedence must be established to provide guidance as to which specification governs in the event of possible conflict.[5]

With the system specification (Type A) being the prime document for *technical* guidance, it is appropriate that the responsibility for its preparation and implementation be assigned as a system engineering task. Care must be taken to ensure that all significant design requirements, applicable at the system level, are included. The requirements must be *integrated*, and meaningful technical performance measures

---

[4]The results of make-or-buy decisions determine whether an item is to be designed and manufactured within the producer's facility or purchased from an outside source. Economic factors and scheduling requirements, combined with the availability of sources of supply are prime considerations in the decision-making process. Make-or-buy (outsourcing) decisions are covered further in Chapter 6 (Section 6.3).

[5]Throughout this section there has been reference to physical "documents" and "specifications." It should be noted that with today's EC technology, this same information may be presented in the form of a database, disk, or via some other electronic means.

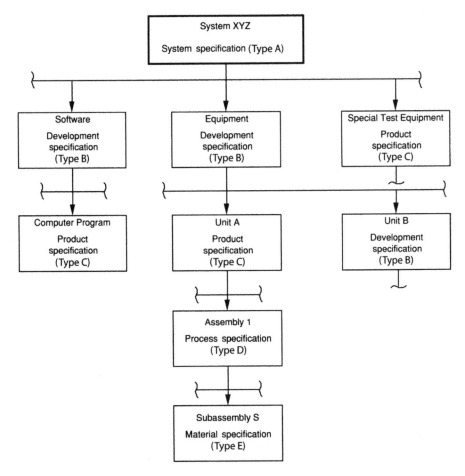

FIGURE 3.3   Sample specification tree (partial).

(TPMs) must be identified. TPMs include those quantitative characteristics of the system that are initially specified, then reflected in the follow-on design, and later used as measures against which the system is evaluated/validated (e.g., speed, range, accuracy, size, weight, capacity, MTBM, MDT, maintenance labor hours per operating hour (MLH/OH), and cost).[6]

An example of the format for a system specification is presented in Figure 3.4. The specification should include a description of the system (system architecture), its major characteristics, some general criteria for design and assembly, major data requirements, logistics and producibility considerations, test and evaluation requirements, maintenance and support requirements, quality assurance provisions, and

---

[6]The system-level TPM requirements, developed through the implementation of a QFD analysis (or equivalent), as discussed in Section 2.6, should be included in the system specification (type A).

System Specification
1.0 Scope
2.0 Applicable Documents
3.0 Requirements
3.1 General Description of System Architecture
3.1.1 System Operational Requirements
3.1.2 Maintenance Concept
3.1.3 Tecnical Performance Measures (TPMs)
3.1.4 Functional Analysis (System Level)
3.1.5 Allocation of Requirements
3.1.6 Functional Interfaces
3.1.7 SOS Interfaces
3.1.8 Environmental Requirements
3.2 System Characteristics
3.2.1 Performance Characteristics
3.2.2 Physical Characteristics
3.2.3 Effectiveness Requirements
3.2.4 Reliability
3.2.5 Maintainability
3.2.6 Usability (Human Factors)
3.2.7 Supportability
3.2.8 Transportability/Mobility
3.2.9 Producibility
3.2.10 Disposability
3.3 Design and Construction
3.3.1 CAD/CAM/CAS Requirements
3.3.2 Materials, Processes, and Parts
3.3.3 Hardware
3.3.4 Software
3.3.5 Electromagnetic Radiation
3.3.6 Interchangeability
3.3.7 Flexibility/Robustness
3.3.8 Workmanship
3.3.9 Safety
3.3.10 Security
3.4 Design Data and Database Requirements
3.5 Logistics
3.5.1 Supply Chain Requirements
3.5.2 Spares, Repair Parts, and Inventory Requirements
3.5.3 Test and Support Equipment
3.5.4 Personnel and Training
3.5.5 Packaging, Handling, Storage, and Transportation
3.5.6 Facilities and Utilities
3.5.7 Technical Data/Information
3.5.8 Computer Resources (Software)
3.6 Interoperability
3.7 Affordability
4.0 Test and Evaluation
5.0 Maintenance and Support (Life Cycle)
6.0 Quality Assurance
7.0 Customer Service

FIGURE 3.4   Example of a system specification (Type A) format.

major customer service activities. Although there may be some variations in format from one program to the next, the intent is to provide a description of the *functional base-line* for the system. This, in turn, constitutes the framework used in the preparation of subordinate specifications (e.g., *development, product, process*, and *material* specifications) by the responsible designer(s) for the numerous components of the system, and it serves as the major technical reference for all program planning documentation.

Finally, the preparation of a good system specification is highly dependent on the abilities of those accomplishing the task relative to their thorough understanding of the system in total, its intended mission, its components and their interrelationships, the various design disciplines required and their interfaces, and so on. It is not sufficient to merely prepare a series of individual write-ups covering each discipline, stapled together and submitted as a specification. This type of output usually results in contradictions, confusion, and inefficiencies throughout all subsequent phases of system design and development. Further, there are likely to be a significant number of errors and inconsistencies in the lower-level specifications if a good baseline is not established from the beginning. Without a good technical baseline, many of the design decisions made later will be suspect. Thus, the realization of a good comprehensive and highly integrated specification is critical from the beginning.[7]

3.3 THE INTEGRATION OF SYSTEM DESIGN ACTIVITIES

Based on the system specification, there may be a variety of design-to requirements, such as those illustrated in Figure 3.5. These requirements may be mutually supportive by nature, or there may be some inherent conflicts in goals. These goals are viewed in terms of relative importance (refer to Figure 2.11), and design optimization is accomplished through trade-off studies with the objective of establishing a mutually satisfactory approach.

In response to the specification and the established goals, certain categories of engineering expertise are identified as being necessary for the design and development of the system in question. These categories, and associated levels of effort, depend on the nature and complexity of the system and the size of the project. For relatively small systems/products such as a radio, an electrical household appliance, or an automobile, the quantity and variety of engineering expertise may be limited. On the other hand, there are many large-scale systems that require the combined input of

---

[7]There have been a number of instances in which the system specification was prepared in a hurry, was incomplete, and included very little in terms of specific and meaningful design criteria. The motivation for such was driven by several factors, expressed as follows: (1) "We don't want to be too specific, and be committed, in the system specification, as we may want to introduce a lot of design changes just before we go into production," or (2) "We are already behind in schedule, we need to get going with the design, and we will worry about the system specification later." The ultimate result: confusion, inconsistencies at the subsystem level and below, and a lot of last-minute costly design changes to correct problems that were induced in the process.

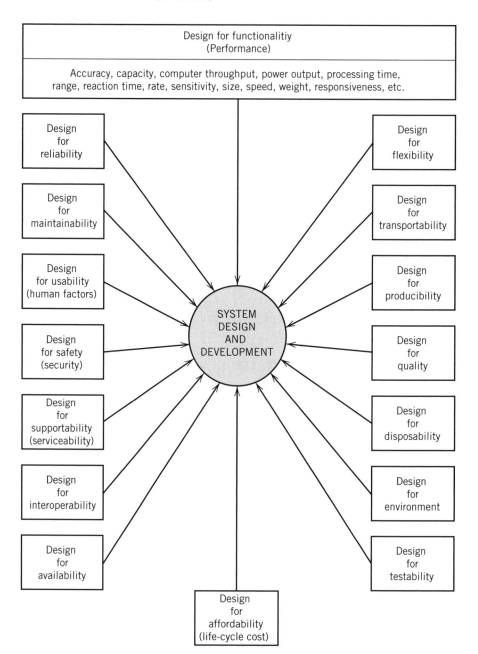

FIGURE 3.5 System design requirements.

specialists representing a wide variety of engineering disciplines. Two examples of relatively large projects are noted:

1. *Commercial aircraft system.* Aeronautical engineers determine aircraft performance requirements and design the overall airframe structure. Electrical engineers design the aircraft power distribution system and ground power requirements. Electronic engineers of various types are responsible for the development of subsystems such as radar, navigation, communications, and data recording and handling. Mechanical engineers are called on in the areas of mechanical structures, linkages, pneumatics, and hydraulics. Metallurgists are needed in the selection and application of materials for the aircraft structure. Reliability and maintainability engineers are concerned with availability, mean time between failure (MTBF), mean corrective maintenance time ($\overline{M}ct$), maintenance labor hours per operating hour (MLH/OH), and system logistic support. Human-factors engineers are interested in the man–machine functions, cockpit and cabin layout, and design of the various operator control panels. System engineering is concerned with the overall development of the airplane as a system and ensuring the proper integration of the numerous aircraft subsystems. Industrial engineers of various types are directly involved in the production of the aircraft itself and its many components. Test engineers are required to evaluate the system to ensure conformance to consumer requirements. Other engineering specialties are employed on a task-by-task basis.

2. *Ground mass-transit system.* Civil engineers are required for the layout and/or design of railroad tracks, tunnels, bridges, cables, and facilities. Electrical engineers are involved in the design of automatic train control provisions, traction power, substations and power distribution, automatic fare collection, digital data systems, and so on. Mechanical engineers are necessary in the design of passenger vehicles and related mechanical equipment. Architectural engineers may provide support in the construction of passenger terminals. Reliability and maintainability engineers would likely be involved in the design for system availability and the incorporation of supportability characteristics. Human-factors engineers are involved in the aspects of lighting, comfort ventilation, boarding access, handicapped accommodations, audio boarding instructions, and comfort stations. Industrial engineers will deal with the production aspects of passenger vehicles and vehicle components. Test engineers will evaluate the system to ensure that all performance, effectiveness, and system support requirements are met. Engineers in the planning and marketing areas will be required to keep the public informed and to promote the technical aspects of the system (i.e., keep the politicians and local citizens happy). Additional engineering specialists of various categories will be required to perform specific project-related tasks on an as-required basis.

Although these examples may not necessarily be all-inclusive, it is apparent that many different engineering disciplines are directly involved. Some of the more traditional disciplines such as electrical engineering and mechanical engineering are fractionated and broken down into specific job-oriented classifications. Engineering requirements for many large projects may include hundreds of individuals (or more, in the case of an aircraft development project or the equivalent) with varied backgrounds assigned to perform engineering functions. These engineers, forming a part of a larger organization, must not only be able to communicate with each other, but must be conversant with such supplemental activities as purchasing, accounting, manufacturing, and legal.

When considering personnel loading for large projects, there are likely to be some fluctuations. Depending on the functions to be performed, some engineers will be assigned to a given project until system development is completed, some will be assigned through the completion of production, and others will be brought in for a short term to perform specific tasks. The requirements, in terms of needed engineering expertise, will also vary from phase to phase because the areas of emphasis change as system development progresses. During the early phases of advance planning and conceptual design, individuals with a broad systems-oriented background are needed in greater quantities than detailed design specialists, whereas the reverse may be true during the detail design and development phase. In any event, the needs for engineering will change as the system evolves through its planned life cycle.

An additional characteristic associated with large projects pertains to the split in the design engineering workload between the major system producer and the suppliers of system components. A great deal of development, evaluation, production, and support of system components is accomplished at supplier facilities located throughout the world (the percentage of supplier activity in terms of total acquisition cost often ranges up to 75%; refer to Chapter 6). In other words, the major producer who is ultimately responsible for the development, integration, production, and support of the total system as an entity is greatly dependent on the results of engineering activities being accomplished at numerous disperse locations.

The project environment for the design and development of a large number of systems today is highly dynamic. There are many individuals with different specialties and backgrounds, rotating "on" and "off" a program at varying times. The need for good communications is essential, as well as a good understanding of the numerous interfaces that exist. The electrical design engineer needs to understand his or her interface relationships with the mechanical designer, the structural engineer, the reliability engineer, and/or the human-factors specialist. The logistics engineer needs to understand the design process and the responsibilities of the electronics engineer. To acquire the necessary design integration, within the context of system engineering, requires this understanding and appreciation, along with good communications.

To enable a better understanding of design requirements overall (with the objective of further promoting the integration process), a few design disciplines have been selected for the purpose of providing additional emphasis. Initially, it is important to address the area of *software engineering* inasmuch as a large part of the makeup of a system today involves the utilization of software. Next, it is important to address reliability, maintainability, human factors, safety, security, producibility, serviceability and supportability, logistics, disposability, environmental quality, value/cost engineering, and a few related disciplines. These areas, by themselves, certainly do not represent the total spectrum of design activity. However, in the past some of these particular requirements have not been adequately addressed or reflected in many of the systems developed, perhaps because of a lack of understanding and appreciation for these areas as they apply to design. As a result, these disciplines (and others) have been addressed independently through specifications and standards, but have not been well integrated into the mainstream design effort. Further, when addressing the individual requirements for each of the disciplines presented, one will find some commonalities. Through a review of these requirements, it is hoped that an even better appreciation of the need for total design integration will occur.

3.4 SELECTED DESIGN ENGINEERING DISCIPLINES

3.4.1 Software Engineering[8]

Software engineering deals with the *process* of bringing software into being. In the system engineering process shown in Figure 1.12, system-level requirements are defined and the accomplishment of functional analysis and allocation leads to the identification of "software" requirements (see Figure 1.13). From this point on, the emphasis may shift to the development and integration of the required software, in conjunction with the development of hardware, facilities, data/information, people requirements, and so on. It must be reemphasized that software is *not* a system in itself, but a major element of a system and must be treated on an *integrated* basis along with the other elements.

Clearly, with the advent of today's technology, software has become a major in gredient in the makeup of many systems being designed and developed. There are estimates that 50 to 80% of the elements of many large systems constitute software. Thus, there has been a great deal of activity in this area, which, in turn, has led to the development of the waterfall model (Figure 1.16), spiral model (Figure 1.17), Vee model (Figure 1.18), and others. Although the nomenclature may vary from one application to the next, the basic overall objectives are similar. The main objective has been to develop a process that will facilitate the design and development of software, which has become a complex issue and a significant challenge in the design of major systems today.

However, in spite of the availability of these recommended guidance-related models, many software developers today view software processes as being unduly rigid, restrictive, and inefficient and would much prefer to progress at their own speed,

---

[8]The object herein is to present an overview of *software* and the acquisition of such, and to relate its importance to the system engineering process. Three good text references are (1) B. W. Boehm, *Software Engineering Economics* (Upper Saddle River, NJ: Prentice-Hall, 1981); (2) R. S. Pressman, *Software Engineering: A Practitioner's Approach*, 4th ed. (New York: McGraw-Hill, 1996); and (3) A. P. Sage and J. D. Palmer, *Software Systems Engineering* (New York: John Wiley & Sons, Inc., 1990). In addition, it is recommended that the reader review the monthly publication *CrossTalk—The Journal of Defense Software Engineering*, published by the Software Technology Support Center (STSC), Hill AFB, Ogden, UT. See Appendix F for additional references.

without any restrictions and independent of the activities associated with the development of the other elements of the system. This, in turn, has resulted in many problems, numerous program/project failures, and high costs.[9]

As a result of these and related experiences, it is believed that adherence to a good and sound process for the development of software (within the context of the overall system engineering process) is essential and must be implemented. As a start, it may be worthwhile to refer to the seven steps identified in the section "The Power of Process" in S. McConnell's book, *Software Project Survival Guide*:[10]

1. Committing all requirements to writing;
2. Using a systematic procedure to control additions and changes to the software requirements;
3. Conducting systematic technical reviews of all requirements, designs, and source codes;
4. Developing a systematic Quality Assurance Plan in the very early stages of the project that includes a test plan, review plan, and defect tracking plan;
5. Creating an implementation plan that defines the order in which the software's functional components will be developed and integrated;
6. Using automated source code control; and
7. Revising cost and schedule estimates as each major milestone is achieved. Milestones include the completion of requirements analysis, architecture, and detailed design as well as the completion of each implementation stage.

The critical issue pertains to the adherence of *disciplined* approach in the development of software.

As a basis for further discussion. Figure 3.6 (which is a simplification of Figure 1.13) is included to show the major hardware–software steps and interfaces. The software development cycle may be described in the following steps:

1. *Software planning.* A description of software requirements as defined through the system engineering process, problem statement, project vision, definition of project scope, sponsorship, proposed targets (schedules), development strategy and acquisition process, organizational approach and personnel strategies, procurement and purchasing philosophy, supplier requirements and interfaces, cost and budgetary requirements, configuration control, measures and measurement control, areas of risk

---

[9]Two good references in this area are (1) "Five Ways to Destroy a Development Project—By Not Adhering to Basic Software-Engineering and Management Principles, Any Project Can Run Aground," by David R. Lindstrom, *IEEE Software: Lessons Learned*, IEEE Computer Society (September 1993). The five ways include "inadequate system engineering, improper requirements management, improper hardware sizing, selection of inappropriate methodologies, and ineffective metrics program"; and (2) "Seven Characteristics of Dysfunctional Software Projects," by M. W. Evans, A. M. Abela, and T. Beltz, *CrossTalk—The Journal of Defense Software Engineering*, vol. 15, no. 4, Software Technology Support Center (STSC), Hill AFB, Ogden, UT, April 2002. One of the seven causes is "failure to implement effective software processes."

[10]S. McConnell, *Software Project Survival Guide* (Redmond, WA: Microsoft Press, 1998), pp. 20–23.

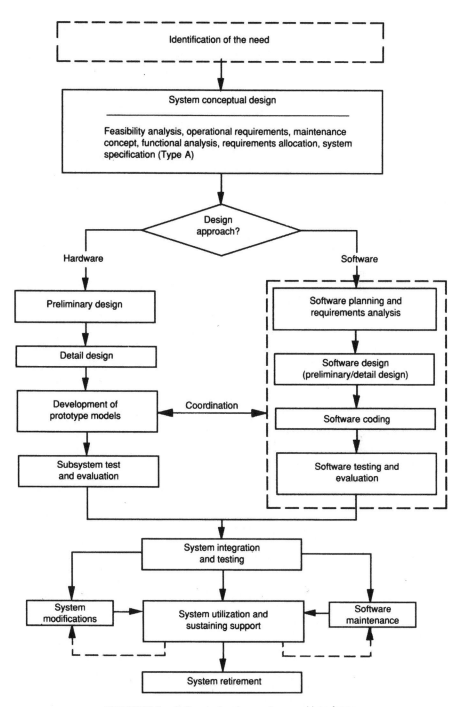

FIGURE 3.6   Software–hardware steps and interfaces.

and risk management, and related management issues, should be included in the *soft-ware plan*.

The software plan should be prepared early in the life cycle and at the time when software requirements are first identified, and the plan must be closely integrated with the System Engineering Management Plan (SEMP).[11]

2. *Requirements analysis.* A definition of system requirements (i.e., operational requirements, maintenance concept, technical performance measures, functional analysis and allocation), an identification of software operational and maintenance functions, top-level software functional block diagrams, performance factors, specific software design criteria/constraints, and so forth, are included. There must be a top-down "traceability" of requirements, and the specific software requirements must be traceable back to one or more system-level functions.[12] Specific design requirements for software should address such attributes (expressed in quantitative terms where possible) as availability, efficiency, flexibility, integrity, interoperability, reliability, maintainability, portability, usability and reusability, testability, and robustness.[13]

3. *Software design.* The development of a software hierarchical structure is accomplished in preliminary design to establish the basic relationships between the functional elements of the software, software modules, and coupling interface requirements. Information/data flow diagrams are developed further, and database requirements are defined. In detail design, the functional flow diagrams are broken down into detailed flowcharts, and the requirements for program design language(s) and application of the appropriate design tools are identified. The key issue here is to thoroughly document (i.e., establish a baseline for) the design, and the evolvement from one configuration to the next should changes be incorporated.[14]

4. *Software coding.* The preparation and writing of programs using structured code, the appropriate code format and code documentation, debugging and error-checking procedures, and so on, are accomplished. When incorporating different

---

[11]J. Cooper et al., *Software Acquisition Capability Maturity Model (SA-CMM)*, Version 1.02 (CMU/SEI-99-TR-002, ADA 362667) (Pittsburgh, PA: Software Engineering Institute (SEI), Carnegie Mellon University, 1999). This model was developed for the purpose of aiding in the acquisition of software.

[12]It should be noted that software development often assumes an *object-oriented* approach whereby "objects" are identified from the bottom up and software modules are developed around these objects. Although this approach may be highly beneficial in many respects, problems often occur when the software developer fails to address the aspects of *functionality* from the top down. Software modules are developed, but when combined and integrated with other elements of the system, there is often a "mismatch." Thus, although the object-oriented approach is certainly feasible, the results must be compatible with the functions defined through the process described in Section 2.7, and the various software modules should fit within some functional block (refer to Figures 2.12–2.18).

[13]K. E. Wiegers, *Software Requirements: Practical Techniques for Gathering and Managing Requirements Throughout the Product Development Cycle* (Redmond, WA: Microsoft Press, 1999), pp. 196–198. This is an excellent text for coverage of software requirements in general.

[14]Properly "documenting" the design baseline is very often an unpopular requirement for the designer, as it is perceived to take too much time and to inhibit progress. However, the absence of such documentation can be very costly, particularly in attempting to trace requirements back from the beginning, adding capability and making changes to the software, and so on.

software packages into the system design, one needs to ensure code compatibility across the board.

5. *Software test and evaluation.* A *Verification* of software design is accomplished to ensure that each of the various individual software products fulfills its specific purpose. The verification process often constitutes a step-by-step approach to testing one program, followed by the testing of another program, given the results of the first, and so on. The verification process is initially iterative, using a *rapid prototyping* approach throughout the preliminary and detail design phases. Later, the software may be verified as part of the overall system test and evaluation process.[15]

6. *Software maintenance.* The maintenance of software may be broken down into two basic categories. The first has to do with *corrective* maintenance and the fixing of problems that have resulted from a failure (or defect). Defects may occur as a result errors in initially defining requirements, errors in design, errors in coding, errors in documentation, and/or errors resulting from bad fixes associated with earlier problems.[16] The second category includes the process of upgrading the system and the incorporation of improvements in the design, and this is often referred to as *adaptive* and/or *perfective* maintenance. Care must be taken to ensure that such maintenance does not cause the introduction of more problems in the process.

In summary, it should be emphasized that software requirements are growing in terms of system applications, and the costs of software development and maintenance continue to increase at a rapid rate. Much of this cost is due to the lack of adhering to a good disciplined process approach in software development and not integrating the requirements for such with the other elements of the system concurrently. Hence, it is essential that these requirements be appropriately *integrated* and *controlled* through the system engineering process.

3.4.2 Reliability Engineering[17]

Reliability, in a generic sense, can be defined as "the probability that a system or product will perform in a satisfactory manner for a given period of time when used under specified operating conditions." The *probability* factor relates to the number of

---

[15]The concept of *rapid prototyping* is discussed further in Chapter 4.

[16]*Software Cost Estimation in 2002*, by C. Jones, *CrossTalk—The Journal of Defense Software Engineering*, vol. 15, no. 6, Software Technology Support Center (STSC), Hill AFB, Ogden, UT, June 2002. According to the author, the "average number of software errors in the United States is about five per function point."

[17]The intent herein is to provide an introductory overview of reliability engineering, in terms of both definitions and program requirements, but not to cover the subject in depth. However, it is highly recommended that the subject area be pursued further. Three good references are: (1) RIAC, *System Reliability Toolkit: A Practical Guide for Understanding and Implementing a Program for System Reliability* (Utica, NY: Reliability Information Analysis Center, 2005); (2) P.D.T. O'Connor, D. Newton, and R. Bromley, *Practical Reliability Engineering*, 4th ed. (Hoboken, NJ: John Wiley & Sons, Inc., 2002); and (3) J. Knezevic, *Reliability, Maintainability, and Supportability* (New York: McGraw-Hill, 1993). Additional references on reliability are included in Appendix F.

times, that one can expect an event to occur in a total number of trials. A probability of 95%, for example, means that (on average) a system will perform properly 95 out of 100 times, or that 95 of 100 items will perform properly.

The aspect of *satisfactory performance* relates to the system's ability to perform its mission. A combination of qualitative and quantitative factors defining the functions that a system is to accomplish, usually presented in the context of the System Specification, is included. These factors are defined under system operational requirements, described in Section 2.4.

The element of *time* is most significant because it represents the measure against which the degree of performance can be rated. A system may be designed to perform under certain conditions, but for how long? Of particular interest is the ability to predict the probability of a system surviving for a designated period of time without failure. Other time-related measures are mean time between failure (MTBF), mean time to failure (MTTF), mean cycles between failure (MCBF), and failure rate (λ).

The fourth key element in the reliability definition, *specified operating conditions*, pertains to the environment in which the system will operate. Environmental requirements are based on the anticipated mission scenarios (or profiles), and appropriate considerations for reliability must include temperature cycling, humidity, vibration and shock, sand and dust, salt spray, and so on. Such considerations must address not only the conditions when the system is operating and in a "dynamic" state, but also the conditions of the system during the accomplishment of maintenance activities, when the system (or components thereof) is being transported from one location to another, and/or when the system is in the storage mode. Experience indicates that the transportation, handling, maintenance, and storage modes are often more critical from a reliability standpoint than the environmental conditions during the periods of actual system utilization.

This definition of reliability is rather basic, and it can be applied to almost any type of system. However, there are instances in which it may be more appropriate to define reliability in terms of some specific mission scenario. In such cases, reliability can be defined as "the probability that a system will perform a designated mission in a satisfactory manner." This definition may, of course, imply the accomplishment of maintenance activities, as long as it does not interfere with the successful completion of the mission. The aspect of maintenance is covered more extensively in subsequent sections of this text.

In applying reliability requirements to a specific system, one needs to relate these requirements in terms of some quantitative measure (or a combination of several figures of merit). The basic reliability function, $R(t)$, may be stated as

$$R(t) = 1 - F(t) \tag{3.1}$$

where $R(t)$ is the probability of success and $F(t)$ is the probability that the system will fail by time t. $F(t)$ represents the failure distribution function.

When dealing with failure distributions, one often assumes average failure rates and attempts to predict the expected (or average) number of failures in a given period of time. To assist in this prediction, the Poisson distribution (which is somewhat analogous to the binomial distribution) can be applied. This distribution is generally expressed as

$$P(x, t) = \frac{(\lambda t)^x e^{-\lambda t}}{x!} \tag{3.2}$$

where λ represents the average failure rate, t is the operating time, and x is the observed number of failures.

This distribution states that if an average failure rate (λ) for an item is known, then it is possible to calculate the probability, $P(x,t)$, of observing 0, 1 2, 3, ..., n number of failures when the item is operating for a designated period of time, t. Thus, the Poisson expression may be broken down into a number of terms:

$$1 = e^{-\lambda t} + (\lambda t)e^{-\lambda t} + \frac{(\lambda t)^2 e^{-\lambda t}}{2!} + \frac{(\lambda t)^3 e^{-\lambda t}}{3!} + \cdots + \frac{(\lambda t)^n e^{-\lambda t}}{n!} \tag{3.3}$$

where $e^{-\lambda t}$ represents the probability of zero failures occurring in time t, $(\lambda t)e^{-\lambda t}$ is the probability that one (1) failure will occur, and so on.

In addressing the reliability objective, dealing with the probability of success, the first term in the Poisson expression is of significance. This term, representing the "exponential" distribution, is often assumed as the basis for specifying, predicting, and later measuring the reliability of a system.[18] In other words,

$$R = e^{-\lambda t} = e^{-t/M} \tag{3.4}$$

where M is the MTBF. If an item has a constant failure rate, the reliability of that item at its mean life is approximately 0.37, or there is a 37% chance that the item will survive its mean life without failure.

Figure 3.7 presents the traditional exponential reliability curve. The basic underlying assumption is that the failure rate is constant. When dealing with failure rates, it is necessary to view them in terms of both time and life-cycle activity. Figure 3.8 presents some typical failure-rate curve relationships. Although somewhat "puristic" in nature, the illustrations are included to support additional discussion of reliability.

In Figure 3.8, the "bathtub" curve will vary somewhat, depending on the type of equipment (whether electronic or mechanical), the degree of system/equipment maturity (new design or production versus state of the art), and so on. Usually, there is an initial "break-in" or "infant mortality" period in which a certain amount of "debugging" or "burn-in" is required in order to reach a stabilized condition. Design and/or manufacturing defects often occur, maintenance is accomplished, and corrective action is taken to resolve any outstanding problems. Subsequently, when stability is

---

[18]It should be noted that many of the assumptions herein are based on an average, or *constant*, failure rate. Although this assumption sometimes simplifies the reliability calculation process, there are instances in which failure rates are constantly changing. In these situations, it may be more appropriate to assume a Weibull distribution (or equivalent) in lieu of the negative exponential.

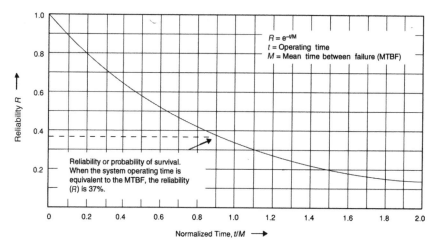

FIGURE 3.7 Traditional reliability exponential function.

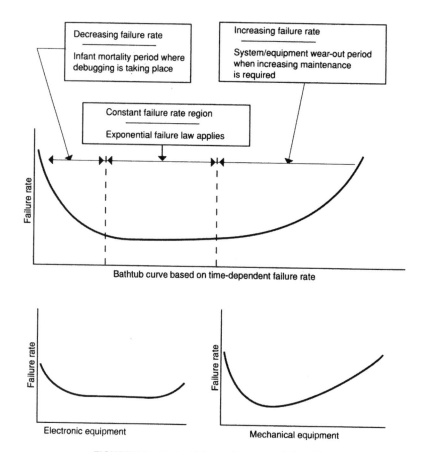

FIGURE 3.8 Typical failure-rate curve relationships.

acquired, the failure rate is relatively constant until a point in time when components begin to wear out, causing an ever-increasing failure rate as time goes on.

The curves presented in Figure 3.8 may also be highly influenced by individual program activities. For example, it is not uncommon for a customer to demand that a system (or components thereof) be delivered earlier than initially scheduled. In responding, the producer may eliminate certain essential quality checks in the production process in order to "get the equipment/software out the door." This usually leads to more initial defects, the consumption of more resources for maintenance and support than initially anticipated, and a higher degree of customer dissatisfaction in the long term. In Figure 3.8, the system becomes operational at the early stages of the bathtub curve before stabilization is attained.

In the world of software, failures may be related to calendar time, processor time, the number of transactions per period of time, the number of faults per module of code, and so on. Expectations are usually based on an operational profile and criticality to the mission. Thus, an accurate description of the mission scenario(s) is required. As the system evolves from the design and development stage to the operational utilization phase, the ongoing maintenance of software often becomes a major issue. Whereas the failure rate of equipment generally assumes the profiles in Figure 3.8, the maintenance of software often has a negative effect on the overall system reliability. The performance of software maintenance on a continuing basis, along with the incorporation of system changes in general, usually impacts the overall failure rate, as shown in Figure 3.9. When a change or modification is incorporated bugs are usually introduced, and it takes a while for these to be worked out of the system.

As indicated in Figure 3.8 and equations (3.2) through (3.4), the failure rate constitutes the number of failures occurring during a specified interval of time, or

$$\lambda = \frac{\text{number of failures}}{\text{total operating hours}} \tag{3.5}$$

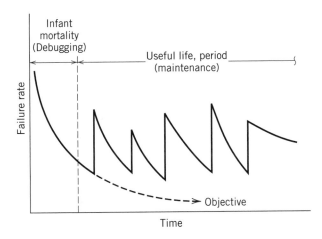

FIGURE 3.9   Failure-rate curve with maintenance (software application).

More specifically, the failure rate may be expressed in terms of failures per hour, failures per million hours, or percent failures per 1,000 hours.[19]

In addition, when defining failures in a pure reliability sense, this refers to "primary" or "catastrophic" failures; that is, instances when the system is not operating in accordance with specification requirements because of an actual component failure stemming from an overstressed condition. A component failure may, in turn, cause other components to fail through a chain reaction of events. Thus, we need to consider both primary catastrophic failures and secondary failures, sometimes known as dependent failures.[20]

With the identification of reliability measures, it is now appropriate to show some applications. System components are functionally related to each other through a series relationship, a parallel relationship, or a combination of these. Figure 3.10 illustrates some examples.

Figure 3.10(a) shows a series network. All components must operate in a satisfactory manner if the system is to function properly. The reliability, or the probability of success, of the system is the product of the reliabilities for the individual components and is expressed as

$$R_s = (R_A)(R_B)(R_C) \tag{3.6}$$

If the system operation is to be related to a specific time period, then, by substituting equation (3.4) into equation (3.6), the overall reliability of the series network is

$$R_s = e^{-(\lambda_A + \lambda_B + \lambda_C)t} \tag{3.7}$$

Figure 3.10(b) illustrates a parallel redundant network with two components. The system will function if either A or B, or both, are working. The reliability expression for this network is

$$R_s = R_A + R_B - (R_A)(R_B) \tag{3.8}$$

Now consider a network with three components in parallel, as shown in Figure 3.10(c). For the system to fail, all three components must fail individually. The reliability of the network is

$$R_s = 1 - (1 - R_A)(1 - R_B)(1 - R_C) \tag{3.9}$$

---

[19]This definition applies primarily to operating equipment. Failure rates may also be expressed in terms of failures per cycle of operation, errors per page of documentation, errors per operator or maintenance task, failures per module of software, and so on.

[20]The overall frequency of unscheduled maintenance considers primary failures, secondary failures, manufacturing defects, operator-induced failures, maintenance-induced failures, defects due to handling, and so on. From a systems engineering perspective, this overall frequency factor needs to be addressed, and is discussed further in Section 3.4.3.

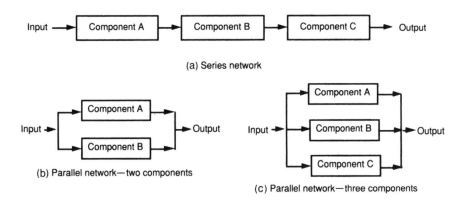

(a) Series network

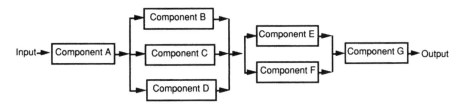

(b) Parallel network—two components

(c) Parallel network—three components

(d) Combined series—parallel network

FIGURE 3.10   Reliability component relationships.

In the event that all three components are identical, the reliability expression in equation (3.9) can be simplified to

$$R_s = 1 - (1 - R)^3$$

For a system with n components, the expression becomes

$$R_s = 1 - (1 - R)^n \tag{3.10}$$

Incorporating redundancy in design helps to improve system reliability. The effects of redundancy on design, presented in a simple generic sense, is illustrated in Figure 3.11. One can also determine the degree of reliability improvement through redundancy by developing some mathematical examples using equations (3.8) and (3.9).

Redundancy can be applied in design at different hierarchical indenture levels of the system. At the subsystem level, it may be appropriate to incorporate parallel functional capabilities, so that the system will continue to operate if one path fails to function properly. The flight control capability (incorporating electronic, digital, and mechanical alternatives) in an aircraft is an example where there are alternate paths in case of a failure in any one. At the detailed piece-part level, redundancy

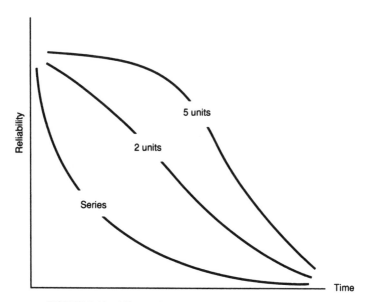

FIGURE 3.11 Effects of redundancy on reliability in design.

may be incorporated to improve the reliability of critical functions, particularly in areas where the accomplishment of maintenance is not feasible. For example, in the design of many electronic circuit boards, redundancy is often built in for the purpose of improving reliability when the accomplishment of maintenance is not practical.

The application of redundancy in design is a key area for evaluation. Although redundancy per se does improve reliability, the incorporation of extra components in a design requires additional space, and the costs are higher. This leads to a number of questions: Is redundancy really required in terms of criticality relative to system operation and the accomplishment of the mission? At what level should redundancy be incorporated? What type of redundancy should be considered (active or standby)? Should maintainability provisions be considered? Are there any alternative methods for improving reliability (e.g., improved part selection, part derating)? In essence, there are many interesting and related concerns that require further investigation.

Given an introduction to series and parallel networks, the next step is to combine these, as shown in Figure 3.10(d). The reliability expression for this network can be derived by applying equations (3.6), (3.8), and (3.9). Thus,

$$R_s = (R_A)[1 - (1 - R_B)(1 - R_C)(1 - R_D)][R_E + R_F - (R_E)(R_F)](R_G) \qquad (3.11)$$

In evaluating combined series-parallel networks, like the one illustrated in Figure 3.10(d), the analyst should first evaluate the parallel redundant elements to obtain the unit reliability, and then combine the unit(s) with other elements of the system in a series format. Overall system reliability is determined by calculating the product of all series reliabilities.

Through various applications of series-parallel networks, a system reliability block diagram can be developed for use in reliability allocation, modeling and analyses, predictions, and so on. The reliability block diagram is derived directly from the system functional analysis described in Section 2.7 (refer to Figures 2.12 to 2.17) and is expanded downward, as illustrated in Figure 3.12. Figure 3.13 shows an expanded block diagram. The block diagram describes the system reliability in terms of various component relationships.

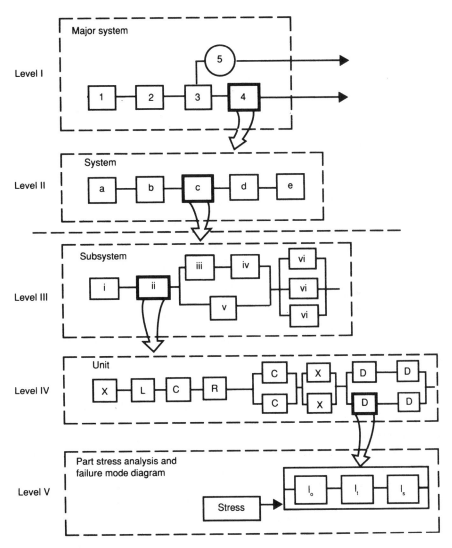

FIGURE 3.12 Progressive expansion of the reliability block diagram. *Source:* MIL-HDBK-338, Military Handbook, *Electronic Reliability Design Handbook* (Washington, DC: Department of Defense, 1975).

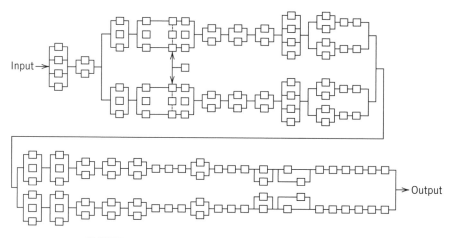

FIGURE 3.13 Expanded reliability block diagram of system.

The material presented in this section is obviously not intended to be a comprehensive text on the subject of reliability, but enough information is included to provide the reader with some overall knowledge of key terms, definitions, and the prime objectives associated with the discipline. Basically, the subject of reliability is being presented as one of the many disciplines requiring consideration within the overall context of system engineering. A general familiarization of the subject area is necessary, as well as an understanding of some of the activities that are usually undertaken in the performance of a typical reliability program. Having covered some key terms and definitions, it is now appropriate to describe related program activities.[21]

In implementing a reliability program for a typical large-scale system, the tasks identified in Figure 3.14 are generally applicable. Although there are variations from one program to the next, the performance of these tasks in terms of overall program phasing is assumed to be in accordance with Figure 3.15. The major program phases, and system-level activities, are derived from the baseline presented in Figure 1.12 (Chapter 1).

In Figure 3.14, the reliability tasks listed can be categorized in three basic areas: (1) program planning, management, and control (tasks 1–5), (2) design and analysis (tasks 6–16), and (3) test and evaluation (tasks 17–20). The first category of tasks must be closely integrated with system engineering activities and reflected in the System Engineering Management Plan (SEMP). The second group of tasks constitutes tools used in support of the mainstream design engineering effort, in response to the reliability requirements included in the system specification and the program plan. The third group, involving reliability testing, must be integrated with system-level testing activities and covered in the Test and Evaluation Master Plan (TEMP).

---

[21] Although specific reliability tasks should be tailored to the system and the associated program needs, the tasks listed in Figure 3.14 are assumed to be typical for the purposes of discussion.

Program Task	Task Description and Application
1. Reliability Program Plan	To develop a reliability program that identifies, integrates, and assists in the implementation of all management tasks applicable in fulfilling reliability program requirements. This plan includes a description of the reliability organization, organizational interfaces, a listing of tasks, task schedules and milestones, applicable policies and procedures, and projected resource requirements. This plan must tie directly into the System Engineering Management Plan (SEMP).
2. Review and control of suppliers of subcontractors	To establish initial reliability requirements and to accomplish the necessary program review, evaluation, feedback, and control of component supplier/subcontractor program activities. Supplier program plans are developed in response to the requirements of the overall Reliability Program Plan for the system.
3. Reliability program reviews	To conduct periodic program and design reviews at designated milestones (e.g., conceptual design review, system design reviews, equipment/software design reviews, and critical design review). The objective is to ensure that reliability requirements will be achieved.
4. Failure reporting, analysis, and corrective-action system (FRACAS)	To establish a closed-loop failure reporting system, procedures for analysis and for determining the cause(s) of failures, and documentation for recording the corrective action initiated.
5. Failure review board (FRB)	To establish a formal review board to review significant or critical failures, failure trends, corrective-action status, and to assure that adequate actions are being taken in a timely manner to resolve any outstanding problems.
6. Reliability modeling	To develop a reliability model for making initial numerical allocations, and for subsequent estimates to evaluate system/component reliability. As design progresses, a reliability block diagram is developed and used as a basis for accomplishing periodic reliability predictions. The reliability block diagram should evolve directly from the system functional flow block diagram.
7. Reliability allocation	To allocate, or apportion, top system-level requirements to lower identure levels of the system (e.g., subsystem, unit, assembly). This is accoplished to the depth necessary to provide specific criteria as an input to design.
8. Reliability prediction	To estimate the reliability of a system (or components thereof) based on a given design configuration. This is accomplished periodically throughout the system design and development processs to determine whether the initially specified system requiremtns are likely to be met given the proposed design at that time.
9. Failure mode, effect, and criticality analysis (FMECA)	To identify potential design weaknesses through a systematic analysis approach considering all possible ways in which a component can fail (the modes of failure), the possible causes for each failure, the likely frequency of occurrence, the criticality of failure, the effects of each failure on system operation (and on various system components), and any corrective action that should be initiated to prevent (or reduce the probability of) the potential problem from occuring in the future. Refer to Case Study C.1, Appendix C.
10. Fault-tree analysis (FTA)	To determine system design weaknesses using a deductive approach involving the graphical enumeration and analysis of different ways in which a system failure can occur. Refer to Case Study C.2, Appendix C.
11. Reliability-centered maintenance (RCM)	To identify alternatives and determine the best overall program for preventive maintenance using life-cycle criteria. Refer to Case Study C.3, Appendix C.
12. Sneak circuit analysis (SCA)	To identify possible latent paths that could cause the occurance of unwanted functions, assuming that all components are functioning properly in the beginning.
13. Electronic parts/circuits tolerance analysis	To examine the effects of parts/circuits electrical tolerances, specified over a range of operations (performance, temperature, etc.), on system reliability. The objective is to asses part drift characteristics, possible tolerance buildup, and identify design weaknesses.
14. Parts program	To establish a procedure for controlling the selection and use of standard and nonstandard parts.
15. Reliability critical items	To identify components requiring "special attention" because of their complexity, their relatively short life, and/or their use in new state-of-the-art technology application. Critical items usually require special maintenance/logistic support provisions.
16. Effects of testing, storage, handling, packaging, transportation, and maintenance	To determine the effects of these activities (i.e., handling, transportation, etc.) on system, or component, reliability.
17. Environmental stress screening	To plan and implement a program where the system (or components thereof) is tested using various environmental stresses (e.g., thermal or temperature cycling, vibration and shock, burn-in, X-ray, etc). The objective is to stimulate potential relevant failures early in the life cycle.
18. Reliability development/ growth testing	To plan and implement a "test-analyze-and-fix" procedure whereby system/component weaknesses can be identified, modifications can be incorporated, and reliability growth can be realized as the system development process evolves. This is an iterative activity, and involves performance testing, environmental testing, accelerated testing, and so on.
19. Reliability qualification test	To plan and implement a program where sequential testing is accomplished, using a preproduction prototype and considering statistical "accept" and "reject" criteria, to measure the reliability MTBF of the system. This occurs prior to entering production.
20. Production reliability acceptance test	To plan and implement a program where testing is accoplished, on a sampling basis, throughout the production process to ensure that degradation has not occured as a result of that process

FIGURE 3.14 Reliability engineering program tasks.

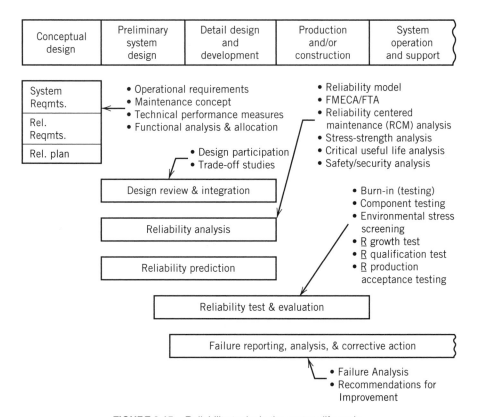

FIGURE 3.15 Reliability tasks in the system life cycle.

Although these tasks are primarily in response to reliability program requirements, there are many interfaces with basic design functions and with other supporting disciplines such as maintainability and logistic support.

Although brief task descriptions are included in Figure 3.14, Seven additional comments covering a select few are noted for the purposes of emphasis:

1. *Reliability program plan.* Although the requirements for a reliability program may specify a separate and dependent effort, it is *essential* that the program plan be developed as part of, or in conjunction with, the SEMP. Organizational interfaces, task inputs-outputs, schedules, and so on, must be directly supportive of system engineering activities. In addition, reliability activities must be closely integrated with maintainability and logistic support functions and must be included in the respective plans for these program areas (which also should be tied directly into the SEMP). The SEMP is introduced in Section 1.5 (refer to Figure 1.26) and is described further in Chapter 6.

2. *Reliability modeling.* This task, along with several others (e.g., allocation, prediction, stress/strength analysis, tolerance analysis), depends on the development of a good reliability block diagram (refer to Figure 3.13). The block diagram should

evolve directly from, and support, the system functional analysis and associated functional flow diagrams (refer to Section 2.7). Further, the reliability block diagram is used for analyses and predictions, the results of which are provided as a major input to maintainability, human factors, logistics, and safety analyses. The reliability block diagram represents a major link in a long series of events and must be developed in conjunction with these other activities.

3. *Failure mode, effect, and criticality analysis* (FMECA). The FMECA is a tool that has many different applications. Not only is it an excellent design tool for determining cause-and-effect relationships and identifying weak links, but it is useful in maintainability for the development of diagnostic routines. It is also required in the accomplishment of supportability analysis (SA) relative to the identification of both corrective and preventive maintenance requirements. The FMECA constitutes a major input to the reliability-centered maintenance (RCM) program. It is used to supplement both the fault-tree analysis and the hazard analysis accomplished in a system safety program. The FMECA is a critical activity, must be accomplished in a timely manner (early during preliminary design and subsequently updated on an iterative basis), and must be directly tied into these other activities. Figure 3.16 points to the application of the FMECA to a package handling system, and Case Study C.1, Appendix C, describes the FMECA process.

4. *Fault-tree analysis* (FTA). The FTA is a deductive approach involving the graphical enumeration and analysis of different ways in which a particular system failure can occur and the probability of its occurrence. A separate fault tree may be developed for every critical failure mode or undesired top-level event. Attention is focused on this top-level event and the first-tier causes associated with it. Each of these causes is next investigated for its causes, and so on. The FTA is narrower in focus than the FMECA and does not require as much input data. Case Study C.2, Appendix C, describes the FTA and its application.

5. *Reliability-centered maintenance* (RCM) analysis. The RCM analysis includes an evaluation of the system/process, in terms of the life cycle, to determine the best overall program for preventive (scheduled) maintenance. Emphasis is on the estab lishment of a cost-effective preventive maintenance program based on reliability information derived from the FMECA (i.e., failure modes, effects, frequency, criticality, and compensation through preventive maintenance). Case Study C.3, Appendix C, describes the RCM analysis and its application.

6. *Failure reporting, analysis, and corrective-action system* (FRACAS). Although this is identified as a reliability program task designed to address recommendations for corrective action as a result of catastrophic failures, the overall task objective relates closely to the system engineering feedback and control loop. Often, as problems arise and corrective action is initiated, the events that take place and the results are not adequately documented. Although it is important to respond to the short-term needs (i.e., correct outstanding problems in an expeditious manner), it is also important to provide some long-term memory through good reporting and documentation. This task should be tied directly with the system engineering reporting, feedback, and control process.

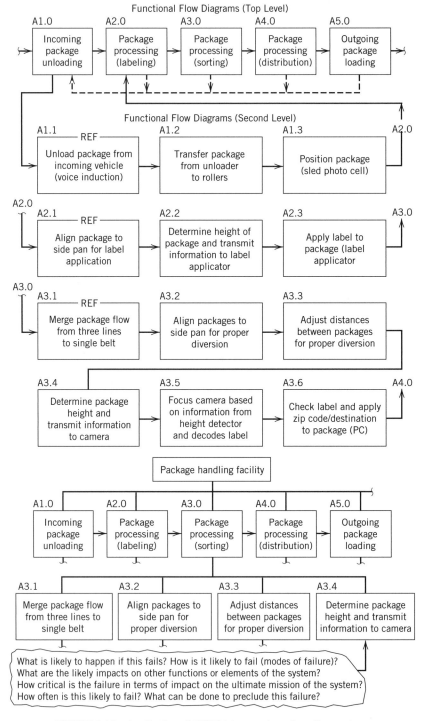

FIGURE 3.16 Application of FMECA to a package handling system.

7. *Reliability, qualification testing.* This task, usually accomplished as part of type 2 testing, should be defined in the context of the *total* system test and evaluation effort (refer to Figure 2.32). The specific requirements will depend on the system complexity, the degree of design definition, the nature of the mission that the system is expected to accomplish, and the TPMs (and their priorities) established for the system. In addition, for this and any other individual test, there are certain expectations and opportunities for gathering information. For instance, the objective of environmental qualification testing is to determine whether the system will perform in a specified environment. In performing this test, it may be possible to gather some reliability information by observing system operating times, failures, and so on. This, in turn, may permit a reduction in subsequent reliability testing. A second example pertains to the gathering of maintainability data during the performance of formal reliability testing. As failures occur during the test, maintenance actions can be evaluated in terms of elapsed times and resource requirements. This, in turn, may allow for some reductions in both maintainability and supportability test and evaluation efforts. In other words, there are numerous possibilities for reducing costs (while still gathering the necessary information) through the accomplishment of an integrated testing approach. Thus, reliability testing must be viewed in the context of the overall system test effort, and the requirements for this must be covered in the TEMP.

In summary, the tasks identified in Figure 3.14 are generally performed in response to some detailed specification or program requirement. For many programs, these are completed on a relatively independent basis. Yet, the interfaces are many, and there are some excellent possibilities for task integration, resulting in reduced costs. Throughout this text, opportunities for integration are discussed further. The intent of this section is to provide an introduction to the requirements associated with many reliability programs.

3.4.3 Maintainability Engineering[22]

Maintainability is an inherent characteristic of system design that pertains to the ease, accuracy, safety, and economy in the performance of maintenance actions. It deals with component packaging, diagnostics, part standardization, accessibility, interchangeability, mounting and labeling, and so on. A system should be designed so that it can be maintained without large investments of time and resources (e.g., personnel, materials, test equipment, facilities, data) and at minimum cost, while still fulfilling its designated mission. Maintainability is the *ability* of an item to be maintained, whereas maintenance constitutes those actions taken to restore an item

---

[22]The objective is to provide an introductory overview of maintainability engineering, including a few definitions and program requirements, but not to cover the subject in depth. However, for more information, two good references are (1) RIAC, *Maintainability Toolkit: A Practical Guide for Designing and Developing Products and Systems*, (Utica, NY: Reliability Information Analysis Center) and (2) B. S. Blanchard, D. Verma, and E. L. Peterson, *Maintainability: A Key to Effective Serviceability and Maintenance Management* (New York: John Wiley & Sons Inc., 1995). Additional references are included in Appendix F.

to (or retain an item in) a specified operating condition. Maintainability is a design parameter, whereas maintenance is the result of design.

Maintainability, defined in the broadest sense, can be measured in terms of a combination of maintenance times, personnel labor hours, maintenance frequency factors, maintenance cost, and related logistic support factors. There is no single measure that will address *all* issues. For instance, an objective may be to shorten the elapsed time for accomplishing maintenance by adding more personnel (and possibly with greater skills). Although such an action may reduce the time requirement, it may cause an increase in personnel requirements and a resultant increase in life-cycle cost. Further, it may be desirable to reduce the frequency of unscheduled maintenance by adding the requirements for more scheduled maintenance. If this is done, there may be an increase in the overall frequency of maintenance and the life-cycle cost may increase as well. In essence, these factors (as applicable) must be addressed on a collective basis, as well as being considered in conjunction with the reliability measures discussed in Section 3.4.2.

One of the most commonly used measures of maintainability is the aspect of "time." In Figure 3.17, the overall time spectrum can be broken down into different applications. *Uptime* pertains to the elapsed time applicable to the system when in operational use, or when in a standby or ready state awaiting for use. On the other hand, *downtime* refers to the total elapsed time required, when the system is not operational, to accomplish corrective maintenance and/or preventive maintenance. These two categories of maintenance are defined as follows:

1. *Corrective maintenance.* The unscheduled actions, initiated as a result of failure (or a perceived failure), that are necessary to *restore* a system to its required level of performance. Such activities may include troubleshooting, disassembly, repair, remove and replace, reassembly, alignment and adjustment, checkout, and so on. In addition, this includes all software maintenance that is not initially planned—for example, *adaptive* maintenance, *perfective* maintenance, and so on.

2. *Preventive maintenance.* The scheduled actions necessary to *retain* a system at a specified level of performance. This may include periodic inspections, servicing, calibration, condition monitoring, and/or the replacement of designated critical items.

In Figure 3.17, total maintenance downtime (MDT) is the elapsed time required to repair and restore a system to full operational status and/or to retain a system in that condition. MDT can be broken down into the following components:

1. *Active maintenance time* ($\overline{M}$). That portion of downtime when corrective and/or preventive maintenancec ativities are being accomplished. This factor is often expressed as

$$\overline{M} = \frac{(\lambda)(\overline{M}ct) + (fpt)(\overline{M}pt)}{\lambda + fpt} \qquad (3.12)$$

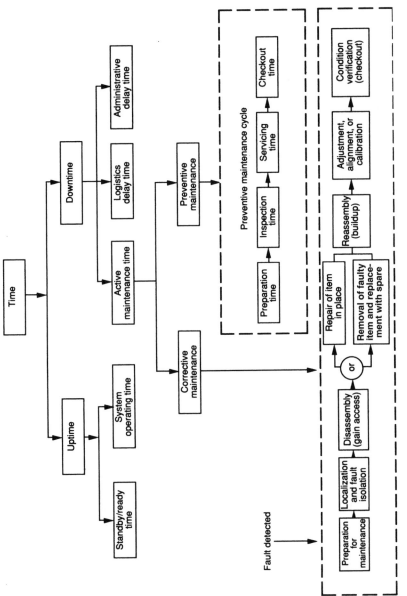

FIGURE 3.17 Time relationships.

where $\overline{M}$ is the mean active maintenance time, $\overline{Mct}$ is the mean corrective maintenance time, $\overline{Mpt}$ is the mean preventive maintenance time, fpt is the frequency of preventive maintenance, and λ is the failure rate (or frequency of corrective maintenance).

2. *Logistics delay time* (LDT). That portion of downtime when the system is not operational because of delays associated with the support capability; for example, waiting for a spare part, waiting for the availability of test equipment, waiting for the use of a special facility.

3. *Administrative delay time* (ADT). That portion of downtime when the necessary maintenance is delayed for reasons of an administrative nature; for example, the unavailability of personnel because of other priorities, organizational constraints, labor strikes.

In looking at these elements of downtime from the design engineer's perspective, it is quite common to address only the *active* maintenance segment (i.e., $\overline{M}$). This is because of being able to directly relate system characteristics such as diagnostic capability, accessibility, and interchangeability to downtime. The producer (i.e., contractor) is responsible for, and usually can control, this element, whereas the LDT and ADT factors are primarily influenced by the consumer (i.e., customer). From the perspective of system engineering, one needs to deal with the *entire* downtime spectrum. There is little point in constraining the design of prime equipment (i.e., an item must be designed so that it can be repaired in 30 minutes) if the support capability is such that it takes three months to acquire the necessary spare part. In essence, the entire spectrum must be considered as reflected in Figure 2.5, and each of these time elements represents an important measure.

By referring to the time relationships presented in Figure 3.17, as well as the factors in equation (3.12), active maintenance time ($\overline{M}$) can be broken down into corrective maintenance and preventive maintenance times. The mean corrective maintenance time ($\overline{Mct}$) is expressed as

$$\overline{Mct} = \frac{\Sigma(\lambda_i)Mct_i}{\Sigma(\lambda_i)} \tag{3.13}$$

where Mct_i represents the time that it takes to progress through the corrective maintenance cycle illustrated in Figure 3.17 (for the ith item), and λ_i is the corresponding failure rate. In the event of a fixed number of maintenance actions, n, then

$$\overline{Mct} = \frac{\sum_{i=1}^{n} Mct_i}{n} \tag{3.14}$$

$\overline{Mct}$, which is a weighted average of repair times using reliability factors, is equivalent to the mean time to repair (MTTR), a measure that is commonly used for maintainability.

The time-dependency relationship between the probability of corrective maintenance and the time allocated for accomplishing corrective maintenance can be expected to produce a probability density function in one of three common forms, as
illustrated in Figure 3.18:

1. *The normal distribution.* Applies to relatively simple and common maintenance actions where times are fixed with very little variation

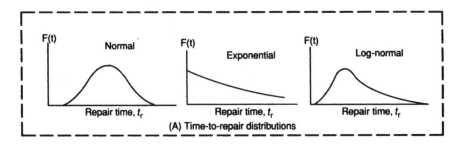

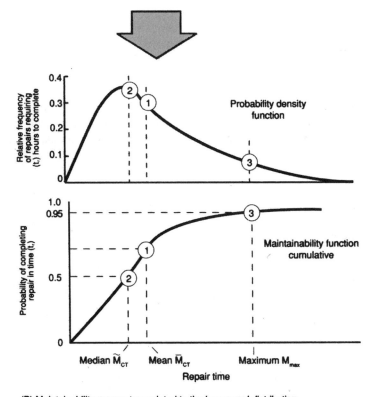

FIGURE 3.18 Maintainability distributions.

2. *The exponential distribution.* Applies to maintenance actions involving part substitution methods of fault isolation in large systems that result in a constant failure rate

3. *The log-normal distribution.* Applies to most maintenance actions involving detailed tasks with unequal frequency and time durations

Experience has indicated that in most instances, the distribution of maintenance times for complex systems follows the log-normal approximation. From Figure 3.18, the key maintainability parameters are the mean time to repair (point 1), the median time to repair (point 2), and the maximum time to repair (point 3). Whereas the "mean" value constitutes the measure that is most commonly used, the median and maximum time values are appropriate measures used in certain applications.

The median active corrective maintenance time ($\tilde{M}ct$) is that value that divides all of the repair-time values so that 50% are less than the median and 50% are greater than the median. For the normal distribution, the median is the same as the mean, and the median in the log-normal distribution is the same as the geometric mean ($MTTR_g$) illustrated in Figure 3.18. The median, represented by Point 2, is calculated as

$$\tilde{M}ct = \text{antilog} \frac{\Sigma(\lambda_i)(\log Mct_i)}{\Sigma(\lambda_i)} = \text{antilog} \frac{\sum_{i=1}^{n} \log Mct_i}{n} \tag{3.15}$$

The maximum active corrective maintenance time (M_{max}) can be defined as that value of downtime below which a designated percent of all maintenance actions can be expected to be completed. This is represented by point 3 in Figure 3.18. Selected points, in the log-normal distribution, at the 90th or 95th percentile are generally used. The maximum corrective maintenance time is expressed as

$$M_{max} = \text{antilog}[\overline{\log Mct} + Z\sigma_{\log Mct_i}] \tag{3.16}$$

where $\overline{\log Mct}$ is the mean of the logarithms of Mct_i, Z is the standard variate at the point where M_{max} is defined (1.65 at 95%, 1.28 at 90%, 1.04 at 85%, and so on); refer to the normal distribution tables in any text on statistics), and σ is the standard deviation of the sample logarithms of average repair times, Mct_i.

In the area of preventive maintenance, both the mean and the median measures are used. The mean preventive maintenance time ($\overline{M}pt$) can be determined by

$$\overline{M}pt = \frac{\Sigma(\text{fpt})(Mpt_i)}{\Sigma(\text{fpt}_i)} = \frac{\sum_{t=1}^{p} Mpt_i}{n} \tag{3.17}$$

where fpt_i is the frequency of the individual (*i*th) preventive maintenance action and Mpt_i is the associated elapsed time to perform the preventive maintenance required.

The median value for preventive maintenance, like the requirement for corrective maintenance specified in equation (3.15), is determined from

$$\widetilde{\text{M}}\text{pt} = \text{antilog} \frac{\Sigma(\text{fpt}_i)(\log \text{Mpt}_i)}{\Sigma(\text{fpt}_i)} \qquad (3.18)$$

Preventive maintenance may be accomplished while the system is in full operation, or the requirements for such may result in downtime. In this instance (and in the case of corrective maintenance), only those actions that are accomplished and result in downtime are considered. Maintenance actions that do not result in system downtime are basically accounted for through the personnel labor-hour and maintenance cost measures of maintainability.[23]

Although the various measures of elapsed time are extremely important, one must also consider the maintenance labor hours expended in the process. In dealing with ease and economy in the performance of maintenance, an objective is to obtain the proper balance between elapsed time, labor hours, and personnel skills at minimum maintenance cost. Personnel time may be expressed in terms of maintenance labor hours per system operating hour (MLH/OH), maintenance labor hours per cycle of system operation (MLH/cycle), maintenance labor hours per maintenance action (MLH/MA), or maintenance labor hours per month (MLH/month). Any of these factors can be presented in terms of mean values, such as mean corrective maintenance labor hours ($\overline{\text{MLH}_c}$), which can be expressed as

$$\overline{\text{MLH}}_c = \frac{\Sigma(\lambda_i)(\text{MLH}_i)}{\Sigma(\lambda_i)} \qquad (3.19)$$

where λ_i is the failure rate of the ith item and MLH_i is maintenance labor hours necessary to accomplish the related corrective maintenance actions.

The aspect of corrective maintenance having been established, the values for mean preventive maintenance labor hours and mean total maintenance labor hours (to include *all* corrective and preventive maintenance actions) can be determined in a similar manner. These factors, predicted for each level of maintenance identified in the system maintenance concept, can be utilized in determining specific maintenance and logistic support requirements and associated costs.

A third measure of maintainability (in addition to the time and labor-hour factors) is maintenance frequency. As indicated in Section 3.4.2, the frequency factors associated with primary and secondary failures are basically reflected through the reliability MTBF and λ measures. These measures are certainly important for determining the overall frequency of unscheduled maintenance; however, there are

---

[23]Although maintainability has already been defined in the broadest context, there are additional definitions that relate to a specific measure. With regard to time, it can be defined as the measure of the ability of an item to be retained in or restored to a specified condition when maintenance is performed by personnel having specified skills, using prescribed procedures and resources, at each prescribed level of maintenance and repair.

additional considerations such as manufacturing defects, operator-induced failures, maintenance-induced failures, and defects due to handling that may be relevant (refer to footnote 20 in this chapter). Moreover, one must consider the aspect of preventive maintenance. With this in mind, it is appropriate to look at the total spectrum of maintenance and the measure of mean time between maintenance (MTBM). This can be calculated as

$$MTBM = \frac{1}{1/MTBM_u + 1/MTBM_s} \tag{3.20}$$

where $MTBM_u$ is the mean interval of unscheduled (or corrective) maintenance, and $MTBM_s$ is the mean interval of scheduled (preventive) maintenance. The reciprocals of $MTBM_u$ and $MTBM_s$ are equivalent to the maintenance rates, or the maintenance actions per hour of system operation. $MTBM_u$ should be equivalent to MTBF, assuming that the possibilities of operator-induced defects, maintenance-induced defects, and so on, have been "designed out" of the system.

Within the overall spectrum of activity represented by the MTBM factor, there are some maintenance actions that result in the removal and replacement of components and the requirement for spare parts. These actions, in response to both corrective and preventive maintenance requirements, can be measured in terms of mean time between replacement (MTBR), a factor of MTBM. In essence, the MTBM factor reflects *all* maintenance actions, some of which result in item replacements.

Figure 3.19 shows a given system where there were 100 unscheduled maintenance actions recorded over a specific segment of time. In all instances, some organizational-level maintenance was accomplished relative to diagnostics and checkout. In 25 cases, it was impossible to verify that a problem existed, as the system appeared to be operating properly when checked. Therefore, no items were removed and replaced. In the other 75 instances, a given component was suspect, resulting in a removal and replacement action. Of the components removed for higher-level maintenance (i.e., intermediate level), a problem was verified in 45 instances and repair was accomplished on-site, 12 components were sent to the factory for higher-level repair, 3 components were condemned (determined to be beyond economic repair), and there were 15 components in which no defect was noted. Of the 12 components sent to the factory, 10 were considered faulty. Through a review of these factors, it can be seen that the MTBM figure must consider all of the 100 maintenance actions, the MTBR figure can be related to the 75 replacements (at the organizational level), and the MTBF measure (as defined in a puristic reliability sense) pertains to the 58 components in which actual catastrophic failures were confirmed. From a *systems* perspective, however, there were 100 failures in total, whether they can be charged to an element of equipment, a module of software, or to a human being.

Given the definitions associated with MTBM, MTBR, MTBF, $\overline{MDT}$, $\overline{M}ct$, $\overline{M}pt$, $\overline{M}$, and so on, it is important to relate some of these figures of merit to a higher-order system parameter. Reliability and maintainability factors, shown in Figure 2.27, are,

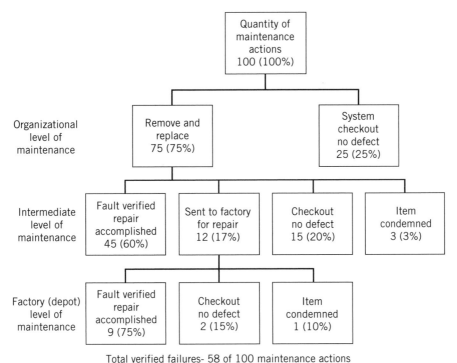

Total verified failures- 58 of 100 maintenance actions

FIGURE 3.19   System XYZ unscheduled maintenance actions.

for example, key inputs in determining system availability, which, in turn, is a major element of system effectiveness. Although the specific measures may vary significantly from one system application to the next, *availability*, is used quite often as a system measure. Availability can be expressed as follows:

$$A_o = \frac{MTBM}{MTBM + MDT} = \frac{uptime}{uptime + downtime} \quad (3.21)$$

where A_o is operational availability. This definition of availability relates to the consumer's operational environment where MTBM reflects *all* maintenance requirements and MDT represents *all* downtime considerations. In instances in which a producer is responsible for designing a system to meet a certain availability requirement, and the producer has no influence or control of the consumer's support structure, it may be appropriate to define availability as

$$A_a = \frac{MTBM}{MTBM + \overline{M}} \quad (3.22)$$

where A_a is achieved availability. It should be noted that the LDT and ADT factors are not considered here. Progressing one step further, there are instances in which availability is defined as

$$A_i = \frac{MTBF}{MTBF + \overline{M}ct} \tag{3.23}$$

where A_i represents inherent availability. Note that preventive maintenance is not included here. Employing this figure of merit as a system measure may be appropriate from a contractual standpoint where the producer is somewhat isolated from the consumer environment. However, in dealing with system engineering requirements, the A_0 factor is more relevant than either the A_a or A_i factors.

Figures 1.4 and 2.27 show two sides of the balance. The reliability and maintainability factors described herein are significant contributors (along with performance) in measuring the *technical* effectiveness of the system. Reliability and maintainability parameters are combined to determine availability, and system availability constitutes a major input in determining system effectiveness. At the other end of the balance is life-cycle cost (LCC). LCC is a function of research and development cost, production/construction cost, operation and support cost, and retirement and disposal cost. The consequences of reliability and maintainability have a direct impact on each of these major cost categories. However, the greatest impact of these design characteristics is on operational and support costs, where the frequency of maintenance and downtime factors are significant in determining the overall support capability for the system. If these characteristics are not appropriately considered in system design, the "iceberg" effect illustrated in Figure 1.7 will likely prevail.

The material presented to this point is intended to provide a familiarization with the terms and definitions associated with maintainability. Maintainability is one of the many disciplines requiring consideration within the overall context of system engineering. A general understanding of the subject is necessary, as well as some familiarity with the activities that are usually undertaken in the performance of a typical maintainability program. Some key terms and definitions have been covered; it is now appropriate to describe related program activities.

In implementing a maintainability program for a typical large-scale system, the tasks identified in Figure 3.20 are generally applicable. Although there are variations from one situation to the next, the performance of these tasks in terms of overall program phasing is assumed to be in accordance with Figure 3.21. The major program phases and system-level activities are derived from the baseline presented in Figure 1.12 (Chapter 1).[24]

In Figure 3.20, the maintainability program tasks listed can be categorized under (1) program planning, management, and control (tasks 1–4), (2) design and analysis (tasks 5–12), and (3) test and evaluation (task 13). The first category of tasks must be closely integrated with system engineering activities and reflected in the SEMP. The

---

[24] Although specific maintainability tasks should be tailored to the system and associated program needs, the tasks listed in Figure 3.20 are assumed to be typical for the purposes of discussion.

Program Task	Task Description and Application
1. Maintainability Program Plan	To develop a maintainability program plan that identifies, integrates, and assists in the implementation of all management tasks applicable in fulfilling maintainability program requirements. This plan includes a description of the maintainability organization, organizational interfaces, a listing of tasks, task schedules and milestones, applicable policies and procedures, and projected resource requirements. This plan must tie directly into the System Engineering Management Plan (SEMP).
2. Review and control of suppliers or subcontractors	To establish initial maintainability requirements and to accomplish the necessary program review, evaluation, feedback, and control component supplier/subcontractor program activities. Supplier program plans are developed in response to the requirements of the overall Maintainability Program Plan for the system.
3. Maintainability program reviews	To conduct periodic program and design reviews at designated milestones; (e.g., conceptual design review, system design reviews, equipemnt/software design reviews, and critical design review). The objective is ensure that maintainability requirements will be achieved.
4. Data collection, analysis, and corrective-action system	To establish a closed-loop system for data collection, analysis, and the initiation of recommendations for corrective action: The objective is to identify potential maintainability design problems.
5. Maintainability modeling	To develop a maintainability model for making initial numerical allocations, and for subsequent estimates to evaluate system/component maintainability. As design progresses, maintainability top-down functional flow block diagrams, logic troubleshooting flow diagrams, and so on, are developed and are used as a basis for accomplishing periodic predictions, logistic support analysis, and testability analysis. These should evolve directly from the system-level maintenance functional flow block diagrams.
6. Maintainability allocation	To allocate, or apportion, top system-level requirements to lower indenture levels of the system (e.g., subsystem, unit, assembly). This is accomplished to the depth necessary to provide specific criteria as an input to design.
7. Maintainability prediction	To estimate the maintainability of a system (or components thereof) based on a given design configuration. This is accomplished periodically throughout the system design and development process to determine whether the initially specified system requirements are likely to be met given the proposed design at that time.
8. Failure mode, effect, and criticality analysis (FMECA)—maintainability information	To identify potential design weaknesses through a systematic analysis approach considering all possible ways in which a component can fail (the modes of failure), the possible causes for each failure, the likely frequency of occurrence, the criticality of failure, the effects of each failure on system operation (and on various system components), and any corrective action that should be initiated to prevent (or reduce the probability of) the potential problem from occuring in the future. The objective is to determine maintainability design requirements as a result of anticipated corrective and/or preventive maintenance needs. Refer to Case Study C.1, Appendix C.
9. Maintainability analysis	To accomplish various design-related studies pertaining to equipment packaging schemes, fault-isolation and diagnostic provisions, built-in test versus external test equipment, levels of repair, component standardization, producibility considerations, and so on. Maintainability mathematical models, level-of-repair analysis models, and life-cycle cost analysis models are utilized as required.
10. Maintenance task analysis (MTA)	To evaluate design data and determine weaknesses relative to the maintainability characteristics incorporated in the design, and to determine the maintenance and support resources required for the system. Refer to Case Study C.4, Appendix C.
11. Level-of-repair analysis (LORA)	To evaluate system components to determine whether it is more economical to repair the item or to discard it in the event of failure. Refer to Case Study C.5, Appendix C.
12. Maintainability data for the detailed maintenance plan and the supportability analysis (SA)	To identify and prepare maintainability data as they apply to the various elements of logistic support—spare and repair parts, test and support equipment, personnel quantities and skill levels, training, facilities, technical manuals, and software.
13. Maintainability demonstration	To plan and implement a program where testing is accomplished (either sequential testing or a "fixed" sample size), using a preproduction prototype and considering statistical "accept" and "reject" criteria, to measure the maintainability characteristics of the system. These characteristics may include $\overline{M}_{ct}$, MLH/OH, $\overline{M}_{pt}$, or equivalent. This test is accomplished prior to entering production.

FIGURE 3.20   Maintainability engineering program tasks.

second group of tasks constitutes tools used in support of the mainstream design engineering effort, in response to maintainability program requirements included in the system specification and the program plan. The third area of activity, maintainability demonstration, must be integrated with system-level testing activities and covered in the TEMP. Although these tasks are primarily in response to maintainability program

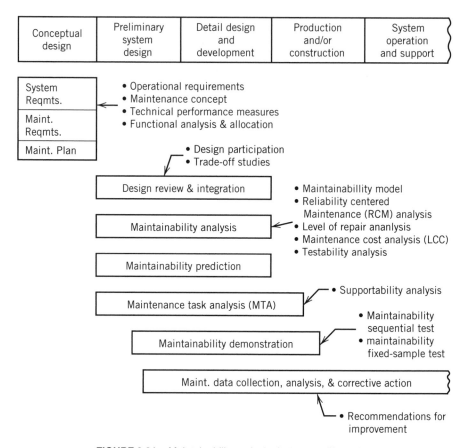

FIGURE 3.21   Maintainability tasks in the system life cycle.

requirements, there are many interfaces with basic design functions and with other supporting disciplines, such as reliability and logistic support.

Although brief task descriptions are included in Figure 3.20, seven dditional comments, as they pertain to a select few, are provided for purposes of emphasis.

1. *Maintainability program plan.* Although the requirements for a maintainability program may specify a separate and independent effort, it is *essential* that the program plan be developed as part, of, or in conjunction with, both the Reliability Program Plan (refer to Figure 3.14, task 1) and the SEMP. Organizational interfaces, task input-output requirements, schedules, and so on, must be integrated with reliability program requirements and must be directly supportive of system engineering activities. Moreover, maintainability activities must be closely integrated with human factors and logistic support functions and must be included in the respective plans for these program areas. The SEMP is introduced in Section 1.5 (refer to Figure 1.26) and is described further in Chapter 6.

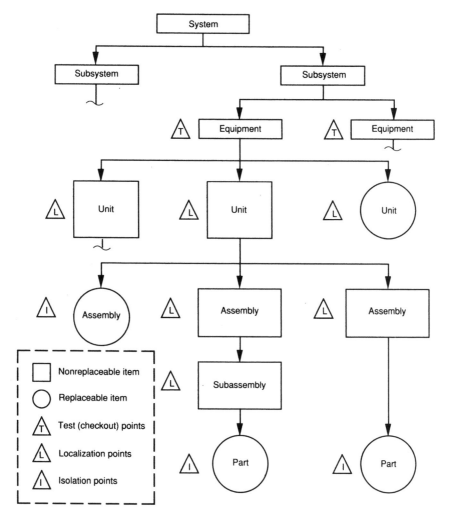

FIGURE 3.22 System/decomposition for maintainability analysis and prediction.

2. *Maintainability modeling.* The completion of this task, along with several others (e.g., allocation, prediction, FMECA, maintainability analysis), depends on the development of functional-level diagrams, similar to the one presented in Figure 3.22. These diagrams should evolve directly from, and must support, the system functional analysis and associated functional flow diagrams described in Section 2.7 (refer to Figures 2.12 to 2.17). The objective is to illustrate system packaging concepts, diagnostic capabilities (depths of localization and fault isolation), items that are repaired in place or removed for maintenance, and so on. The results of this task constitute a major input to the maintenance task analysis (MTA) and the supportability analysis (SA) and must be provided in a timely manner.

3. *Failure mode, effect, and critically analysis* (FMECA). FMECA, as it applies to maintainability, is primarily used as an aid in the development of system packaging schemes and diagnostic routines and is employed to assist in determining critical preventive maintenance requirements. This task should be closely integrated with reliability and logistics activities, because the FMECA is also a required task in these program areas. Case Study C.I, Appendix C describes the FMECA process.

4. *Maintainability analysis.* This includes the accomplishment of many different design-related studies dealing with system functional packaging concepts, levels of diagnostics, levels of repair, built-in versus external test, and so on. It must be accomplished in conjunction with the FMECA and maintainability modeling, and it must be coordinated with logistic support analysis (LSA) requirements. The LSA also requires a level-of-repair analysis and life-cycle cost analysis in fulfilling the requirements related to the design for supportability. Case Study C.6, Appendix C, describes an evaluation of alternative design configurations accomplished in support of a maintainability analysis effort.

5. *Maintenance task analysis* (MTA). This includes a detailed analysis and evaluation of the system to (a) assess a given configuration relative to the degree of incorporation of maintainability characteristics in design and compliance with the initially specified requirements and (b) determine the maintenance and logistic support resources required to support the system throughout its planned life cycle. Such resources may include maintenance personnel quantities and skill levels, spares and repair parts and associated inventory requirements, tools and test equipment, transportation and handling requirements, facilities, technical data, computer software, and training requirements. Such an evaluation may be accomplished during the preliminary and detail design phases utilizing available design data as the source of information and/or through a review and assessment of an existing item using checklists as an aid. An MTA may be conducted on a commercial off-the-shelf (COTS) item in the event that the maintenance resource requirements have not already been identified. This task should be closely coordinated with human-factors activities (i.e., the operator task analysis and the development of operational sequence diagrams) and with logistics activities (i.e., the MTA is an integral part of the logistic support analysis effort). Case Study C.4, Appendix C, includes an abbreviated example of the results of an MTA.[25]

6. *Level-of-repair analysis* (LORA). This includes an evaluation of various system components to determine whether it is economically feasible to repair an item or to discard it in the event of failure. If repair is to be accomplished, should the component be repaired at the intermediate level or at the factory (i.e., depot)? A LORA may be performed initially, in the development of the system maintenance concept, to provide design guidelines for packaging, diagnostics, and so on, and later in the evaluation of a given design configuration to determine maintenance resource

---

[25] A more in-depth presentation of the MTA, its content, and the procedure for accomplishing such is included in B. S. Blanchard, *Logistics Engineering and Management*, 6th ed. (Upper Saddle River, NJ: Pearson Prentice-Hall, 2004).

requirements. The LORA should be performed in conjunction with the MTA and as part of the logistic support analysis effort. Case Study C.5, Appendix C, includes an example of the LORA process.

7. *Maintainability demonstration.* This task, usually performed as part of type 2 testing, should be defined in .the context of the *total* system test and evaluation effort. The objective of maintainability demonstration is to simulate different maintenance task sequences, record the associated maintenance times, and verify the adequacy of the resources required to support the demonstrated maintenance activities (e.g., spare/repair parts, support equipment, software, personnel quantities and skills, and data). The results from this activity should not only determine whether maintainability requirements have been met, but should also help to determine whether the supportability objectives have been met in response to logistic support requirements. Maintainability demonstration requirements must be covered in the TEMP.

In summary, the tasks identified in Figure 3.20 are generally performed in response to some detailed specification or program requirement. Like reliability tasks, these tasks are completed on a relatively independent basis for many programs. Yet the interfaces are numerous, and there are some excellent opportunities for task integration, resulting in reduced program costs. Figure 3.23 conveys an example of

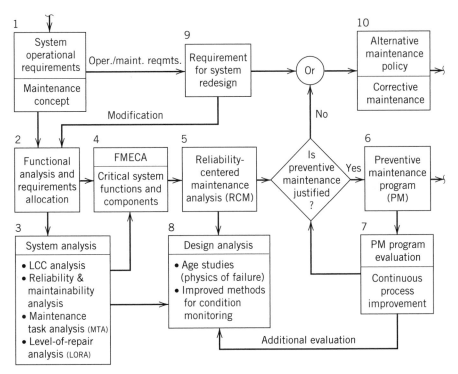

FIGURE 3.23 Example of the relationships between selected reliability and maintainability tools.

the relationships between selected reliability and maintainability tools. As one progresses further through this text, the opportunities for integration will become even more apparent. The intent of this section is to provide an introduction to the requirements associated with most maintainability programs.

3.4.4 Human-Factors Engineering[26, 27]

Quite often in the development of a system, the emphasis is on the design of hardware and software and the *human* element tends to be ignored. For a system to be complete, the human being and the interfaces between the human and the other elements of the system (e.g., equipment, software, facilities, data, elements of support) must be addressed. Optimum hardware or software design alone will not guarantee effective results.

The requirements for the "human" (i.e., operator, maintainer, supporting personnel) stem from the functional analysis, along with the requirements for hardware, software, and so on (see Figure 1.13). From this point, operational and maintenance functions are broken down into job operations, duties, tasks, subtasks, and task elements, as illustrated in Figure 3.24. Through subsequent analyses, the various activities and tasks to be performed by the human are combined and related in terms of personnel types, quantities, skill levels, and proposed assigned workstations. This, in turn, leads to the definition of training requirements and the development of training support (e.g., simulators, equipment, software, facilities, data/information). As the design evolves through the steps identified in Figure 3.24, it is essential that the proper level of integration be accomplished with the development of hardware, software, and so on, as the interfaces are many, and continuous.

In the development of a system for human beings, specific considerations in design must include four factors:

1. *Anthropometric factors.* Anthropometry deals with the measurement of the dimensions and the physical characteristics of the human body (e.g., standing height, sitting height, arm reach, breadth, buttock–knee length, hand size, and weight). When establishing basic design requirements involving the human being (for work space application, work surface design, control panel layout), one obviously must take into consideration the physical dimensions of the human body. Both "structural" dimensions (when the body is fixed and in a static state) and "functional" dimensions (when the body is engaged in some physical activity and in a dynamic state) must be measured and used in designing for the performance of operational functions and

---

[26]The objective is to provide an introduction to human factors (or human engineering), but not to cover the subject in depth. However, for more information, three good references are (1) A. Chapanis, *Human Factors in Systems Engineering* (Hoboken, NJ: John Wiley & Sons, Inc., 1996); (2) G. Salvendy, ed., *Handbook of Human Factors and Ergonomics*, 3rd ed. (Hoboken, NJ: John Wiley & Sons, Inc., 2006); and (3) M. S. Sanders and E. J. McCormick, *Human Factors Engineering and Design*, 7th ed. (New York: McGraw-Hill, 1993). Additional references are included in Appendix F.

[27]Although the term *human factors* is used throughout this text, other terms often applied in covering the same material include *ergonomics and human engineering*, and there are other variations of these.

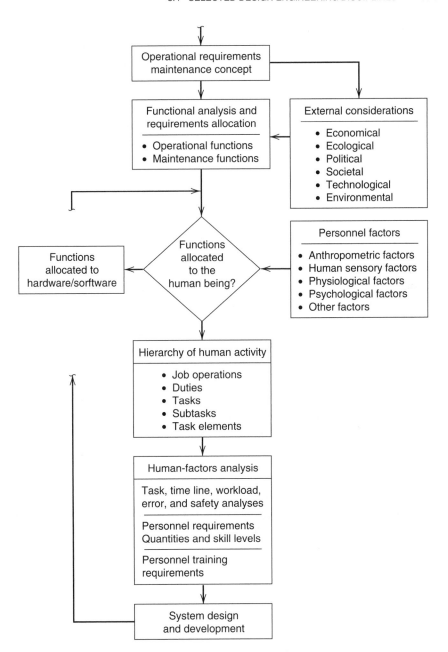

FIGURE 3.24   Human-factors requirements.

maintenance functions. Further, the design engineer must consider both *male* and *female* dimensions, with the appropriate ranges of variability (usually from the 5th to the 95th percentiles). For instance, the height of a male may range from 63.6 inches (5th percentile) to 68.3 inches (50th percentile) or 72.8 inches (95th percentile), and the height of the female from 59.0 inches (5th percentile) to 62.9 inches (50th percentile) or 67.1 inches (95th percentile). Although the average values may be used, the design of work spaces, surfaces, and so on, must consider possible variations for both male and female operators and maintainers; for example, from the 5th percentile female to the 95th percentile male. For specific design criteria, the reader should refer to additional sources.[28]

2. *Human sensory factors.* This category relates to the human sensory capacities, particularly sight or vision, hearing, feel or touch, smell, and so on. In the design of workstations, surfaces, operator consoles, and panels, the engineer must be cognizant of the human's capability relative to sight as it pertains to vertical and horizontal fields of view, angular fields of view, the detection of certain objects from different angles, the detection of certain colors and varying degrees of brightness from different angles, and so on. The placement of panel displays and controls as a function of use and the employment of different color combinations to facilitate the accomplishment of manual tasks require knowledge of the human being's capability for seeing. In addition, the designer needs to understand the human's capacity for hearing in terms of both frequency and intensity (or amplitude). The design of work areas for oral communications and/or the use of auditory displays requires knowledge relative to the effects of noise on the performance of work. For instance, as the noise level increases, a human begins to experience discomfort and both productivity and efficiency decrease. If the noise level approaches 120 to 130 dB, then a physical sensation in some form, or pain, will likely occur. In essence, the system designer needs to integrate the capabilities of the human into the final product.[29]

3. *Physiological factors.* Although the study of physiology is obviously well beyond the scope of this text, it is appropriate to recognize the effects of environmental stresses on the human body during the performance of manual tasks. *Stress* refers to any type of external activity, or environment, that acts on an individual in such a manner as to cause a degrading impact. Some typical causes of stress are (1) high and low temperatures, or temperature extremes, (2) high humidity, (3) high levels of vibration, (4) high levels of noise, and (5) large amounts of radiation or toxic

---

[28]Anthropometry data are included in National Aeronautics and Space Administration (NASA), *Anthropometric Source Book.* Vol. 1; *A Handbook of Anthropometric Data*, Vol. 2; and *Annotated Bibliography*, Vol. 3; NASA Reference Publication 1024, 1978. Also refer to Kroemer and Kroemer (2001) and Sanders and McCormick (1993) in Appendix F.

[29]Human sensory factors are covered further in H. P. Van Cott and R. G. Kinkade, eds., *Human Engineering Guide to Equipment Design* (Washington, DC: U.S. Government Printing Office, 1972); and M. S. Sanders and E. J. McCormick, *Human Factors in Engineering Design*, 7th ed. (New York: McGraw-Hill, 1993).

substances in the air. To varying degrees, these environmental effects will negatively impact on human performance; that is, physical fatigue will occur, motor response will be slower, mental processes will slow down, and the likelihood of error will increase. These externally related stress factors will normally result in individual human "strain." *Strain* may, in turn, have an impact on any one or more of a human's biological functions (e.g., the circulatory system, digestive system, nervous system, and respiratory system). Measures of strain may include parameters such as blood pressure, body temperature, pulse rate, and oxygen consumption. These factors of strain, caused by external stresses, will definitely have an impact on the performance of human operator and maintenance functions if the design fails to consider the physiological effects on the human.

4. *Psychological factors.* This category relates to the factors that pertain to the human mind; that is, the emotions, traits, attitudinal responses, and behavioral patterns as they relate to job performance. All other conditions may be perfect relative to completing a task in an effective manner. However, if the individual operator (or maintenance technician) lacks the proper motivation, initiative, dependability, self-confidence, communication skills, and so on, the likelihood of performing the task in an effective manner is extremely low. Generally, a person's attitude, initiative, motivation, and so on, are based on the needs and expectations of the individual. This, in turn, is a function of system design and the organizational environment within which the individual performs. If the tasks to be completed are perceived as being too complex, the individual may become frustrated, a poor attitude may develop, and errors will occur. On the other hand, if the tasks are too simple and routine, there is little challenge, boredom prevails, and errors will occur as a result of attitude. Further, as an external factor, the management style of the supervisor may cause an attitudinal problem. In any event, it is appropriate to consider the possible psychological effects on the human in the design and development of a system.[30]

In addition to considering the aforementioned general characteristics associated with the human, it is necessary to have some understanding of the human's ability to deal with and process information. Whether a function should be automated or accomplished by the human and, if accomplished by the human, to what extent, is dependent on the human's ability to detect, react, and process information. Figure 3.25 portrays a simple information-processing model, which includes four basic subsystems. The *sensing* subsystem responds to specific types of energy identified through the human senses (i.e., vision, hearing, feeling, smelling). This provides the stimulus to initiate some form of action. The *information-processing* subsystem addresses the human's capacity to receive and process information. Of particular interest is the type and amount of information the human can transmit (often expressed in terms of

---

[30] Additional information on human behavioral characteristics, psychological factors, motivation, attitude, leadership characteristics, and so on, may be found in most texts dealing with organizational theory, organizational dynamics, behavioral science, and related subjects.

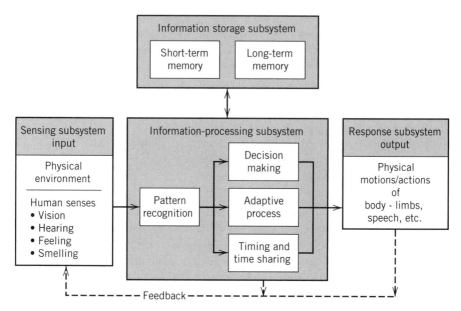

FIGURE 3.25 The processing of information and subsequent human response (simplified).[31]

"bits") and the rate at which he or she can transmit it. The *storage* subsystem refers to the human memory and its capacity, or the ability to retrieve data and facilitate the information-processing activity. Finally, there is the *response* subsystem, which allows the accomplishment of some function/task through a combination of physical motions (i.e., the output from the model). Inherent within this model is the *feedback* loop, which helps to verify that the responses are accurate in terms of the original input.

In the implementation of a human-factors program for a typical large-scale system, the tasks identified in Figure 3.26 are generally applicable. There are (1) program planning, management, and control tasks (tasks 1–3), (2) design and analysis tasks (tasks 4–13), and (3) test and evaluation tasks (tasks 14 and 15). In addition, some of these tasks have been presented, in terms of the life cycle, in Figure 3.27. Although brief task descriptions are included in Figure 3.26, some additional comments pertaining to a few are provided here for emphasis:

- *Human-factors program plan.* Although the requirements for a human-factors program may specify a separate and independent effort, it is *essential* that the program plan be developed as part of, or in conjunction with, the reliability

---

[31]Figure 3.25 constitutes a modified version of Figure 2.1 in H. P. Van Cott and R. G. Kinkade, *Human Engineering Guide to Equipment Design*, rev. ed. (Washington, DC: U.S. Government Printing Office, 1972).

Program Task	Task Description and Application
1. Human-factors program plan	To develop a human-factors program plan that identifies, integrates, and assists in the implementation of all management tasks applicable in fulfilling human-factors engineering requirements. This plan includes a description of the human-factors organization, organizational interfaces, a listing of tasks, task schedules and milestones, applicable policies and procedures, and projected resource requirements. This plan must tie directly into the System Engineering Management Plan (SEMP).
2. Review and control of suppliers or subcontractors	To establish initial human-factors requirements and to accomplish the necessary program review, evaluation, feedback, and control of component supplier/subcontractor program activities. Supplier program plans are developed in response to the requirements of the overall human-factors program plan for the system.
3. Human-factors program reviews	To conduct periodic program and design reviews at designated milestones (e.g., conceptual design review, system design reviews, equipment/software design reviews, and critical design review). The objective is to ensure that human-factors requirements will be achieved.
4. System analysis (mission analysis)	To determine the overall capabilities and the performance requirements for the system, and to develop appropriate mission scenarios identifying basic activity sequences. This should be accomplished as part of the system requirements definition process in conceptual design.
5. Functional analysis	To identify the major functions that the system is to perform (based on operational requirements), and to develop functional flow block diagrams defining system design requirements in functional terms. This task must track the system-level functional analysis.
6. Function allocation	To conduct trade-off studies, evaluate, and determine the resources required in accomplishing the functions identified through the Functional Analysis activity (i.e., determining the "hows" versus the "whats"), particularly in situations where there are human-machine interfaces.
7. Detailed operator task analysis	To evaluate functions that are to be accomplished by the human, and to establish a hierarchical breakdown to the lowest level where human activity exists (i.e., job operation, duty, task, subtask, and task element). Personnel quantity and skill-level requirements are identified through analysis.
8. Operational sequence diagrams	To identify the human-machine interfaces, and to develop a sequential flow of information, decisions, and actions through the generation of operational sequence diagrams (OSDs).
9. Time line analysis	To select and evaluate critical task sequences, and to verify that the necessary events can be performed and that they are compatible in terms of allocated time; Can the tasks be performed within the appropriate time allotted for accomplishing the mission?
10. Workload analysis	To evaluate human-operator activities throughout a given mission scenario (or through a number of designated scenarios) to determine the workload level (e.g., the relationship between the maximum time allowed and the actual time for task performance).
11. Error analysis	To systematically determine the various ways in which errors can be made by the human, and to make design recommendations to reduce the likelihood of such errors occurring in the future. This task is comparable to the reliability FMECA, except that the system/equipment failures are the result of *human* errors.
12. Safety analysis	To systematically evaluate, through cause-and-effect analysis, the effects of system/equipment failures on safety. Although safety pertains to both personnel and equipment, the aspect of *personnel* safety is emphasized herein. This task ties in directly with the reliability FMECA and the human-factors error analysis.
13. Models and/or mock-ups	To develop a three-dimensional physical model or a mockup of the system (or a component thereof) to demonstrate human-machine interfaces, spatial relationships, equipment layouts, panel displays, accessibility provisions for maintenance, and so on.
14. Training program requirements	To plan and implement a formal training program. This includes the determination of personnel training requirements (quantity of personnel and the skill levels desired as an output), categories of training, training equipment, training data, training facilities, mockups and models, special training aids, and so on. The plan should include a description of the training organization, a listing of tasks, task schedules and milestones, policies and procedures, and projected resource requirements.
15. Personnel test and evaluation	To plan and implement a program to physically demonstrate human-machine interfaces, task sequences, task times, personnel quantity and skill-level requirements, the adequacy of operating procedures, the adequacy of personnel training, and so on. This test and evaluation activity is accomplished prior to entering production.

FIGURE 3.26   Human-factors engineering program tasks.

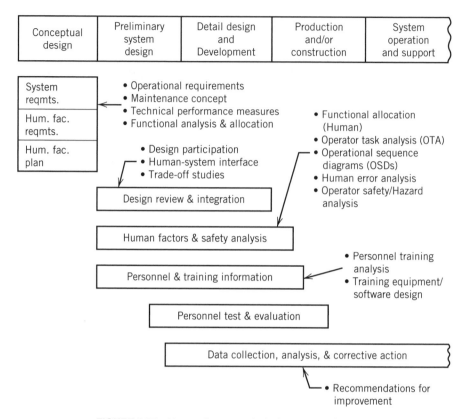

FIGURE 3.27 Human-factors tasks in the system life cycle.

program plan (Figure 3.14, task 1), the maintainability program plan (Figure 3.20, task 1), and the SEMP. Many of the activities in each of the plans are mutually supportive and require integration in terms of task input-output requirements, schedules, and so on.

- *Functional analysis.* The purpose of a functional analysis (in this context) is to identify those functions that are to be performed by the human being and where there is a human–machine interface. This activity should evolve directly from, and must support, the system functional analysis and associated functional flow diagrams described in Section 2.7 (refer to Figures 2.12–2.17).

- *Detailed operator task analysis.* This part of the overall human-factors analysis effort constitutes the expansion of major functions from the system functional analysis into job operations, duties, tasks, and so on. Ultimately, this will lead to the definition of operator and maintenance personnel requirements, in terms of quantities and skill levels, and the subsequent development of training program requirements (Figure 3.26, task 14). With the identification of personnel and training requirements, close coordination must be established with reliability, maintainability, and logistics program activities, as there are common interests in this area.

- *Operational sequence diagrams.* As part of the human-factors design analysis effort, operational sequence diagrams (OSDs) are developed to show various groups of activities involving the human–machine interface. An example of an OSD is presented in Figure 3.28, where a communications sequence between operators and workstations is illustrated. Through a symbolic presentation, different actions are shown that, in turn, lead to the identification of specific design requirements. Of significance is the requirement that OSDs must evolve from the functional analysis.

- *Personnel test and evaluation.* The purpose of this task is to demonstrate selected human activity sequences to verify operating/maintenance procedures and to ensure compatibility between the human and other elements of the system. Demonstrations are conducted using a combination of analytical computer simulations, physical mock-ups (wooden, metal, and/or cardboard), and preproduction prototype equipment. Computerized simulations may include the insertion of a 5th percentile female or a 95th percentile male into a work space, in a sitting or standing position, in order to evaluate activity sequences and space requirements. A great deal of information can be acquired through use of the appropriate computer graphics employing a three-dimensional database. Type 2 testing, using preproduction prototype equipment, may include the use of personnel, trained as recommended from the results of task 14, in the performance of selected operator and/or maintenance task sequences accomplished in accordance with approved procedures. The conductance of such tests should not only allow for the evaluation of critical human–machine interfaces, but should provide reliability information pertaining to operator functions, maintainability data when maintenance tasks are performed, verification and validation of information in formal technical manuals/procedures, verification of the adequacy of the training program for operator and maintenance personnel, and so on. Basically, this activity must be coordinated with other testing requirements and must be covered in the TEMP.

In summary, many of the tasks identified in Figure 3.26 (and the tools/techniques used in accomplishing them) are interrelated, the interfaces are many, and they feed on one another. Figure 3.29 provides an example showing the relationships between the functional analysis, the operator task analysis (OTA), the development of operational sequence diagrams (OSDs), the development of training requirements, and the appropriate feedback loop. In addition, note that a safety/hazard analysis has been included, as personnel safety is a major issue in the design for human factors.[32]

---

[32]The safety/hazard analysis is discussed further in Section 3.4.5.

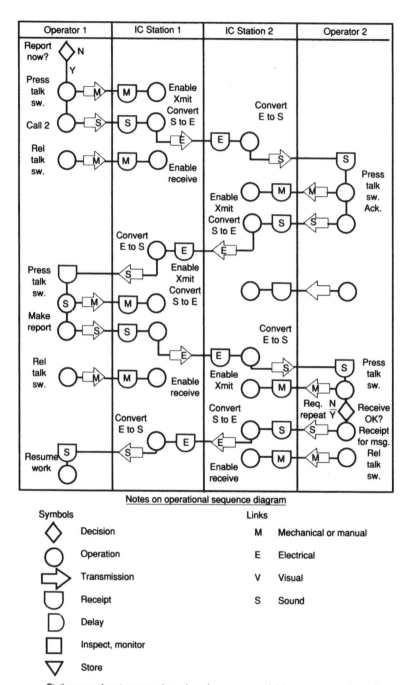

Notes on operational sequence diagram

Symbols		Links	
◇	Decision	M	Mechanical or manual
○	Operation	E	Electrical
⇨	Transmission	V	Visual
▽	Receipt	S	Sound
D	Delay		
□	Inspect, monitor		
▽	Store		

Stations or subsystems are shown by columns; sequential time progresses down the page.

FIGURE 3.28 Example operational sequence diagram. *Source:* MIL-H-46855, Military Specification, "Human Engineering Requirements for Military Systems. Equipment and Facilities" (Washington, DC: Department of Defense).

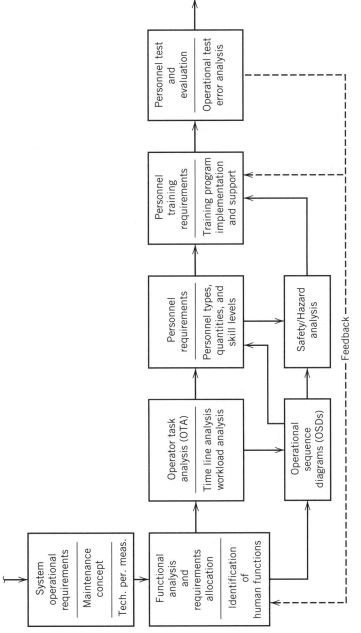

FIGURE 3.29 The application and relationships of selected tools/methods used for human factors in design.

3.4.5   Safety Engineering[33]

Safety is a system design characteristic. Certain materials selected for the design and construction of a system element may produce harmful toxic effects on the human; the placement and mounting of components may cause injuries to the operator and/or the maintainer; the use of certain fuels, hydraulic fluids, and/or cleansing liquids may result in an explosive environment; the location of certain electronic components close together may cause the generation of an electrical hazard; the performance of a series of strenuous tasks during the operation or maintenance of the system may cause personal injury; and so on.

Safety is important, both from the standpoint of the human operator and/or maintainer and from the standpoint of the equipment and other elements of the system. Through faulty design, one can create problems that may result in human injury. Moreover, problems can be created that result in damage to other elements of the system. In other words, the concerns in design deal with both personal safety and equipment safety.

Relative to the system design and development process, safety engineering requirements are comparable to those described for reliability, maintainability, and human factors (Sections 3.4.2, 3.4.3, and 3.4.4, respectively). Figure 3.30 provides a listing of safety program tasks for a typical large-scale system. There are (1) program planning, management, and control tasks (tasks 1–3), (2) design and analysis tasks tasks (4–7), and (3) test and evaluation tasks (tasks 8 and 9). There are three basic tasks shown in Figure 3.30 that require additional comment:

1. *System Safety Program Plan.* Although the requirements for this task may specify a separate and relatively independent effort, it is essential that the program plan be developed as part of, or in conjunction with, the reliability program plan (Figure 3.14, task 1), the maintainability program plan (Figure 3.20, task 1), the human-factors program plan (Figure 3.26, task 1), and the System Engineering Management Plan (SEMP). Tasks 4 and 5 of the safety program (fault-tree analysis and hazard analysis) are closely related to the reliability FMECA, the maintainability analysis (diagnostics and testability analysis), and the human-factors safety analysis. Task 7 should tie in with the reliability FRACAS and the maintainability task 4 (data collection and analysis). Task 8 (training program) should be related to the human-factors task 14. Task 9 (testing) should be coordinated with reliability tasks 18–20, maintainability task 13, and human-factors task 15. Many of the activities in each of the plans are mutually supportive and require integration in terms of task input-output requirements, schedules, and so on.

2. *Fault-tree analysis (FTA).* This is an ongoing top-down analytical process, using deductive analysis and Boolean methods, for determining system events that will, in turn, cause undesirable events, or hazards. Further, these events are ranked in terms

---

[33]Two good references for a more in-depth coverage of the subject are (1) H. E. Roland and B. Moriarity, *System Safety Engineering and Management*, 2nd ed. (Hoboken, NJ: John Wiley & Sons, Inc., 1990); and (2) N. J. Bahr, *System Safety Engineering and Risk Assessment: A Practical Approach* (New York: Taylor & Francis, 1997).

Program Task	Task Description and Application
1. System Safety Program Plan	To develop a system safety program plan that identifies, integrates, and assists in the implementation of all management tasks applicable in fulfilling safety engineering require-ments. This plan includes a desription of the safety engineering organization, organizational interfaces, a listing of tasks, task schedules and milestones, applicable policies and procedures, and projected resource requirements. This plan must tie directly into the System Engineering Management Plan (SEMP).
2. Review and control of suppliers or subcontractors	To establish initial system safety requirements and to accomplish the necessary program review, evaluation, feedback, and control of component supplier/subcontractor program activities. Supplier program plans are developed in response to the requirements of the overall System Safety Program Plan for the system.
3. System safety program reviews	To conduct periodic program and design reviews at designated milestones (e.g., conceptual design review, system design reviews, equipment/software design reviews, and critical design review). The objective is to ensure that safety engineering requirements will be achieved.
4. Fault-tree analysis (FTA)	To accomplish a fault-tree analysis (FTA) for determining system events that may cause undesireable events (or hazards), and to establish a ranking of these undesireable events. Fault-tree diagrams are developed from early hazard analyses, critical paths are identified, and probable causes are noted (a top-down approach). This task is closely related to the reliability FMECA. Refer to Case Study C.2, Appendix C.
5. Hazard analysis	To accomplish an analysis of the system with the objective of (a) identifying all major hazards and the anticipated probability of occurance, (b) identifying the "cause" factors that will result in a hazard, (c) evaluating the impacts (effects) on the system in the event that hazards occur, and (d) categorizing the identified hazards (i.e., catastrophic, critical, marginal, negligible). This task is closely related to the reliability FMECA and the Human-Factors Safety Analysis.
6. Risk analysis	To initiate a risk management program for the evaluation and control of the probability of occurrance and the consequences of hazardous events. Risk analysis, risk assessment, and risk abatement activities are included.
7. Data collection, analysis, feedback, and corrective action	To plan and implement a data collection and reporting capability for identifying and evaluating potential areas of risk. Participate in failure analysis activity and in accident investigations as appropriate. Recommendations for corrective action are initiated in areas when potential risk exists.
8. Safety training program	To plan and implement a training program covering the procedures and steps necessary to ensure that operator and maintenance personnel are properly trained in the performance of all system functions. This includes consideration of the requirements for training materials and data, training equipment, training aids, training facilities, and so on.
9. Safety test and evaluation	To plan and implement a program to test the system (and its components) to ensure that it can be safely operated and maintained, and that all necessary safety precautions have been taken. This test and evaluation activity is accomplished prior to entering production.

FIGURE 3.30   Safety engineering program tasks.

of their influence in causing the potential hazards. Fault-tree logic diagrams are de-veloped commencing with the top event and proceeding downward through succes-sive levels of causation steps, determining at each level what the next set of events will be. Fault-tree analysis is closely related to both reliability and maintainability analysis, particularly in considering possible symptoms and frequencies of failure, diagnostic and test routines, and so on. Case Study C.2, Appendix C, describes the FTA approach.

3. *Hazard analysis.* The objective of this task is to evaluate the design and deter-mine possible events that may result in hazards at the system level. By simulating failures, critical activities, and so on, at the component level, one can (through a cause-and-effect analysis) identify possible hazards, anticipated frequency of occur-rence, and classification in terms of criticality. Recommendations for design change are made where appropriate. This task, with regard to methodology and objectives, is very closely related to the reliability FMECA (which also categorizes events in terms of criticality) and the human-factors safety analysis.

In summary, the tasks identified in Figure 3.30 are generally performed in response to some detailed program requirement and are often completed on an independent basis. However, the interfaces are numerous, and it is essential that these requirements be appropriately integrated into the overall system engineering process.

3.4.6 Security Engineering[34]

Although not usually included within the class of the more traditional disciplines associated with engineering and the design of systems, the issue of *security* has certainly assumed a high priority in view of the continuing threats of terrorism and the terrorist acts that are taking place in today's world. Thus, there is an added dimension that must be addressed within the overall spectrum of system engineering: the *design for security*. The question at this point is, How does one design a system to preclude the planned introduction of faults/failures that will cause the system (or any portion thereof) to be completely destroyed, resulting in the damage of material, facilities, and/or the loss of life? The objective, of course, is to prevent an individual (or group of individuals) from intentionally sabotaging a system for one reason or another.[35]

Although such a problem may be caused intentionally versus inadvertently, the goal here is similar to the design objectives specified within the disciplines of human-factors engineering and safety engineering. In human-factors engineering, one of the objectives is to design a system to preclude the introduction of faults by the operator (or maintainer) that will result in the system's not being able to perform its mission. In safety engineering, an objective is to design a system such that faults cannot be introduced that will result in system damage and/or personal injury/death. In both cases, the major concern is related to the possibility of inducing problems in the process of performing system functions during the accomplishment of a mission, in the performance of a maintenance task, and/or in the accomplishment of a support activity. The assumptions in this case relate to the possibility that such problems may occur though some unintentional act or series of acts.

In designing for security, it is necessary to go one step further by addressing the issue of intent. The question is, What characteristics should be incorporated in the design of a system that will prevent (or at least deter) one or more individuals from

---

[34]Two references for a more in-depth coverage are R. Anderson, *Security Engineering: A Guide to Building Dependable Distributed Systems* (Hoboken, NJ: John Wiley & Sons, 2001); and D. S. Herrmann, *A Practical Guide to Security Engineering and Information Assurance* (New York: CRC Press/Auerbach, 2001).

[35]Subsequent to 9/11, there has been a great deal of emphasis on security and the design for security. In the defense sector, in particular, an added requirement in the development of new (and the modification of existing) systems has been the inclusion of the necessary characteristics in design to counter the threat of terrorism.

intentionally inducing faults that will destroy the system, cause harm to personnel, and/or have an impact that will endanger society and the associated environment? In response, the design should consider the following:

1. The development and incorporation of an external security alarm capability that will detect the presence of unauthorized personnel and prevent them from operating, maintaining, and/or gaining access to the system and its elements, and one that will ultimately lead to the prevention of an "outsider" from inducing a problem that will result in system damage or destruction.

2. The incorporation of a "condition-based monitoring" capability that will enable one to check the status of the system and its elements on a continuing basis. To accomplish this requires the appropriate sensors, readout devices, inspection methods, and the like, be included that will verify that the system and its components are in the condition intended and that the appropriate diagnostics be incorporated that will lead to the correction of any problem that may exist. An objective is to initially determine (through inspection and/or test methods) that the system is in satisfactory condition and to provide the necessary subsequent controls that will ensure that this condition will continue to exist.[36]

3. The incorporation of a built-in capability (mechanisms) that will detect and initiate an alarm in the event that problem is detected and a design that will, in the event of a problem, prevent a subsequent chain reaction of failures leading to system damage or destruction.

In other words, the designer must address such issues as (1) preventing unauthorized personnel from gaining access to the system in question, (2) being able to initially determine the condition of the system and the follow-on monitoring of its components at all times, and being able to control the processing of these components as they progress through the *forward* and *reverse* flow of activities identified in Figure 1.20, and (3) being able to both detect and subsequently prevent any failures that are induced through incorporation of the appropriate characteristics in the system design.[37]

At this point (and in summary), it should be emphasized that it is certainly easier to define a problem than to arrive at a proposed solution, and much remains to be accomplished to ensure better system security in the future. Hence, it is anticipated

---

[36] A major challenge for the future is to develop the appropriate sensors and inspection methods that will allow for the proper condition-verification of *all* of the materials, cargo containers, and related items that are being transported both internally and worldwide. The current absence of such a capability constitutes a potential threat.

[37] An objective in system design is to determine the cause-and-effect relationships among the various system elements/components, and the effects of a system failure on the mission being accomplished. Some failures, of a more catastrophic or critical nature, will ultimately result in system damage, destruction, and/or personal injury. The goal is to design the system to prevent these failures from occurring. An excellent tool that may be utilized to facilitate this objective is the FMECA; see Case Study C.1, in Appendix C.

that a great deal of research and design effort will be expended from here on to arrive at better solutions for the problem at hand.

3.4.7 Manufacturing and Production Engineering[38]

The role of manufacturing/production may take several forms, including the construction of a single one-of-a-kind system entity and the production of a quantity of similar items. In the first case, there is an obvious strong interface between the design activity and the follow-on construction of the system, which, in turn, is based on the recommended design configuration. In the second situation, one needs to (1) design the product that is to be manufactured for *producibility* and (2) design the manufacturing/production capability to be both effective and efficient in producing that product. A major goal in the application of system engineering requirements is to address these various life-cycle activities and their interfaces, as conveyed in Figure 1.10.[39]

In regard to a product and its design configuration, a key objective is to *design for producibility*. "Producibility" is a measure of the relative ease and economy of producing an item. The characteristics of design must be such that the item can be produced easily and economically, using conventional and flexible manufacturing methods and processes without sacrificing function, performance, effectiveness, or quality. Four major objectives are as follows:

1. The quantity and variety of components utilized in system design should be held to a minimum. Common and standard items should be selected where possible, and there should be a number of different supplier sources available throughout the planned life cycle of the system.

2. The materials selected for constructing the system should be standard, available in the quantities desired and at the appropriate times, and should possess the characteristics for easy fabrication and processing. The design should preclude the specification of peculiar shapes requiring extensive machining and/or the application of special manufacturing methods.

3. The design configuration should allow for the easy assembly (and disassembly as required) of system elements; that is, equipment, units, assemblies, and modules. Assembly methods should be simple, repeatable, and economical and should not require the utilization of special tools and devices or high personnel skill levels.

4. The design configuration should be simple, to the extent that the system (or product) can be produced by more than one supplier, using a given data

---

[38] A good reference that presents some of the current trends in manufacturing is P. M. Swamidass, *Innovations in Competitive Manufacturing* (Boston, MA: American Management Association (AMACOM), 2002). Refer to Appendix F for additional references.

[39] When referring to a *product*, the assumption is that we are dealing with a relatively large repairable entity versus a smaller nonrepairable commercial consumable item.

package and conventional manufacturing methods/processes. The design should be compatible with the application of computer-aided design (CAD)/computer-aided manufacturing (CAM) technology where appropriate.

Figure 3.31 presents a simplified step-by-step approach addressing some important considerations in design. Referring to the eighth block in the figure, the design review checklist in Appendix D (Item 21), or something equivalent, may be utilized to provide additional guidance in this area.

In considering the design characteristics of the manufacturing/production capability itself, there are a number of goals and objectives that are important, particularly in view of the current trends pertaining to increased globalization and greater international competition, the need for producing a wide variety of products in shorter time frames, the need to reduce product costs, and so on (refer to Section 1.1 and Figure 1.4). More specifically, there has been a great deal of emphasis on *flexibility* and *agility*. The central theme in "agile manufacturing" is to develop a capability that can react quickly in producing a wide variety of high-quality products, with continuously changing configurations, in a short time frame, with rapid response and maximum customer satisfaction as the goal. Another key objective relates to *lean production*, which emphasizes the elimination of waste in the utilization of all resources, including people and time. At the same time, there has been a great deal of activity related to improving all of the functions within the supply chain (e.g., purchasing, materials handling, transportation and distribution, customer service), as well as modernizing some of the business processes necessary in the manufacture of products. The development of electronic commerce (EC) methods has enabled the integration and rapid processing of information and data packages supporting key business operations. For example, the advent of the *Enterprise Resource Planning* (ERP) approach has enabled the integration of manufacturing operations and other functions of a given firm with suppliers and customers.

Although the aforementioned areas of activity are primarily dedicated to improving the operations of a manufacturing/production capability, one must also address the life-cycle issues associated with the maintenance and support of this capability. There have been a number of instances where a relatively high percentage of the cost of a product has been be attributed to the maintenance costs associated with the equipment in the factory that is used to manufacture/produce that product, with such costs being amortized and assigned to the product. Thus, in a highly competitive environment, one must consider not only the operational issues but the maintenance and support issues as well.[40]

---

[40]Refer to Section 1.4.4 and a description of the concept of *total productive maintenance* (TPM). This concept was first introduced in 1971, primarily because of the low level of effectiveness in manufacturing products and the resulting high costs of maintenance experienced in many factories at the time. Subsequently, implementation of the principles and concepts of TPM have become popular internationally and have been adopted by many factories throughout the world today. For additional information, refer to the bibliography in Appendix F, Section F.5 Maintainability Engineering and Maintenance.

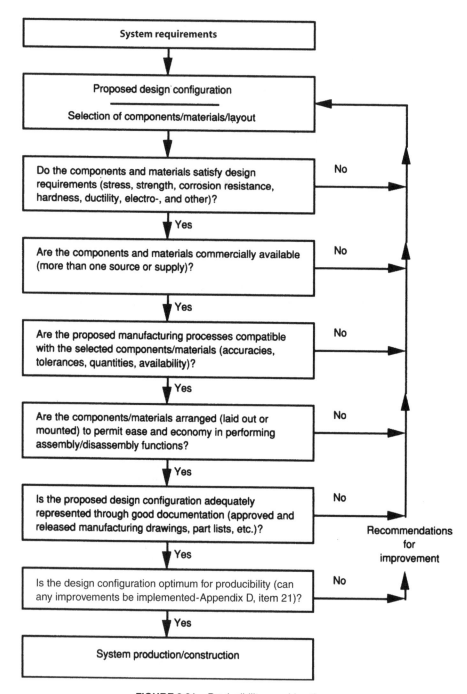

FIGURE 3.31 Producibility considerations.

3.4.8 Logistics and Supportability Engineering[41]

As shown in the system "operational and maintenance" flow diagram in Figure 1.20, there are a wide variety of activities conducted throughout the system life cycle (see Sections 1.4.3 and 1.4.4). Included in the *forward* flow of activities (i.e., from the supplier to the consumer/user) are the functions identified in Figure 1.21—purchasing, materials processing and handling, inventory management, packaging and transportation, warehousing and storage, distribution, customer service, information flow, and all of the related business practices that are necessary to support the effective and efficient implementation of the supply chain. Although the product *design* and *maintenance and support* interfaces are generally not addressed within the bounds of supply chain management (SCM), there has been much progress in recent years in modernizing the physical supply and distribution channels in the interest of improving the competitive position of firms worldwide.

Included in the *reverse* flow of activities shown in Figure 1.20 (i.e., from the consumer/user to the applicable maintenance facility and back) are the maintenance and support functions identified throughout the infrastructure illustrated in Figure 1.22, along with the required resources, which include the following general categories.[42]

1. *Manpower and personnnel.* Includes all personnel required in the installation, checkout, operation, handling, and sustaining maintenance of the system throughout its planned life cycle. Maintenance personnel considerations cover activities at all levels of maintenance, operation of test equipment, operation of facilities, and so on.

2. *Training, training equipment, and devices.* Includes the initial training of all system operator and maintenance personnel and the follow-on "replenishment" training to cover attrition and replacement personnel. Training equipment, training simulators, mock-ups, training data and manuals, special facilities, special devices and aids, and software to support personnel training operations are also included.

3. *Supply support.* Includes all spares (units, assemblies, modules, etc.), repair parts, consumables, special supplies, and related inventories needed to support prime mission-oriented equipment, software, test and support equipment, transportation and handling equipment, training equipment, and facilities. Provisioning documentation, procurement functions, warehousing, distribution of material, and personnel

---

[41]To gain a complete perspective of the field of logistics (presented in a broad context), it is recommended that additional study in this area be pursued. Four good references are (1) B. S. Blanchard, *Logistics Engineering and Management*, 6th ed. (Upper Saddle River, NJ: Pearson Prentice Hall, 2004); (2) J. J. Coyle, E. J. Bardi, and C. J. Langley, *The Management of Business Logistics*, 7th ed. (Mason, OH: South-Western, 2003); (3) E. H. Frazelle, *Supply Chain Strategy: The Logistics of Supply Chain Management* (New York: McGraw-Hill, 2002); and (4) *Journal of Business Logistics* (Oak Brook, IL: Council of Supply Chain Management Professionals (CSCMP)). Additional references are included in Appendix F.

[42]B. S. Blanchard, *Logistics Engineering and Management*, 6th ed. (Upper Saddle River, NJ: Pearson Prentice Hall, 2004).

associated with the acquisition and maintenance of spare/repair part inventories at all support locations are also included in this category.

4. *Test and support equipment.* Includes all tools, special condition monitoring equipment, diagnostic and checkout equipment, metrology and calibration equipment, maintenance stands, and servicing and handling equipment required to support operation, transportation, and scheduled and unscheduled maintenance actions associated with the system or product. Both "peculiar" (newly developed) and "standard" (existing and already in the inventory) items must be covered.

5. *Packaging, handling, storage, and transportation.* Includes all special provisions, materials, containers (reusable and disposable), and supplies necessary to support packaging, preservation, storage, handling, and/or transportation of prime mission-oriented equipment, test and support equipment, spares and repair parts, personnel, technical data, and mobile facilities. In essence, this category covers the initial distribution of products and the transportation of personnel and materials for maintenance purposes.

6. *Facilities.* Includes all special facilities needed for system operation and the performance of maintenance functions at each level. Physical plant, real estate, portable buildings, housing for personnel, intermediate maintenance shops, calibration laboratories, and special depot or overhaul facilities must be considered. Capital equipment and utilities (heat, power, energy requirements, environmental controls, communications, etc.) are generally included as part of facilities.

7. *Technical data.* Includes system installation and checkout procedures, operating and maintenance instructions, inspection and calibration procedures, overhaul procedures, modification instructions, facilities information, drawings and specifications, and associated databases that are necessary for the performance of system operation and maintenance functions. Information processing requirements (networks and equipment) are also included in this category.

8. *Computer resources.* Includes all software, computer equipment, tapes/disks, databases, and accessories necessary in the performance of system maintenance functions at each level. This covers condition monitoring requirements and maintenance diagnostic aids.

These basic elements of logistics and the maintenance and support infrastructure (also identified in Figure 1.23) must be completely integrated and viewed in the context of the "system" as an entity—that is, the combining and integration of all of the activities identified in Figures 1.21 and 1.22. Otherwise, there is no guarantee that system requirements will be met should a failure occur. Further, considering past experience and the downstream costs associated with system support (and the cause-and-effect relationships—refer to Figures 1.7 and 1.8), the ultimate requirements for these elements must be addressed in terms of the entire system life cycle, with emphasis in the early phases of design and development. More specifically, (1) the prime mission-related elements of the system must be *designed for supportability*, and (2) the logistics and maintenance support infrastructure must be designed so that it will provide for effective and efficient support through the system's planned life cycle. Thus, it is essential that these requirements be included and inherent within the system engineering process (refer to Figure 3.5).[43]

This life-cycle approach, with emphasis on system design, has been recognized in the defense sector and applied in the development of relatively large-scale defense systems through the introduction of the concept of "acquisition logistics."[44] *Acquisition logistics* can be defined as a

> multifunctional technical management discipline associated with the design, development, test, production, fielding, sustainment, and improvement modifications of cost-effective systems that achieve the user's peacetime and wartime readiness requirements. The principal objectives of acquisition logistics are to ensure that support considerations are an integral part of the system's design requirements, that the system can be cost-effectively supported throughout its life cycle, and that the infrastructure elements necessary to the initial fielding and operational support of the system are identified and developed and acquired.[45]

Inherent within the spectrum of acquisition logistics are a number of program activities, including initial planning; a variety of design-related tasks throughout the system development process; the identification, procurement, processing, distribution, and installation of the required elements of support at the appropriate consumer/user's operational sites; and the ongoing sustaining customer service and maintenance support of the system throughout its planned life cycle. An abbreviated discussion of key activities follows.[46]

1. *Integrated logistic support plan (ILSP)*. An ILSP (or a planning document of an equivalent nature) is usually initiated during the conceptual design phase and updated in preliminary system design; it covers all planning activities, design activities, procurement and acquisition activities, and sustaining support activities. Often included are individual subplans covering the different elements of the maintenance and support infrastructure and related life-cycle activities—for example, detailed maintenance concept/plan (including applicable logistics performance factors); reliability and maintainability plan (interface requirements); supportability analysis plan; supply support plan; test and support equipment plan; personnel training plan;

[43] Although the term *supportability* is primarily used through this text, similar terms such as *serviceability* and *sustainability* are also used interchangeably; for example, the former primarily in the commercial sector and the latter, assuming some recent emphasis, in the defense sector. Independent of such, the objective is to design the system so that it can be supported effectively and efficiently throughout its programmed life cycle.

[44] DOD 5000.2-R, *Mandatory Procedures for Major Defense Acquisition Programs (MDAPS) and Major Automated Information System (MAIS) Acquisition Programs* (Washington, DC: Office of the Secretary of Defense, April 5, 2002), Section C5.2.3.5.4.

[45] MIL-HDBK-502, *Acquisition Logistics* (Washington, DC: Department of Defense, May 30, 1997), Section 4.1.

[46] While many of the terms and definitions (such as ILS, PBL, SA, etc.) are more commonly related to major *defense* systems, the same type of requirements may be applicable to any system, including large nondefense systems as well. The important point here is to view the objectives of each item and apply the principles accordingly.

technical data plan; packaging, handling, storage, and transportation plan; facilities plan; distribution and user support plan (customer service); postproduction support plan; information systems plan; and system retirement plan.

The ILSP includes a description of logistics concepts, research results, and acquisition strategy; logistics organization, supplier requirements, and organizational interfaces; a listing of program tasks, task schedules, major milestones, and applicable policies and procedures; projected resource requirements; and areas of program risk. Basically, the ILSP must cover all of the applicable logistics and related activities identified by forward and reverse flows in Figure 1.20. The ILSP must tie directly into the SEMP, particularly in regard to those tasks dealing with logistics engineering (see Figure 1.26; the SEMP is covered further in Chapter 6).

2. *Logistics engineering.* Logistics engineering commences with the definition of specific design-to requirements evolving from the development of system operational requirements, the maintenance concept, and the identification and prioritization of technical performance measures (refer to Sections 2.4, 2.5, and 2.6). These requirements are further delineated through the accomplishment of the functional analysis and requirements allocation process (refer to Sections 2.7 and 2.8). From this point on, there are requirements pertaining to the day-to-day design participation process, including the initial establishment of design-to criteria, conductance of trade-off analyses, accomplishment of a supportability analysis (SA), review of supplier activities, participation in formal design reviews, participation in test and evaluation (validation) activities, and so on. In essence, the area of logistics and system support must be represented and included as a "member" of the design team and must be involved in the ongoing design integration activities (refer to Sections 2.9, 2.10, and 2.11).

3. *Performance-based logistics (PBL) and associated design-to requirements.* As indicated in Section 2.6, a QFD analysis approach is utilized to aid in the identification and prioritization of specific quantitative design-to goals for the system; an example identifying the results of such an analysis is presented in Figure 2.11. Although the factors (requirements) included in the table primarily pertain to the prime mission-related elements of the system and its design for supportability, there is a need to further delineate these requirements down to the maintenance and support infrastructure (conveyed in the illustration presented in Figure 2.10). While specific requirements will vary for each system, Figure 3.32 provides an example of some of the measures/metrics that may apply to each of the major elements of support. If one is to ultimately realize the objectives that have been emphasized throughout this text, specific design-to requirements (established from the beginning) must be applied to *all* of the elements of the system, not limited to just those elements that are directly involved in accomplishing a given mission scenario.[47]

---

[47]PBL is addressed in DOD 5000.2-R, *Mandatory Procedures for Major Defense Acquisition Programs (MDAPS) and Major Automated Information System (MAIS) Acquisition Programs* (Washington, DC: Office of the Secretary of Defense, April 5, 2002), Section C2.8.3.

Maintenance and Support Infrastructure
- Effectiveness (reliability) of support capability
- Logistics responsiveness (response times)
- Efficiency of support (cost per support action)

Supply Support
- Spares/repair parts demand rates(s)
- Mean time between replacement (MTBR)
- Spares/repair parts processing time
- Inventory item location time
- Probability of system success with spares
- Probability of spares availability
- Inventory stockage level(s)
- Inventory turnover rate(s)
- Economic order quantity (EOQ)
- Cost/supply support action

Test, Meas., Handling, and Support Equip.
- Utilization rate (period of usage)
- Utilization time (test station processing time)
- Availability of equipment
- Reliability (MTBF, λ)
- Maintainability (MTBM, $\overline{M}$ct, $\overline{M}$pt, MDT)
- Calibration rate and cycle time
- Cost/test action
- Cost/hour of utilization

Maintenance Facilities
- Number of items processed/period
- Item process times(s)
- Item turnaround time (TAT)
- Waiting line (length of queue)
- Materials consumption rate
- Utility consumption/Maintenance action
- Utility consumption/period
- Cost/Maintenance action

Maintenance and Support Personnel
- Personnel quantities and skill levels
- Personnel attrition rate (turnover rate)
- Maintenance labor hours/maint. action
- Personnel error rate(s)
- Cost/Person/Organization

Training and Training Support
- Quantity of personnel trained/Period
- Number of personnel training days/Period
- Frequency and duration of training
- Training program input/output factors
- Training data/Student
- Training equipment/Program
- Training software/Program
- Cost/person trained

Packaging, Handling, Stor., and Trans.
- Transportation mode, route, distance, Frequency, time, and cost
- Packaging materials/Items shipped
- Container utilization rate
- Effectiveness of transportation (reliability)
- Successful delivery rate
- Package damage rate

Computer Resources
- Software reliability/Maintainability
- Software complexity (Lang./Code level)
- Number of software modules/syst. element
- Cost/Element of software

Technical Data/Information Systems
- Number of data items/system
- Data format and capacity
- Data access time(s)
- Database size
- Information processing time
- Change implementation time

FIGURE 3.32 Selected technical performance measures for the support infrastructure.

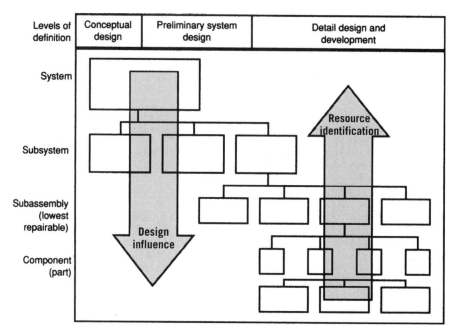

FIGURE 3.33 Supportability analysis emphasis. *Source:* Adapted from AMC Pamphlet 700-22, USAMC Material Readiness Support Activity, Lexington, KY.

4. *Supportability analysis (SA)*. A supportability analysis is an ongoing iterative analytical process, included within the context of the overall system analysis activity, with the basic objectives of initially *influencing design* and subsequently *determining logistic support resource requirements*, based on an assumed design configuration. In Figure 3.33, these objectives are best accomplished through the integration and application of various analytical techniques/methods utilized to ensure that logistics and supportability requirements are considered in the design process. Basically, the SA (which is inherent within the system engineering process):

(a) Aids in the initial establishment of PBL metrics and supportability requirements during conceptual design through the evaluation of system operational requirements, alternative technology applications, and alternative logistics and maintenance support concepts. Through the development and prioritization of system requirements, design criteria are established for the logistics and maintenance support infrastructure and are included in the appropriate specifications.

(b) Aids in the evaluation of alternative system, or equipment/software, design configurations (e.g., alternative material applications, repair policies, packaging schemes, diagnostic routines, and the selection of components). This includes the ongoing process of synthesis, analysis, and design optimization, utilizing trade-off studies to arrive at a recommended approach for supportability.

(c) Aids in the evaluation of a given design configuration (whether final or interim) to determine the specific logistic support resource requirements for that configuration. Resource requirements include personnel quantities and skill levels, training, spares/repair parts and related inventories, test and support equipment, packaging and transportation, facilities, maintenance software, and data/information. The maintenance task analysis (MTA), supplemented through the utilization of other models, constitutes the database for the determination of these requirements (refer to Appendix C, Case Study C.4).

(d) Aids in the ultimate measurement and evaluation (i.e., assessment) of an operating system being utilized by the consumer in the user's environment. Field data are collected, analyzed, and utilized to update the SA, which was initially based on design data. The objective is to determine the *true* effectiveness of the system, the *true* effectiveness of the logistics and maintenance support infrastructure, and so on, and to provide the appropriate feedback and any feasible recommended changes for system improvement (refer to Section 2.11.5 covering system modifications).

The SA includes the evaluation of many alternatives, following the basic analysis steps illustrated in Figure 2.28. Inherent within this activity is the utilization of such tools as the life-cycle cost analysis (LCCA); level-of-repair analysis (LORA); maintenance task analysis (MTA); reliability-centered maintenance (RCM) analysis; failure mode, effects, and criticality analysis (FMECA); testability and diagnostic analysis; and so on. In essence, the application of reliability and maintainability analysis techniques/methods are inherent in the completion of SA requirements. Further, such analytical techniques such as simulation, linear and dynamic programming, queuing analysis, accounting methods, and control theory may be employed in solving a wide variety of problems.[48]

5. *Sustaining system support.* Given that a system design configuration has been established, there is a series of logistics activities to be performed, including the selection of suppliers, provisioning and procurement of materials and services, movement of items through the production process, and the transportation and distribution of products to the consumer's operational sites. As the system is introduced and delivered to the ultimate user, there may be some customer service requirements in the form of training and assistance in the performance of operational and maintenance tasks. Subsequently, there are those activities necessary for the sustaining maintenance and support of the system throughout its planned life cycle. The system engineering role here is that of *assessment* (data collection, analysis, and feedback) and verification that the system is in compliance with the initially specified requirements. The ultimate objective is, of course, to ensure complete customer satisfaction.

---

[48]Many of the principles and concepts pertaining to the SA have been implemented in the past under such titles as *logistic support analysis* (LSA), *maintenance engineering analysis* (MEA), *maintenance level analysis* (MLA), *maintenance engineering analysis record* (MEAR), *maintenance analysis data* (MAD), and similar titles. Although the titles have changed through the years, the intent, objectives, and methods of implementation have basically remained the same.

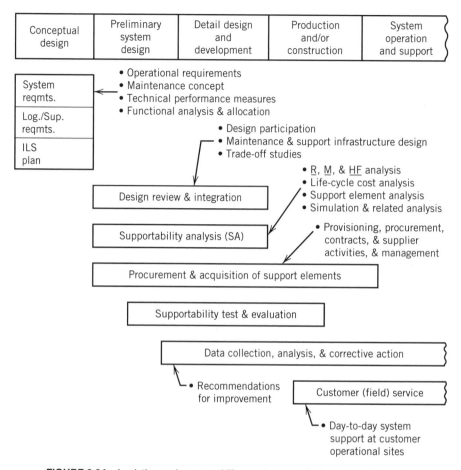

FIGURE 3.34 Logistics and supportability requirements in the system life cycle.

In summary, Figure 3.34 is included to show the various logistics- and supportability-related activities in the context of the system life cycle. The major program phases and system-level activities are derived from the baseline presented in Figure 1.12 (Chapter 1).

3.4.9 Disposability Engineering

Referring to Section 1.4 (and Figure 1.20), there is both a "forward" flow and a "reverse" flow of activities. While the emphasis thus far has been on the forward flow and the activities and resources associated with system design and development, production/construction, operation, and sustaining support of the system in accomplishing its intended mission, one must also deal with the reverse flow and the resources required and associated with such activities.

In Section 2.14 (and Figure 2.36), this reverse flow is amplified to include system retirement and the various system components that are phased out of the operational inventory for one reason or another. This includes components that are retired because: (1) they are obsolete as a result of technology upgrades and the incorporation of modifications for system improvement; (2) there is a reduction in inventory as a result of changes in mission requirements for the system; and (3) failures have occurred and the applicable faulty components must be recycled for repair and/or rework. In each situation, there are logistics requirements and the expenditure of maintenance support resources to complete the activities involved. That is, there are personnel needs, transportation requirements, facility and inventory requirements, special tools and test equipment requirements, data and information needs, and so on. Much of this can be included within the broad spectrum of what is often referred to as *reverse logistics.*[49]

As components are phased out of the operational inventory, they must be either recycled for other applications or disposed of in some way as to preclude any degradation to the environment. An objective is to make those products, which cannot be recycled and must be processed for disposal, out of completely biodegradable materials (or equivalent) so that they can be easily decomposed at the end of their service life without negatively impacting the environment. As is the case when specifying other system objectives, this objective must be addressed early in the system design process. Figure 3.35, which is an extension and expansion of Figure 2.36, conveys the process that should be implemented early.

As shown in Figure 3.5, the *design for disposability* becomes another major consideration in the early system design and development process. With the growing concern(s) associated with the dwindling of resources and the environment worldwide, addressing this issue will become even more important in the future and, thus, must be inherent within the overall system engineering process.

3.4.10 Quality Engineering[50]

In today's context, the word *quality* has come to mean more than it did in the past. Basically, it pertains to meeting or exceeding the requirements, expectations, and needs

---

[49] Many of the supply chain and maintenance support activities shown in Figures 1.21 and 1.22 are also applicable in the system retirement and material recycling/disposal phase. Addressing the subject of reverse logistics is increasing in importance, particularly in the commercial sector, as available material resources are dwindling worldwide and environmental impact concerns are increasing. Two references are H. Dyckhoff, R. Lackes, and J. Reese, eds., *Supply Chain Management and Reverse Logistics* (New York: Springer, 2003); and D. F. Blumberg, *Introduction to Management of Reverse Logistics and Closed Loop Supply Chain Processes* (New York: CRC Press/St Lucie, 2004).

[50] Selected references covering various facets of *quality, quality assurance, quality control,* and so on, are identified in the bibliography in Appendix F. Specifically, the reader is encouraged to review some of the writings of Crosby, Deming, and Juran to gain a better insight as to background, basic principles, and concepts. This section emphasizes some of the principles of *total quality management* (TQM); that is, total customer satisfaction, individual participation, continuous improvement, robust design, variability control, supplier integration, and management responsibility. More recently, emphasis in these areas has continued through the introduction of Six Sigma methods.

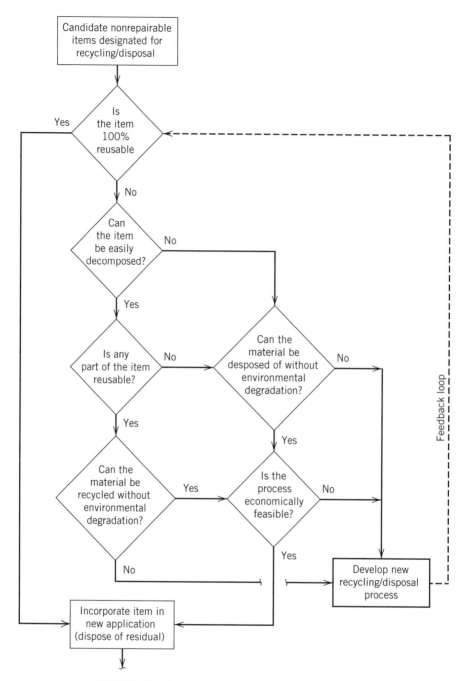

FIGURE 3.35   The material reuse/recycling/disposal process.

of the consumer (customer). The prime motivator is that of "survival" in a highly competitive international environment. In general, the availability of cost-effective, high-quality systems/products from international sources has been increasing, and competition is encouraging industries to do a better job in the design and production of systems and their components. As a result, the field of quality, although not new, is undergoing a continual change in emphasis.

In the past, the fulfillment of quality objectives has primarily been accomplished in the production and/or construction phase of the life cycle through the implementation of formal quality control (QC) or quality assurance (QA) programs. Statistical process control (SPC) techniques, incoming and in-process inspection activities, closely monitored supplier control programs, periodic audits, and selective problem solving methods have been implemented with the objective of attaining a designated level of system quality. In addition, the advent of such techniques as Six Sigma, applying Baldridge criteria for the purposes of evaluation, and the ever-increasing application and strengthening of ISO standards (e.g., ISO 9000 and ISO 14001) have aided in the maintenance of high-quality programs in many firms today. However, these efforts (although very effective in their application) have, for the most part, been accomplished "after the fact" and the overall results have been somewhat questionable.[51]

Recently, the aspect of quality has been viewed more from a top-down life-cycle perspective, and the concept of *total quality management* (TQM) has evolved. TQM can be described as a *total integrated management approach that addresses system/product quality during all phases of the life cycle and at each level in the overall system hierarchical structure.* It provides a before-the-fact orientation to quality, and it focuses on system design and development activities, as well as production, manufacturing, assembly and test, construction, product support, and related functions. TQM is a unification mechanism linking human capabilities to engineering, production, and support processes. It provides a balance between the "technical system" and the "social system." Specific characteristics of TQM include the following:

1. Total customer satisfaction is the primary objective, as compared with the practice of accomplishing as little as possible in conforming to the minimum requirements. The customer orientation is important (versus the "What can I get away with?" approach).

2. Emphasis is placed on the iterative practice of "continuous improvement" as applied to engineering, production and support processes, functions, and the like. The objective is to seek improvement on a day-to-day basis, as compared

---

[51]Three good references that include some discussion of the more recent quality-related methods/techniques, their applications, and results are (1) Y. Fasser and D. Brettner, *Management for Quality in High-Technology Enterprises* (Hoboken, NJ: John Wiley & Sons, Inc., 2002); (2) P. Swamidass, *Innovations in Competitive Manufacturing* (Boston, MA: American Management Association (AMACOM), 2002); and (3) A. Gaal, *ISO for Small 9001:2000 Business: Implementing Process-Approach Quality Management* (New York: CRC Press/St. Lucie, 2001).

with the often-imposed last-minute single thrust initiated to force compliance with some standard. The Japanese practice this approach through the implementation of a process known as *kaizen*.

3. In support of item 2, an individual understanding of processes, the effects of variation, the application of process control methods, and so on, is required. If individual employees are to be productive relative to continuous improvement, they must be knowledgeable of various processes and their inherent characteristics. Variability must be minimized (if not eliminated).

4. TQM emphasizes a total organizational approach, involving every group in the organization and not just the quality control function. Individual employees must be motivated from within and should be recognized as being key contributors to meeting quality objectives.

Included within the broad spectrum of TQM are the very important aspects of engineering and the *design for quality*; that is, quality engineering. The projected life cycles illustrated in Figure 1.10 (Section 1.3.1) must be considered in *total*. A system is conceived, designed, produced, utilized, and supported throughout its planned life cycle. As part of the initial system design effort, consideration must be given to (1) the design of the process that will be utilized to produce the system and (2) the design of the support configuration that will be utilized to provide the necessary ongoing maintenance and support for that system. As the interactions between these various facets of program activity are numerous, it is important that these areas be addressed on an integrated basis from the start.

These program relationships have been recognized through initiation of the concept of "concurrent engineering" (see Section 1.4.1), which is defined as "a systematic approach to the integrated, concurrent design of products and their related processes, including manufacture and support. This approach is intended to cause the developers, from the outset, to consider all elements of the product life cycle from conception through disposal, including quality, cost, schedule, and user requirements." The objectives of concurrent engineering include (1) improving the quality and effectiveness of systems/products through a better integration of requirements and (2) reducing the system/product development cycle through a better integration of activities and processes. This, in turn, should result in a reduction in the total life-cycle cost for a given system.

From the perspective of this text, the primary thrust is *quality engineering* and its role as a part of the system engineering process. In a relationship similar to that expressed for logistics (refer to Section 3.4.8), there is the larger concern for TQM and there are some specific concerns associated with quality as it pertains to engineering design. The following activities are thus considered appropriate in regard to system engineering:

- *Quality planning.* The development of a TQM plan (or equivalent) must be accomplished during conceptual design and updated during preliminary and detail design as required. Inherent within this overall plan are quality engineering activities including the (a) determination of engineering design requirements using a QFD, "house of quality," or equivalent approach (refer to Section 2.6); (b) evaluation and design of manufacturing and assembly processes in response to design technology decisions; (c) participation in the evaluation and selection of system components and supplier sources; (d) preparation of product, process, and material specifications as required (types C, D, and E); (e) participation in on-site supplier reviews; and (f) participation in formal design reviews. These and related activities should also be included in the SEMP.[52]

- *Quality in design.* This area of activity, viewed in the broad context, pertains to many of the issues discussed throughout the earlier sections of this chapter. Emphasis is directed toward design simplicity, flexibility, standardization, and so on. Of a more specific nature are the concerns for *variability*, whereby a reduction in the variation of the dimensions for specific component designs, or tolerances in process designs, will likely result in an overall improvement. Taguchi's general approach to "robust design" is to provide a design that is insensitive to the variations normally encountered in production and/or in operational use. The more robust the design, the less the support requirements, the lower the life-cycle cost, and the higher the degree of effectiveness. Overall design improvement is anticipated through a combination of careful component evaluation and selection, the appropriate use of statistical process control (SPC) methods, and application of experimental testing approaches, applied on a continuous basis.[53]

The subject of quality pertains to both the technical characteristics of design and the human aspects in the accomplishment of design activities. Not only is there a concern relative to the selection and application of components, but the successful fulfillment of quality objectives is highly dependent on the behavioral characteristics of those involved in the design process. A thorough understanding of customer requirements, good communications, a team approach, a willingness to accept the basic principles of TQM, and so on, are all necessary. In this respect, the objectives of quality engineering are inherent within the scope of system engineering.

3.4.11 Environmental Engineering

Although the previous sections in this chapter dealt primarily with some of the more tangible considerations in design, it is essential that one also address the aspect of

---

[52]*House of quality* refers to a basic methodology used to implement a "quality function deployment" (QFD) program. QFD focuses on planning and communications, using a cross-functional team approach. It provides a framework for assessing product attributes and for transforming them into engineering design requirements. Refer to J. R. Hauser and D. Clausing, "The House of Quality," *Harvard Business Review* (May–June 1988): 63–73.

[53]Genichi Taguchi developed some mathematical techniques relative to the evaluation of design variables, with the objective of reducing variability through continuous process improvement. Refer to (1) Y. Fasser, and D. Brettner, *Management for Quality in High-Technology Enterprises* (Hoboken, NJ: John Wiley & Sons, Inc., 2002), Section 13.2, pp. 245–248; and (2) P. J. Rose, *Taguchi Techniques for Quality Engineering* (New York: McGraw-Hill, 1988).

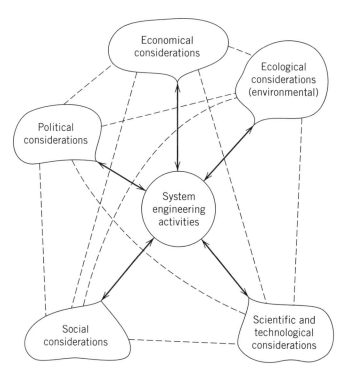

FIGURE 3.36   Environmental influences considered in system design and development.

design for the environment (DFE), "Environment," in this context, refers to the numerous external factors that must be dealt with in the overall system design and development process. In addition to the *technical* and *economic* factors discussed earlier (see Figure 1.24), one must deal with *ecological, political*, and *social* considerations as well. The system being developed must be compatible with, acceptable in, and ultimately must exist within an environment that addresses the many factors illustrated in Figure 3.36. A requirement within the spectrum of system engineering is to ensure that the system being developed will be socially acceptable, compatible with the political structure, technically and economically feasible, and will not cause any degradation to the environment overall.

Of particular interest here are the ecological considerations. *Ecology* generally pertains to the study of the relationships between various organisms and their environment. This includes consideration of plant, animal, and human populations in terms of rate of population growth, food habits, reproductive habits, and ultimate death. In other words, one is concerned with the conventional biological process as viewed in the broad context.

In recent decades, the world population growth, combined with the many technological changes associated with our living standards, has resulted in greater consumption of our natural resources, thereby causing potential shortages which, in turn, have stimulated shifts toward establishing other means for accomplishing our objectives. Concurrently, the amount of waste has increased significantly. The net effects of this trend have been alterations to the basic biological process, and to some extent these alterations have been harmful. Of particular concern are those problems resulting from the following:

- *Air pollution.* Any gaseous, liquid, toxic, or solid material suspended in air that can result in health hazards to humans. Air pollutants may fit into a number of categories: participate matter (small particles of substances in air resulting from fuel combustion, incineration of waste materials, or industrial processes), sulfur oxides, carbon monoxide, nitrogen oxides, and hydrocarbons.
- *Water pollution.* Any contaminating influence on a body of water brought about by the introduction of materials that will adversely affect the organisms living in that body of water (measure of dissolved oxygen content).
- *Noise pollution.* The introduction of industrial noise, community noise, and/ or domestic noise that will result in harmful effects on humans (e.g., loss of hearing).
- *Radiation.* Any natural or human-made energy transmitted through space that will result in harmful effects on the humans.
- *Solid waste.* Any garbage and/or refuse (e.g., paper, wood, cloth, metals, plastics, etc.) that will result in a health hazard. Roadside dumps, piles of industrial debris, junk car yards, and so on, are good examples of solid waste. Improper solid waste disposal may be a significant problem in view of the fact that flies, rats, and other disease-carrying pests are attracted to areas where there are solid wastes. In addition, there may be a significant impact on air pollution if windy conditions prevail or on water pollution if the solid waste is located near a lake, river, or stream.

Thus, in the development of systems and in the selection of components, the designer must be sure that the materials selected will not have a negative impact on any one (or more) of these ecological areas of concern, either when the system is operational and responding to a specific mission requirement or when the system is undergoing some form of maintenance. Throughout the utilization phase of the life cycle, there may be numerous instances when faulty components (residual material) will be removed and discarded in the accomplishment of system maintenance functions. Further, when obsolescence occurs and the system is ultimately retired from the inventory, there will be additional challenges relative to the disposal of its components. A prime objective is to design components so that they can be directly *reused* in similar applications where possible. If there are no opportunities for reuse, then the component should be designed so that it can be easily decomposed, with the residual elements being *recycled* and converted into materials that can be remanufactured for other purposes. In addition, the *recycling process* itself should not create any detrimental effects on the environment (see Figure 3.35).

In summary, all of the factors identified in Figure 3.36 need to be addressed in the design and development of systems, and on an integrated basis. The basic questions are, how will the introduction and operation of this new system capability impact the political, social, economical, and ecological infrastructure? How will the accomplishment of system maintenance and support activities influence this infrastructure? One can develop the best "technical" solution in the world, but it may not be feasible from a political perspective, or socially acceptable, or economically justifiable. An objective in system engineering is to achieve the proper balance among all of these factors.

3.4.12 Value/Cost Engineering (Life-Cycle Costing)[54]

The material presented thus far has primarily emphasized the *technical factors* associated with the system, as referenced in Figure 1.24. These factors, which include performance, reliability, maintainabiiity, human factors, supportability, and quality, represent only one side of the overall spectrum. The other side of the spectrum pertains to *economic factors*, and a proper balance between the two must be attained.

In the system evaluation process, these technical and economic factors are often combined in such a manner as to provide a measure of effectiveness (MOE) for a given system. Although these effectiveness figures of merit (FOMs) will vary from one application to the next, a few examples are noted:

$$\text{Effectiveness FOM} = \frac{\text{Performance} \times \text{Availability}}{\text{Life-cycle cost}} \tag{3.24}$$

$$\text{Effectiveness FOM} = \frac{\text{System capacity}}{\text{Revenues} - \text{Cost}} \tag{3.25}$$

$$\text{Effectiveness FOM} = \frac{\text{Life-cycle cost}}{\text{Facility space}} \tag{3.26}$$

$$\text{Effectiveness FOM} = \frac{\text{Supportability}}{\text{Life-cycle cost}} \tag{3.27}$$

In regard to the *economic* side of the balance, both *revenues* and *costs* must be considered, as conveyed in Figure 3.37, particularly in the commercial sector where the loss of revenues often represents a major segment of cost. However, the emphasis in this section is on cost; that is, the total cost of all activities throughout the system

---

[54]Value engineering, cost engineering, life-cycle costing, and related areas are covered further in (1) B. S. Blanchard and W. J. Fabrycly, *Systems Engineering and Analysis*, 4th ed. (Upper Saddle River, NJ: Prentice-Hall, 2006); (2) G. J. Thuesen, and W. J. Fabrycky, *Engineering Economy*, 9th ed. (Upper Saddle River, NJ: Prentice-Hall, 2001); and (3) Department of Defense Regulation 5000.2-R, *Mandatory Procedures for Major Defense Acquisition Programs (MDAPS) and Major Automated Information System (MAIS) Acquisition Programs* (Washington, DC: Office of the Secretary of Defense, latest edition). Additional references are noted in Appendix F. The life-cycle cost analysis (LCCA) process is covered in detail in Appendix B, and is illustrated through a case study in Appendix C.

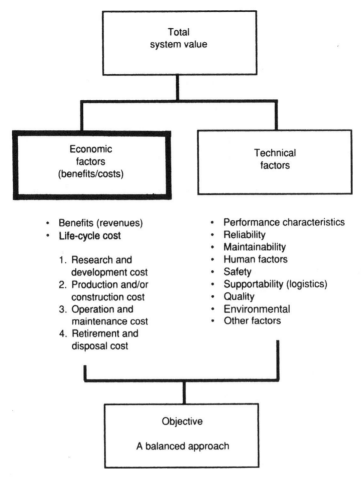

FIGURE 3.37 System evaluation factors.

life cycle. *Life-cycle cost* (LCC) includes the consideration of *all* future costs associated with research and development (i.e., design), construction and/or production, distribution, system operation, maintenance and support, retirement, and material disposal and/or recycling. It involves the costs of all technical and management activities throughout the system life cycle; that is, producer activities, contractor and supplier activities, and consumer or user activities. In addition, costs are often related to "functions" accomplished over the long term, as compared with the rather short-term perspective conveyed through the traditional accounting structure for most organizations. With this in mind, one may pose the following questions:

1. Do you know the total costs associated with each of the functions being accomplished within your company or organization?

2. Do you know what functions constitute the high-cost contributors over the long term? For a given system, what are the high-cost elements? What are the high-cost "drivers?"

3. Are you aware of the "cause-and-effect" relationships and the criticalities as they relate to the accomplishment of a given mission (or operational scenario)?

4. Can you identify the high-risk areas or elements of the system in question?

The answers to these and related questions are not easily attained. Yet individual design and management decisions are often based on some smaller aspect of cost (e.g., initial purchase price or acquisition cost) without first assessing the consequences of these decisions in terms of *total* cost. As conveyed in Section 1.3 (Chapter 1), many of the decisions made in the early stages of system design will have a large impact on the costs of downstream activities such as production, operations, maintenance and support, and retirement and material disposal. Although some of these early decisions may be necessary, the decision maker is remiss unless they are made in the context of total life-cycle cost. *Full-cost visibility* is essential if the risks associated with the decision-making process are to be properly assessed.

Life-cycle cost analyses, in one form or another, are performed throughout system design and development, during construction/production, and/or for the purposes of assessment while the system is being utilized in the field. The completion of such an effort generally requires that one follow certain steps, such as those presented in Figure 3.38.

In Figure 3.38, one of the first steps in the process is to describe the system in *functional* terms, and then to construct a functional flow diagram covering all of the activities in the system life cycle, evolving from the identification of need through retirement and material disposal (refer to Section 2.7). Given this requirement, it is necessary to develop a *cost breakdown structure* (CBS), such as shown in Figure 3.39. The CBS constitutes a vehicle for including all costs and is broken down to the depth required to provide the appropriate level of visibility for determining the costs of various functions, processes, and/or elements of the system over time. The CBS serves as a structure that will allow for the initial allocation of cost targets in a design-to-cost application (refer to Figure 2.23) and for the subsequent collection of costs in a *life cycle cost analysis*. Costs are estimated for each year in the system life cycle, inflationary and other influencing factors are included, costs profiles are developed, and costs are summarized by category in the CBS. The high-cost contributors are noted, cause-and-effect relationships are established, a sensitivity analysis is performed, feasible alternatives are evaluated, and recommendations are made based on the results.

A life-cycle cost analysis may serve many purposes, and the possible applications are varied, as conveyed in Figure 3.40. Of particular note is the use of LCC analysis in the evaluation of different design configurations in the early stages of system development, the evaluation of different commercial off-the-shelf (COTS) alternatives, and the evaluation of an existing system configuration with the objective of identifying the high-cost contributors leading to possible recommendations for product/

1. Describe the system configuration being evaluated in *functional* terms, and identify the appropriate technical performance measures (TPMs) or applicable metrics for the system.
2. Describe the system life cycle and identify the major activities in each phase as applicable (system design and development, construction and/or production, utilization, maintenance and support. retirement and disposal).
3. Develop a work breakdown structure (WBS), or cost breakdown structure (CBS), covering <u>all</u> activities and work packages throughout the life cycle.
4. Estimate the appropriate costs for each category in the WBS (or CBS), using activity-based costing (ABC) methods, or equivalent.
5. Develop a computer-based model to facilitate the life-cycle cost analysis process.
6. Develop a cost profile for the "baseline" system configuration being evaluated.
7. Develop a cost summary, identifying the high-cost contributors (i.e., high cost drivers).
8. Determine the "cause-and-effect" relationships, and identify the causes for the high-cost areas.
9. Conduct a sensitivity analysis to determine the effects of input factors on the analysis results. and identify the high-risk areas
10. Construct a Pareto diagram and rank the high-cost areas in terms of relative importance and requiring immediate management attention.
11. Identify feasible alternatives (potential areas for the improvement), construct a lifecycle cost profile for each, and construct a break-even analysis showing the point in time when a given alternative assumes a point in preference.
12. Recommend a preferred approach, and develop a plan for system modification and improvement (this may entail a modification of equipment or software, a facility change, and/or a change in some process). This constitutes an ongoing iterative approach for *continuous process improvement.*

FIGURE 3.38   The basic steps in a life-cycle cost analysis.

process improvement. In each application, the steps identified in Figure 3.38 and the process illustrated in Figure 3.41 are followed.

Figure 3.42 provides an example of LCC analysis applications in the system design and development process. Cost targets may be established initially in conceptual design through the development of TPMs (refer to Section 2.6). Trade-off studies are performed during the preliminary and detail design phases to support design and procurement decisions. During the latter stages of detail design and throughout the construction/production and system utilization phases, LCC analyses may be conducted for assessment of the overall cost-effectiveness of the system. Computer-based models are used to facilitate the analysis process (as shown in Figure 2.28). Figure 3.43 shows the LCC analysis as it may be applied throughout the system life cycle. For a more in-depth discussion of life-cycle costing, the analysis process, and its benefits, refer to Appendix B.

3.5 SOS INTEGRATION AND INTEROPERABILITY REQUIREMENTS

One of the most challenging areas in the design of any given system pertains to the many external interfaces that may exist between (1) the system in question and other

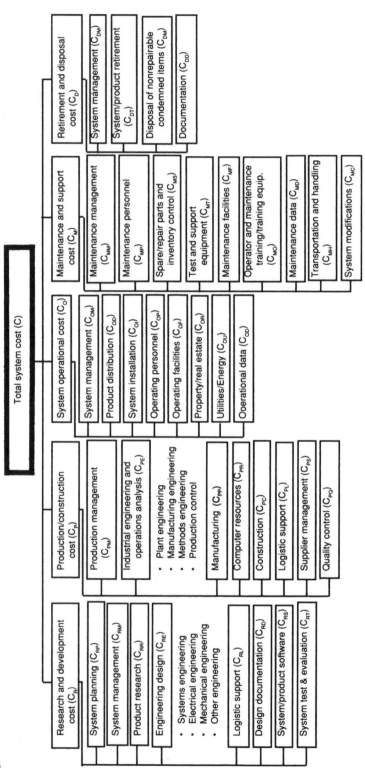

FIGURE 3.39 Sample cost breakdown structure (CBS).

206

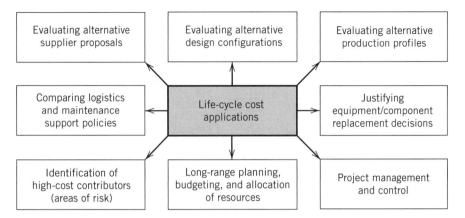

FIGURE 3.40   Life-cycle cost applications.

systems within the same system-of systems (SOS) configuration, and (2) with other systems operating independently and in the same environment. Although much of the discussion in the earlier sections of this chapter deal primarily with the design of a given system as a unique entity, projecting this system design configuration in the context of "other operating systems" is a critical final step in the overall design process.

Referring to Figure 3.44 (page 211) the first case illustrated in part (a), which is derived from Figure 1.3, presents an *aircraft system* as an element of an *air transportation system*, which is within an overall higher-level *transportation system* capability. The functional analysis and allocation requirements for the newly designed system, described in Sections 2.7 and 2.8, must be properly integrated within the overall vertical hierarchical structure, both upward and downward as appropriate. Further, it is important to ensure that the system of interest will not in any way be degraded as a result of external impacts evolving from other elements within the same SOS structure.

In the second case, illustrated in Figure 3.44(b), one must address the interrelationships and interaction effects between the newly designed system and other systems operating independently and in the same general geographical area. Thus, the *design for interoperability* assumes a degree of importance along with the design for reliability, design for maintainability, design for the environment, and so on. The questions are as follows:

- Will the newly designed system be able to operate effectively and efficiently when deployed and utilized?
- What are the external effects (i.e., impact) of the newly designed system on the operation of others systems in the user environment?
- What is the impact(s) of these other external systems on the new system?

A design objective is, of course, to preclude (i.e., or eliminate) any negative impacts from these various external system capabilities.

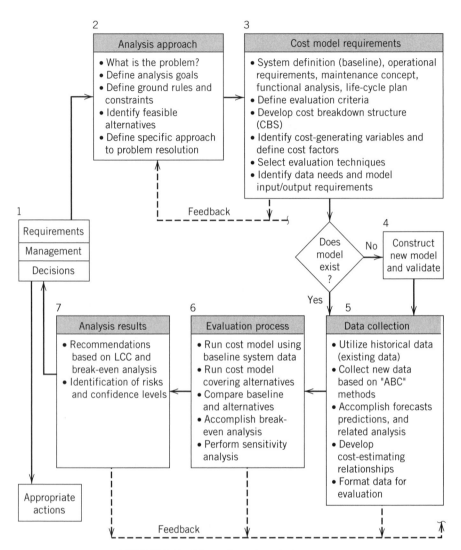

FIGURE 3.41 Life-cycle cost analysis process.

3.6 SUMMARY

Inherent within the system engineering process described in Chapter 2 are the requirements for reliability, maintainability, supportability, quality, and the like. A few design disciplines such as these have been selected for discussion in Section 3.4 of this chapter. In each instance, there are certain steps that are followed in order to meet the objectives as specified. Initially, the requirements for reliability, maintainability, and so on, must be established in defining operational requirements and the

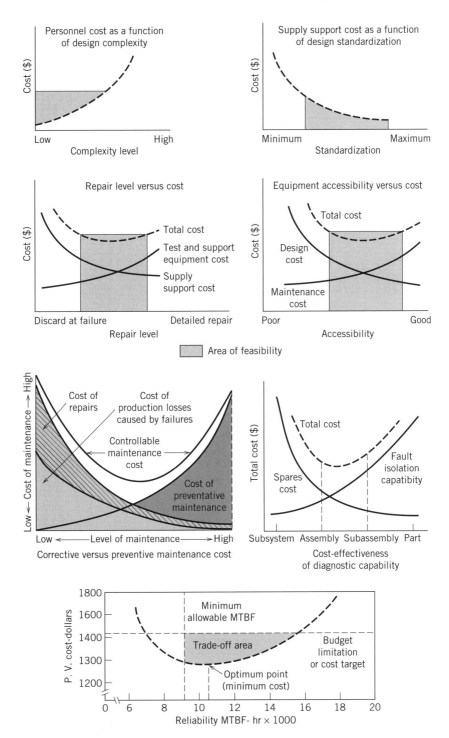

FIGURE 3.42 Examples of cost-effectiveness analysis applications.

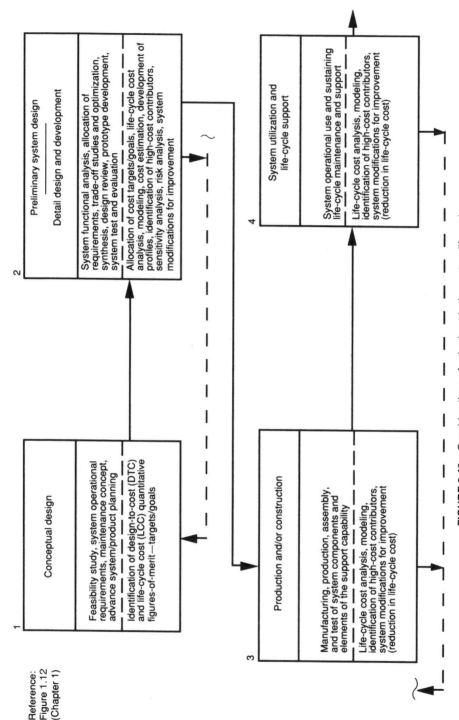

FIGURE 3.43 Considerations of value/cost in the system life cycle.

Reference Figure 1.3

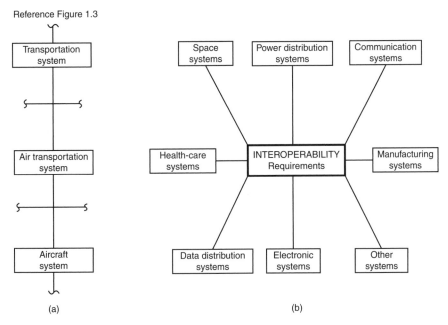

(a)   (b)

FIGURE 3.44   System-of-systems (SOS) integration and interoperability requirements.

maintenance concept for the system. Functional analyses and the allocation of these requirements are necessary to identify input criteria for design. Analyses and trade-off studies are accomplished in the design optimization process. Finally, the initially specified requirements are verified through system test and evaluation. These steps, which are characteristic in each instance, are illustrated in Figure 3.45.

Although the design disciplines in this chapter have been introduced as separate individual requirements, there is a certain degree of interdependence among them. Maintainability requirements are based on reliability, supportability requirements are dependent on reliability and maintainability data, safety factors are based on human factors, and so on. These disciplines not only build on the basic design (i.e., electrical design, mechanical design, etc.), but they build on each other. An attempt is made to show these relationships through the order of material presentation in Section 3.4.

Finally, with the objectives of system engineering in mind, it is essential that the appropriate level of communications be established among these disciplines. This communication must be reflected throughout the individual respective program plans, and there must be a free exchange of design-related data in order to fulfill the various analyses and design support functions. The necessity to integrate these activities into a total effective engineering design effort is a major aspect of system engineering.

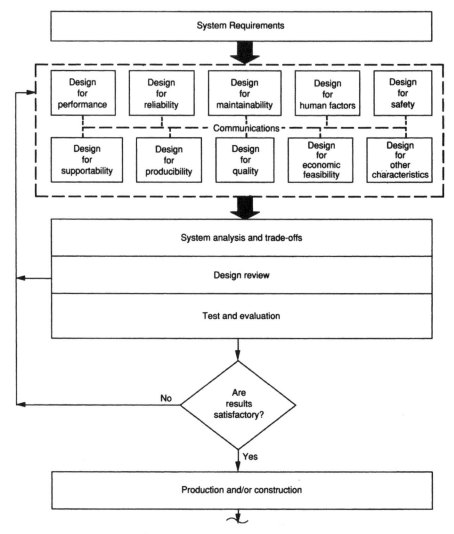

FIGURE 3.45 The design process.

QUESTIONS AND PROBLEMS

1. Describe the steps involved in defining the quantitative and qualitative design criteria for a system.
2. Select a system of your choice and develop a detailed outline of a type A specification for this system.
3. Define *reliability*. Provide an example of some reliability measures/metrics for a typical system, and describe the bases for these.

4. One hundred (100) parts were tested for 10 hours, and 10 failures occurred during the test. The times when the failures occurred are 1, 3, 6, 2, 3, 6, 8, 9, 2, and 1 hour, respectively. What is the failure rate?

5. Field data have indicated that unit A has a failure rate of 0.0004 failure per hour. Calculate the reliability of the unit for a 150-hour mission.

6. A system consists of four subassemblies connected in series. The individual subassembly reliabilities are A = 0.98, B = 0.85, C = 0.90, and D = 0.88. Determine the overall system reliability.

7. A system consists of three subsystems in parallel. Subsystem A has a reliability of 0.98, subsystem B has a reliability of 0.85, and subsystem C has a reliability of 0.88. Calculate the overall system reliability.

8. In Figure 3.46, the component reliabilities are A = 0.95, B = 0.97, C = 0.92, D = 0.94, E = 0.90, and F = 0.88. Determine the overall network reliability.

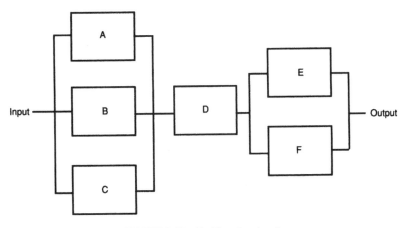

FIGURE 3.46   Problem 8 network.

9. Develop the overall reliability expression (R_n) for the network shown in Figure 3.47.

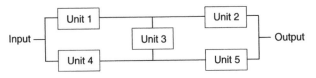

FIGURE 3.47   Problem 9 network.

10. There are a variety of tools/techniques that can effectively be utilized in the design process to help meet the objectives of reliability engineering. Briefly describe the objective and the application of each of the following (what is it? how and when can it be applied? what are the anticipated results?): *reliability modeling, reliability allocation, reliability prediction, FMECA, FTA, RCM.*

11. Define *maintainability*. Provide an example of some maintainability measures/metrics for a typical system and describe the bases for these.

12. The following corrective maintenance task times were observed:

    (a) What is the range of observations?

    (b) Using a class interval width of 4, determine the number of class intervals. Plot the data and construct a curve. What type of distribution is indicated by the curve?

    (c) What is the $\overline{M}ct$?

    (d) What is the geometric mean of the repair times?

    (e) What is the standard deviation of the sample data?

    (f) What is the $\overline{M}_{max}$ value (assume 90%)?

Task time (min)	Frequency	Task time (min)	Frequency
41	2	37	4
39	3	25	10
47	2	35	5
36	5	31	7
23	13	13	3
27	10	11	2
33	6	15	8
17	12	29	8
19	12	21	14

13. There are a variety of tools/techniques that can be effectively utilized in the design process to help meet the objectives of maintainability engineering. Briefly describe the objective and the application of each of the following (what is it? how and when can it be applied? what are the anticipated results?): *maintenance concept, maintainability allocation, maintainability analysis, maintainability prediction, FMECA (as it applies to maintainability), LORA, MTA.*

14. Calculate as many of the following parameters as you can with the given information: Determine

$$A_i \qquad MTBM$$
$$A_a \qquad MTBF$$
$$A_o \qquad \overline{M}$$
$$\overline{M}ct \qquad MTTR_g$$
$$M_{max}$$

Given:

$\lambda = 0.004$

Total operation time $= 10,000$ hours

Mean downtime $= 50$ hours

Total number of maintenance actions $= 50$

Mean preventive maintenance time $= 6$ hours

Mean logistics plus administrative time $= 30$ hours

15. Define *human factors*. Provide an example of some human measures/metrics for a typical system and describe the bases for these.

16. Identify and briefly describe some of the characteristics that must be considered in the design for the human.

17. Describe the objective and application of each of the following (what is it? how and when can it be applied? what are the anticipated results?): *functional analysis and allocation, operator task analysis, error analysis, OSD*.

18. Describe the steps involved in defining the requirements for personnel training. What is included?

19. Describe the steps involved in defining the requirements for system safety and system security. What are the relationships between the two? Identify and briefly describe some of the tools that are utilized in helping to meet the objectives of safety and security engineering.

20. Define *logistics*. What is meant by *supply chain management* (SCM)? Describe how the two relate to each other (if at all). What are the elements of logistics? Briefly describe each. What is meant by *logistics engineering*? Define *supportability*?

21. Describe some of the measures/metrics of logistics, and describe how they might be applied for a typical system.

22. What is the *supportability analysis*? What are the input requirements, and what is the expected output in terms of type of information and application?

23. How are the requirements for software determined? What are some of the measures/metrics for software? How is software reliability measured? How is software maintainability measured?

24. Define *TQM*. What is meant by *SPC*? Describe quality engineering. How does it relate to system engineering?

25. Describe concurrent engineering. How does it relate to system engineering?

26. Describe what is meant by *agile manufacturing*, *lean production*, and *enterprise resource planning* (ERP).

27. What is meant by *disposability engineering* and *environmental engineering*? How do they relate to each other? How do they differ?

28. Describe what is meant by *total productive maintenance* (TPM). How can it be measured?

29. Define *life-cycle cost* (LCC). How does LCC relate to value? How are economic factors considered in the system design process?

30. Describe the steps involved in accomplishing a life-cycle cost analysis (LCCA).

31. What is the CBS? What is included/excluded? How does the CBS relate to the functional analysis?

32. Describe some of the more commonly used cost estimating methods. Under what conditions should they be applied?

33. What is *activity-based costing* (ABC)? Why is it important (if at all)?

34. What is meant by *cost-estimating relationships* (CERs)? How are they determined?

35. In the evaluation of alternative design configurations, an individual cost profile is developed for each. These individual profiles must be reviewed and evaluated in terms of some form of "equivalence." Briefly describe the steps you would follow in accomplishing such.

36. Calculate the anticipated life-cycle cost for your personal automobile.

37. Select a system of your choice, and accomplish a life-cycle cost analysis (LCCA) in accordance with the steps identified in Figure 3.38 and the process described in Appendix B.

38. Why is the accomplishment of life-cycle costing important? Please explain.

39. Describe the overall process (i.e., steps) that you would apply in the design and development of a SOS configuration. What tools/techniques would you apply as an aid in the accomplishment of such?

40. Define *interoperability*. Briefly describe the requirements in the *design for interoperability*. Describe some of the problems that could be encountered in meeting the objectives of such.

4

ENGINEERING DESIGN
METHODS AND TOOLS

Throughout the system engineering process described in Chapter 2, there is a series of design activities accomplished with the objective of ultimately providing a system that will fulfill a designated customer (consumer) need. The successful completion of these activities, amplified to some extent through the specific requirements covered in Chapter 3, is dependent on the proper application of selected design methods and practices. This, in turn, is strongly influenced by the technology and tools available to and utilized by the responsible design engineer(s).

For years, the basic design process had involved a series of activities, accomplished on an individual-by-individual basis using step-by-step manual procedures. Ideas were generated, conceptually oriented layout or arrangement drawings were prepared and approved, system components were evaluated and selected from design standards documentation, detailed drawings and parts lists were developed and reviewed, mock-ups and models were constructed, and so on. In essence, the design process involved a long series of activities, usually requiring a great deal of time and often not very well coordinated.

With the advent of computer technology, the design process has changed significantly. Through the introduction of computer graphics in the late 1950s and early 1960s, user-input devices in the late 1960s and early 1970s (keyboards, light pens, joysticks), and the current development of sophisticated design workstations, system/product design includes the application of many new innovations. Computer technology is available to facilitate the generation of graphics material, the accomplishment of various types of analyses, and the fulfillment of data management activities. More recently, the development of information technology (IT), electronic data interchange (EDI), and electronic commerce (EC) techniques/methods has enabled the design process to become much more extensive, more efficient, and

217

accomplished in less time and with reduced cost. It is not uncommon today to establish a central design database (using various CAD/CAM techniques), to communicate using Web-based technology, and to have individual designers from various parts of the world providing their respective inputs from remote locations and in a timely manner. In essence, the design process has changed significantly through the past several decades.

It is within this context (i.e., the current design environment) that the subject of system engineering needs to be addressed. The purpose of this chapter is to briefly highlight some of these relatively recent concepts in design, and to discuss system engineering objectives as they relate to current design methods.

4.1   CONVENTIONAL DESIGN PRACTICES[1]

For most projects dealing with small- and large-scale systems, the major steps in system development include conceptual design, preliminary design, and detail design, as illustrated in the flow process presented in Figure 1.12. This flow process should, of course, be tailored to the specific requirement. Although the basic steps are applicable in the development of all systems, the level of effort and the duration of the project will vary from one situation to the next.

Inherent within the fundamental steps identified in Figure 1.12 are many different activities, all related to the prime objective of designing a system to meet a customer need. For relatively large projects, such as the commercial aircraft system and the ground mass-transit system referenced in Section 3.3, there may be a requirement for technical expertise representing many different disciplines; for example, electrical engineers, mechanical engineers, structural engineers, materials engineers, aerospace engineers, civil engineers, and reliability engineers. In support of the responsible design engineers in the various fields, there is a need for draftspersons, technical illustrators, component part specialists, laboratory technicians, computer programmers, test technicians, purchasing and contracts specialists, legal experts, and so on. There are many different and varied levels of specialization required for most projects, and each assigned specialty contributes to the design.

To review the steps in design further, Figure 4.1 is presented as an amplification of Figure 3.1. As design progresses, actual definition is accomplished through documentation in the form of plans and specifications (already discussed), procedures, drawings, materials and parts lists, reports and analyses, computerized databases, and so on. The design configuration may be the best possible in the eyes of the designer; however, the results are practically useless unless properly documented so that others can first understand what is being conveyed and then be able to translate the output into a producible entity. '

---

[1]With the introduction of new technologies evolving on a day-by-day basis, the methods by which many of these conventional design practices are accomplished have and will continue to change. However, what is presented in this section represents practices in the past, and forms a baseline for further discussion and leading to subsequent sections in this chapter.

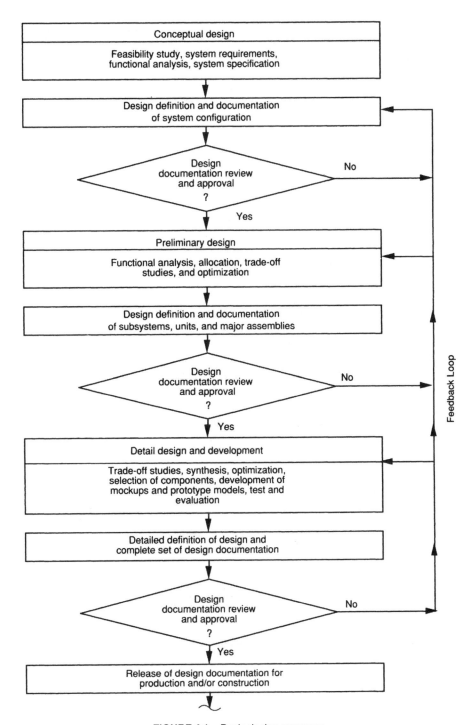

FIGURE 4.1 Basic design sequence.

In addressing the aspect of documentation, emphasized in Figure 4.1 to define the various levels of system development, the results of design have generally been conveyed through a combination of the following:

- *Design drawings:* Assembly drawings, specification control drawings, construction drawings, installation drawings, logic diagrams, piping diagrams, schematic diagrams, interconnection diagrams, wiring and cable harness diagrams.[2]
- *Materials and parts lists:* Parts lists, materials lists, long-lead-item lists, bulk-item lists, provisioning lists.
- *Analysis reports:* Trade-off study reports supporting design decisions, finite-element analysis reports, reliability and maintainability analyses and predictions, safety reports, logistic support analysis records, configuration identification reports, computer software documentation.

Based on the steps in design definition (shown in Figure 4.1), ideas are generated and converted to drawings (or equivalent), drawings are reviewed by various interested disciplines and/or organizations, recommended changes are initiated and incorporated as appropriate, and approved drawings (designated by drawing "sign-off") are released for production. For the most part, the steps in this informal day-to-day process have been completed in series and often require a great deal of time. For instance, an electrical engineer may start the process with a proposed layout of components on a circuit board, a mechanical engineer may then provide the necessary structural and cooling requirements, a reliability engineer may follow with a prediction and evaluation of the selected components, and so on. These responsible individuals may be located in different geographical locations, and the data processing and communications often become quite lengthy. Further, this procedure takes on an additional degree of complexity as the number of drawings and drawing change notices increases.

In essence, this day-to-day design activity, particularly for large-scale systems, may include the generation and processing of hundreds of drawings and drawing change notices (DCNs). Many different design disciplines may be represented, with engineering personnel located throughout the producer's organization and, in some instances, at various remote supplier facilities. The communications, both in terms of verbal discussions pertaining to design approaches and the handling of design data documentation, are marginal, and the elapsed times are often extensive. In some cases, it may require a month, or more, to route a single drawing through the review steps for approval.

These somewhat conventional practices of the past have created some major challenges. In numerous instances, procedures have been bypassed, changes have been implemented without the proper approvals and necessary coordination, and

---

[2]It should be noted that many of these categories of drawings have been replaced with more current ways of presenting the same type of information (e.g., three-dimensional sketches, geometric solid models, isometric drawings, and supporting computer-aided graphic presentations showing various features).

the appropriate configuration controls have not been practiced. In essence, the basic objectives of system engineering have not been followed.

4.2 ANALYTICAL METHODS

Inherent within the overall spectrum of system engineering is an ongoing analytical process in which the design engineer(s) is engaged in some form of synthesis, analysis, evaluation, and design optimization activities (refer to Section 2.9, Chapter 2). When known solutions fail to hold sufficient promise for future application, more promising opportunities are sought. This leads to the identification of various possible future courses of action and the ultimate selection of a preferred approach. In other words, the design engineer(s) is often faced with many different decisions in progressing from the early identification of a conceptual approach to the development of a well-defined product output. The accomplishment of various types and levels of trade-off analyses is necessary in reaching the ultimate objective, and some familiarization with decision theory and available analytical techniques is required in reaching this goal.[3]

Although it is not intended herein to provide comprehensive coverage of the techniques (or models) utilized in system analysis, the following list presents five areas in which some familiarization is highly recommended:

1. *Probability theory and analysis.* An inherent knowledge of statistics and probability theory is required as a prerequisite for understanding the concepts and principles of reliability and maintainability and for the application of selected program management methods (e.g., risk analysis). Recommended probability distribution models should include the uniform, binomial, Poisson, exponential, normal, lognormal, and Weibull distributions.

2. *Economic analysis.* A basic knowledge of economic concepts and principles, interest and interest formulas, and ability in determining economic equivalence for various design alternatives and in conducting break-even evaluations are required in the performance of life-cycle cost (LCC) and related analysis activities.

3. *Optimization methods.* A basic familiarity with classical optimization theory, constrained and unconstrained optimization, and linear and dynamic programming is required in determining optimum system equipment life, recommended components replacement policies (frequencies), equipment packing schemes, preferred transportation routes, and so on.

---

[3] An understanding of some of the basic tools/methods utilized in operations research (or operations management) is essential for the successful implementation of system engineering objectives in this area. Three good references are (1) B. S. Blanchard and W. J. Fabrycky, *Systems Engineering and Analysis*, 4th ed. (Upper Saddle River, NJ: Pearson Prentice Hall, Part III, 2006); (2) F. S. Hillier and G. J. Lieberman, *Introduction to Operations Research*, 8th ed. (New York: McGraw Hill, 2005); and (3) H. A. Taha *Operations Research: An Introduction*, 8th ed. (Upper Saddle River, NJ: Pearson Prentice Hall, 2006).

4. *Queuing theory and analysis.* A basic understanding of queuing theory (e.g., the arrival mechanism, the waiting line, the service mechanism, and their associated distributions), single-channel queuing models, multiple-channel queuing models, and the application of Monte Carlo analysis is required in the design of facilities and those functions involving the processing of various items through some "channel" activity. Establishing the appropriate number of logistics and/or repair channels in the design of a maintenance facility is a good example.

5. *Control concepts and techniques.* A basic familiarity with statistical process control (SPC) techniques, the application of various types of control charts (e.g., *x-bar, R, p,* and *c* charts), determining optimum policy control approaches, quality control methods, and the application of control networks for project/program scheduling is essential for the implementation of total quality management (TQM) and selected program management requirements.

Although a detailed knowledge of these and related *operations research* techniques is not expected, some familiarity with these models, and when and how they can be applied, is essential.

4.3 INFORMATION TECHNOLOGY, THE INTERNET, AND EMERGING TECHNOLOGIES[4]

The term *information technology* (IT) refers to the infrastructure that fosters the integration of all the various mechanisms for converting, storing, protecting, processing, transmitting, and retrieving information. The Internet is a worldwide network of interconnected computer networks (e.g., commercial, academic, and government) that operates using a standardized set of communications protocols and TCP/IP (transmission control protocol/Internet protocol). The evolution of the Internet has had, and continues to have, a significant effect on information technology processes by increasing productivity, reducing resources consumed, and greatly enhancing communications. Information flows have changed drastically and rapidly as the Internet has become widely available and put to use in a variety of models that are rational, flexible, and capable of taking advantage of distributed work practices and the volume of available information and resources. The term *workflow* is most often used to describe the tasks, organizations, information, and tools involved in a business process. The increase in computer programming developed to capture and develop human-to-machine interaction enhances workflow by giving end users the ability to easily manage and process complex data in order to reach their specific goals. More recently, significant effort is being put into defining workflow patterns in order to

---

[4]It should be noted that the terms and definitions are continually in a state of change, with the ever-increasing growth in the various facets of *technology*. In this instance, my thanks to Lisa McCade, McCade Design, Charlottesville, VA, for her assistance in preparing this section.

provide a conceptual basis for process technology. By examining and evaluating various workflow patterns, the results can be used for examining the suitability of a particular process or system for a particular project.

The Internet has transitioned from a series of isolated information silos to sources of content and functionality, thus becoming computer platforms serving web applications to end-users. A Web application (Webapp) is an application that is accessed via the Internet. Webapps allow access and use of software on an unlimited number of computers without distribution or installation of software. A very significant advantage of building and using Webapps is that they are designed to perform regardless of the operating system on any given client. One of the most significant movements affecting the creation, deployment, and use of Webapps, and many other software developments, is the open-source program initiative. An open-source program has its source code freely available to anyone interested in using or working with it. This method of development allows anyone (often programmers) to modify, enhance, and redistribute the program based on specific requirements. Many supporters of the open source movement believe that by allowing distributed peer review and transparency of process, the end result will be better quality, higher reliability, more flexibility, and lower costs.

The emergence of a new category of software, collaboration platforms, provides facilities to monitor and manage the collaborative work process. To meet business objectives and maintain competitive advantage, companies have recognized the need for cost-effective secure solutions that allow its distributed teams to work together as if they were located in one office. There is also a need to coordinate resources and financial support and to track the status of all projects within an area of responsibility. Finally, investors and customers often share project information as the projects evolve and real time communication with them is vital. Collaborative technologies are designed to capture the efforts of many into a managed and meaningful end-product by allowing team members to communicate, coordinate, and manage the process. With collaborative tools, it becomes possible to harness data and knowledge leading to a greater capacity for innovation and organizational agility. Collaboration tools are being introduced in supply chain and manufacturing environments where organizations need to collaborate and negotiate on product design and other issues.

A more specific example of how the Internet and the emergence of e-commerce technologies have affected business processes is its use in supply chain management (SCM). Referring to Section 1.4.3, SCM pertains to the integrated flow of materials from various sources of supply, through the production process, and in the form of finished products distributed and installed at the various user operational sites (refer to Figure 1.21). SCM is inherently information intensive, as it involves communication and collaboration between multiple parties (e.g., production teams, suppliers, distribution centers, customers). The capability of the Internet to transfer massive amounts of information with minimal time and effort has enabled efficient and effective sharing of information among business partners. E-commerce technologies allow the opportunity for businesses to develop and streamline their supply chains. Currently, complex software systems with Web applications are providing part or all of the SCM service for an increasing number of companies. A number of large corporations have achieved great success by developing comprehensive and global supply chains characterized by low inventory levels, lean production, demand collaboration, and deferred assembly. This has resulted in lower costs and greater responsiveness to changing market conditions.

It is essential for those involved in the implementation of system engineering requirements to be familiar with and take advantage of the latest technologies and how they can be effectively applied toward meeting the objectives described throughout this text. The key to acquiring real value from these information-related technologies is to motivate people to use them on a frequent and consistent basis, as these technologies work only when the proper incentives and training are provided early in a given program.

4.4 CURRENT DESIGN TECHNOLOGIES AND TOOLS

With the advancements in information technology, many new tools and techniques have been developed and adapted to streamline the design process. The development of software and associated computer models has increased exponentially, and the results have provided the design engineer (and the manager) with a wide variety of tools intended to enhance productivity.

Given the appropriate computerized design aids, the engineer is able to accomplish a performance analysis employing simulation methods (dynamic versus static, stochastic versus deterministic, continuous versus discrete) in a relatively short time frame. Mathematical programming methods (linear programming, quadratic programming, dynamic programming) can be used in solving resource allocation and assignment problems. Statistical tools are available for plotting distributions and for determining related characteristics (e.g., mean, standard deviation, range, maximum value). Project management aids are used to plot scheduling networks (e.g., the Program Evaluation and Review Technique/PERT and the Critical Path Method/CPM) and cost projections. Database management models are employed extensively for data acquisition and storage, information processing, and report generation. Finally, there are a wide variety of specialized engineering tools that can be effectively utilized by the designer to help solve specific problems (e.g., the design of a digital filter, the layout of components on a circuit board, and the accomplishment of a reliability analysis for system XYZ).

Relative to application in the design and development process, the availability of these tools offers three distinct advantages:

1. The combination of personal computers (PCs), included as part of individual design workstations located throughout an entire project design area, with the connections to a central workstation and a mainframe computer (or equivalent), has created an excellent data communications network. Not only is it possible to transmit many different categories of data, in varying formats, to all applicable workstations on a concurrent basis, but this can be accomplished rapidly and efficiently. Design data packages can be developed by individual

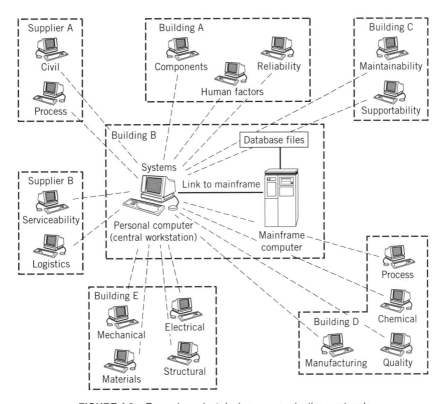

FIGURE 4.2 Example project design communications network.

designers, transmitted to many others simultaneously, and reviewed with recommended changes submitted to the designer in a very short period of time. This capability minimizes the requirement for completing tasks in series, as described in Section 4.1, and allows for a reduction in overall system development time. Figure 4.2 illustrates this concept.[5]

2. The versatility and variety of software packages/models provide the designer with many new tools, not readily available in the past. For instance, in systems design, the use of simulation methods early in the conceptual phase enables the designer to do a better job in determining operational requirements and accomplishing a performance analysis. Three-dimensional computer models may be developed in order to evaluate a variety of possible system configurations, to study the interrelationships among system components, to investigate space allocations, to study the performance of human task sequences, and so on. The use of mathematical/statistical models allows the designer to investigate many more alternatives (as compared with what has been possible in the

---

[5]One of the objectives through concurrent engineering is to reduce the length of the system acquisition process by appropriately applying new technologies in design and development.

past), involving numerous calculations and the processing of large amounts of data in a short period of time. In essence, the design engineer of today has many more tools available, and the capabilities of these tools allow for a more in-depth analysis and investigation of design alternatives. This capability, applied early in the system life cycle, helps to reduce the risks associated with design decision making by eliminating nonfeasible options in the beginning.

3. The data handling capabilities, provided through computer technology, allow for the acquisition, processing, storage, and retrieval of greater varieties and larger amounts of data than in the past. Data storage and retrieval methods are simpler, and data processing times are much less. Not only is it easy to capture certain design information on drawings, but the generation and distribution of parts/materials lists can be handled expeditiously. Further, reports and technical publications can be produced automatically by employing a combination of graphics and word processing methods.

In regard to the system engineering objectives defined in Chapter 1, the advent of computer technology (along with the many tools that are currently available) can have a very beneficial impact. Specifically:

- A more in-depth and complete analysis of system requirements can be accomplished during an early stage in design when the life-cycle impacts are the greatest. This, in turn, tends to foster more emphasis on a top-down, life-cycle approach in system development.

- An improved communications process can evolve as a result of (a) being able to disseminate data rapidly and effectively, (b) being able to transmit information to multiple individuals and/or organizations concurrently, and (c) being able to incorporate design changes expeditiously. These improvements in communications help to provide the necessary integration of the many different disciplinary areas involved in the design process.

By contrast, adapting to the new computerized technology is not without some concerns and challenges. First, the development and implementation of a standardized approach for converting design information into a digital format are required. All designers, supporting personnel, producer and consumer organizations, and suppliers must utilize the same language and data format, follow identical practices, and so on.[6]

Second, the methods for linking personal computers (PCs) throughout a company via a local-area network (LAN), between companies via a wide-area network (WAN), and/or the linking of PCs to mainframe computers, are critical to the communications process. Figure 4.2, illustrating a "star topology," shows one approach,

---

[6]The Department of Defense recognized this requirement and attempted to provide some standardization through the application of MIL-D-28000, Military Specification. "Digital Representation for Communication of Product Data: IGES Application Subsets" (Washington, DC: Department of Defense). This standard includes the definition of a series of application-specific subsets of the initial graphics exchange specification (IGES), the popular name for American National Standard ANSI Y14.26M, "Digital Representation for Communication of Product Definition Data."

with the objective of allowing for the transmission of design information to multiple workstations concurrently. Another approach, often utilized, is a circular configuration (i.e., ring topology), in which data must pass through various workstations in sequence prior to arriving at the centralized computer. This configuration generally does not allow for the transmission to all interested designers simultaneously, and the process assumes a series of actions progressing from one remote workstation to another.

In any event, the design of the protocols and the various networks (e.g., LANs, WANs) must be such as to promote the rapid and effective communication of system/product design information. Further, the hardware and software packages to be utilized must be fully compatible. The various workstations must be able to "talk to each other," and the appropriate levels of communications within a network, and between networks, must be possible. In other words, methods and procedures must be established to allow for the rapid and efficient transmission of design data between the various suppliers and the producer (i.e., the contractor), and between the producer and the consumer (i.e., the customer).[7]

In considering the system engineering objectives highlighted in Chapter 1, these must be viewed in the context of the overall design environment. The objectives relate to the application of a total, top-down, integrated, life-cycle approach in the development of a system that will meet the needs of the consumer, both effectively and efficiently. The successful fulfillment of these objectives is, of course, dependent on the activities accomplished throughout the system development process. This, in turn, is highly influenced by the design capability and environment that exist within the producer/supplier organization. If, on the one hand, the appropriate tools are available and properly utilized, the realization of system engineering objectives may be relatively easy. On the other hand, if the environment is not conducive to the accomplishment of *good* design, then additional efforts will be required to ensure that the desired system engineering objectives are met.

Thus, not only will the requirements for system engineering vary as a function of the nature and complexities of the system being developed, but they must also include consideration of the capability designated to support the design process. The degree to which computerized technology is being utilized may have a significant impact on the system engineering process.

4.4.1 The Use of Simulation in System Engineering[8]

Simulation is the process of designing and utilizing an operational model of a system to conduct experiments for the purpose of either understanding the behavior of the

---

[7]The development of the initial graphics exchange specification (IGES) was initiated to allow for the transfer of design data between networks, CAD/CAM systems, and so on. The capability is being supplemented by the product data exchange specification (PDES), which will encompass the complete set of data elements that defines a system/product for all applications over its expected life cycle.

[8]Two good references are (I) *Systems Engineering Fundamentals* (Fort Belvoir, VA: Defense Acquisition University (DAU) Press, December 2000), Chapter 13; and (2) S. V. Hoover and R. F. Perry, *Simulation: A Problem Solving Approach* (Reading, MA: Addison-Wesley, 1990). The subject of simulation is also covered in a number of texts dealing with operations research methods (refer to Appendix F).

system or evaluating alternative strategies and/or system design configurations. The objective is to construct a simplified representation of a system or a process in order to facilitate the analysis, synthesis, and/or evaluation of the system/process. The use of simulation is particularly appropriate in the early stages of system development prior to the point when the various physical elements of the system are available for evaluation; that is, during the "analytical" stage specified in Figure 2.34.

Simulation methods can be applied in the development of three-dimensional computer-aided design (CAD) models to show the overall system configuration and its components, their location, accesses, interrelationships, and so on. A design engineer can visually see the system configuration, as an entity, during the early stages of preliminary design. This will, of course, enable the designer to evaluate different alternatives, identify potential problems, and accomplish an early synthesis of the design. Examples of applications include the simulation of an airplane and the lay-out of its components, an aircraft cockpit with crew, an automobile with driver, a submarine with installed components, a manufacturing facility with capital equipment installed, and so on. Simulation methods can also be used to illustrate process flow rates (e.g., the flow of materials through a factory along with stoppages), different transportation routings, reliability failure patterns, maintenance and support policy alternatives, and so on.

As conveyed in earlier chapters, one of the objectives of system engineering is to gain as much visibility as possible in the early stages of system development. The intent is to investigate all feasible approaches to design, identify and eliminate potential problems, select a preferred design configuration, and reduce (if not eliminate) the possibility of risk. The use of simulation techniques allows the designer to investigate many different potential design solutions prior to the procurement of equipment, the development of software, and the acquisition of other physical elements of the system. This, in turn, can lead to significant reductions in cost.

4.4.2   The Use of Rapid Prototyping[9]

In the area of software development in particular, designers are oriented toward the building of "one-of-a-kind" software packages. The issues in software development differ from those in other areas of engineering in that mass production is not the normal objective. Instead, the goal is to develop software that accurately portrays the features that are desired by the user; that is, the interfaces with the customer. For instance, in the design of a complex workstation display, the user may not at first comprehend the implications of the proposed command routines and data format on the screen. When the system is ultimately delivered, problems occur, and the "user interface" is not acceptable for one reason or another. Changes are then recommended and implemented, and the costs of modification and rework are usually high.

---

[9]Two pertinent references are B. W. Boehm, *Software Engineering Economics* (Upper Saddle River, NJ: Prentice-Hall, 1981); and B. Thome, ed., *Systems Engineering, Principles and Practice of Computer-Based Sytems Engineering* (New York: John Wiley & Sons, Inc., 1992).

The alternative is to develop a prototype early in the system design process, design the applicable software, involve the user in the operation of the prototype, identify areas that need improvement, incorporate the necessary changes, involve the user once again, and so on. This iterative and evolutionary process of software development, accomplished throughout the preliminary and detail design phases, is referred to as *rapid prototyping*. Rapid prototyping is a practice that is often implemented and is inherent within the system engineering process, particularly in the development of large software-intensive systems.

4.4.3 The Use of Mock-ups

Although much information may be acquired through the use of simulation methods in early design, it may be desirable to construct a three-dimensional scale model or physical *mock-up* of the system, or an element thereof, during the preliminary or detail design phase to provide a realistic replica of a proposed equipment/facility configuration. These models, or mock-ups, can be produced to any desired scale and to varying degrees of detail, depending on the level of emphasis required. Mock-ups may be constructed of heavy cardboard, wood, metal, or a combination of materials. Mock-ups can be developed relatively inexpensively and in a short period of time. The utilization of mock-ups can provide many benefits, including the following eight applications:

1. They provide the design engineer with an opportunity to experiment with different facility layouts, packaging schemes, panel displays, and so on, prior to the preparation of formal design data.

2. They provide the reliability/maintainability/human-factors engineer with an opportunity to accomplish a more effective review of a proposed design configuration for the incorporation of supportability characteristics. Problem areas readily become evident.

3. They provide the maintainability/human-factors engineer with a tool for use in the accomplishment of predictions and detailed task analyses. It is often possible to simulate operator and maintenance tasks to acquire task sequence and time data.

4. They provide the design engineer with an excellent tool for conveying the final design approach during a formal design review.

5. They serve as an excellent marketing tool.

6. They can be employed to facilitate the training of system operator and maintenance personnel.

7. They are utilized by production and industrial engineering personnel in developing fabrication and assembly procedures and in the design of factory tooling and associated test fixtures.

8. At a later stage in the system life cycle, they may serve as a tool for the verification of a modification kit design prior to the preparation of formal data and the development of kit hardware.

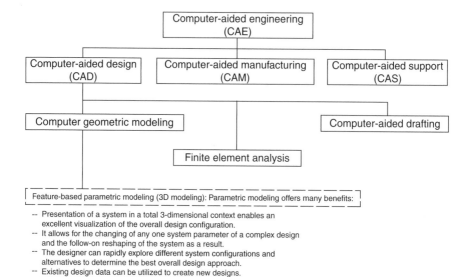

FIGURE 4.3   Computer-aided engineering (CAE) and related features (example).

In general, mock-ups are extremely beneficial. They have been used effectively in facility design, aircraft design, and the design of smaller systems/equipment.

4.5   COMPUTER-AIDED DESIGN (CAD)[10]

Within the overall spectrum of *computer-aided engineering (CAE)*, there are a wide variety of applications to include what we might classify under the terms of *computer-aided design (CAD), computer-aided manufacturing (CAM)*, and *computer-aided support (CAS)*. The interrelationships among these areas of activity are illustrated (for example) in Figure 4.3, although the specific terms and definitions may vary somewhat from one application to the next.

Computer-aided design (CAD), defined in a broad sense, refers to the application of computerized technology to the design process. With the availability of computer tools that can be appropriately utilized in the performance of certain design functions, the designer can accomplish more, at a faster pace, and earlier in the life cycle. These tools, including supporting software, incorporate graphics capabilities (vector and raster graphics, line and bar charts, x–y plotting, scatter diagrams, three-dimensional displays), analytical capabilities (mathematical and statistical programs for analysis and evaluation), and data management capabilities (data storage and retrieval, data processing, drafting, and reporting). These capabilities are usually combined in

---

[10]Other terms often used to define this area of activity are *computer-aided engineering* (CAE) and *computer-aided systems engineering* (CASE).

integrated packages for application to solve a specific design problem. Three examples of design applications follow:

1. The utilization of graphics, combined with word processing and database management capabilities, enables the designer to do the following:

(a) Lay out components on electrical/electronic circuit boards, design routing paths for logic circuitry, incorporate diagnostic provisions on microelectronic chips, and so on. CAD capabilities are being used extensively in the design of large-scale integration (LSI) and very large scale integration (VLSI) electronic modules and standard packages.

(b) Lay out individual components for sizing, positioning, and space allocation purposes through the use of three-dimensional displays.

(c) Develop solids models enabling assemblies, surfaces, intersections, interferences, and so on, to be clearly delineated via the automatic generation of isometric and exploded views of detailed dimensional and assembly drawings. Component surface areas, volumes, weights, moments of inertia, centers of gravity, and other parameters can be determined automatically. CAD capabilities are being used extensively in the development of solids models for large systems; e.g., airplanes, surface ships, submarines, ground vehicles, facilities, bridges, dams, and highways. Many of these models allow the designer to view the system as an entity while providing a top-down hierarchical breakout of system components at different levels. Both two- and three-dimensional displays can be presented, using color graphics as an enhancement.

(d) Develop three-dimensional models of facilities, operator consoles, maintenance work spaces, and the like, for the purpose of evaluating the interface relationships between humans and other elements of the system. CAD tools are being utilized to simulate both operator and maintenance task sequences.

2. The utilization of analytical methods, combined with word processing, spreadsheet, and database management capabilities, enables the designer to do the following:

(a) Accomplish system requirements and performance analyses in support of design trade-off studies (e.g., finite-element analysis, structural analysis, stress-strength analysis, thermal analysis, weight/loads analysis, materials analysis).

(b) Perform reliability analyses in support of design (e.g., allocations, predictions, failure mode, effect, and criticality analysis (FMECA), fault-tree analysis (FTA), sneak-circuit analysis, critical-useful-life analysis, environmental stress analysis).

(c) Perform maintainability analyses in support of design (e.g., allocations, predictions, FMECA, diagnostic and test requirements analysis, maintenance task analysis (MTA).

(d) Perform human-factors analyses (e.g., functional analysis, operator task analysis, operational sequence diagrams (OSDs), training requirements analysis).

(e) Perform safety analyses (e.g., hazard analysis, fault-tree analysis).

(f) Perform supportability analyses (e.g., maintenance requirements analysis, level-of-repair analysis, spares requirements analysis, transportation requirements analysis, test equipment requirements analysis, facilities requirements analysis).

(g) Perform value/cost analyses (e.g., value engineering analysis, life-cycle cost analysis).

3. The utilization of database management, combined with graphics, spreadsheet, and word processor capabilities, enables the designer to do the following:

(a) Develop functional flow diagrams, information/data flow diagrams, dependency diagrams, reliability block diagrams, action diagrams, decision trees and tables.

(b) Develop and maintain a database that includes historical design data, parts lists, materials lists, supplier information, technical reports. The purpose is to be able to store standard data on common items and to be able to retrieve (or recall) such information quickly and reliably for future applications.

(c) Develop a management information system (MIS) that enables the accomplishment of project review and control functions (e.g., data communications, personnel loading projections, cost projections, technical performance measures (TPMs) "tracking," PERT/CPM reporting, project reporting requirements).

Through a review of these areas of application, one can see that the utilization of computer technology is widespread and growing at a rapid rate. However, although there are many uses of CAD technology, they are often not very well integrated.

From the overall life cycle illustrated in Figure 4.4, CAD tools are being applied throughout the design and development process, the results of which feed directly into the computer-aided manufacturing (CAM) and computer-aided support (CAS) capabilities. The application of CAD tools has been evolving since the 1960s and 1970s, primarily following a "bottom-up" approach.[11]

Unlike the integrated network concept illustrated in Figure 4.2, individual design workstations have been developed on a somewhat independent basis. A workstation constitutes a grouping or arrangement of equipment (e.g., graphics terminal, computer, keyboard), combined and laid out in such a manner as to help the designer in completing selected tasks. The capabilities of a design workstation will vary, depending on (1) the specific design functions to be performed, (2) the nature and complexity of the system being developed (and its components), (3) the education and vision of the responsible designer, in terms of his or her interest and willingness to adapt relative to the utilization of new computer technologies, (4) the degree of management support relative to modernizing the design capability, and/or (5) the availability of the necessary budgetary resources to acquire and support the capital equipment items that are incorporated as part of the design workstation.

---

[11]Through the years, different terms have been used to describe computer-aided applications applied to logistics and logistic support, including *computer-aided support* (CAS), *computer-aided logistics* (CAL), and *computer-aided logistic support* (CALS). Although the terms vary with usage, the principles and concepts remain the same.

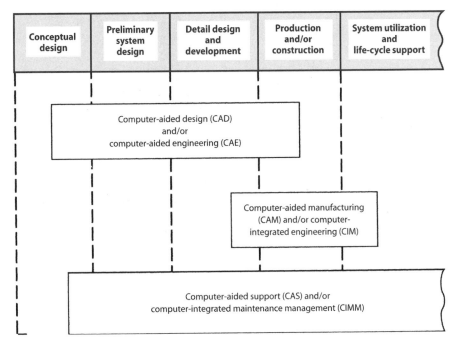

FIGURE 4.4   Application of CAD/CAM/CAS.

In many instances, adapting to computerized technologies in design requires a *cultural* change, or reorientation, pertaining to the methods used in task accomplishment. Further, whereas experience has indicated that the use of CAD methods is very cost-effective in the long term, the initial acquisition cost of the required capital assets may be perceived as being too high. In any event, these apparent obstacles must be overcome if progress is to be made in this area.

The capabilities of design workstations have evolved from a configuration in which the user was able to design a specific component part, or perform a detailed analysis of some rather isolated function of the system, to a comprehensive configuration integrating many of the diverse technologies discussed earlier. Initially, a graphics terminal was used to design a part, analyze stresses, study and evaluate a mechanical action, and so on. Now, a design workstation can be utilized to effectively accomplish many of the functions noted earlier; for example, the layout of components on circuit boards, the development of three-dimensional solid models, and the utilization of sophisticated analytical methods. Relative to the future, much more can be accomplished in terms of design integration and the incorporation of new capabilities, and the potential for additional growth is great.

The total integrated flow concept illustrated in Figure 4.5 reflects a long-range goal. At the beginning, a comprehensive design workstation can be developed and constructed, along with the appropriate software, to reflect the capabilities inferred from the integrated approach presented in Figure 4.2. In addition to the requirements

234

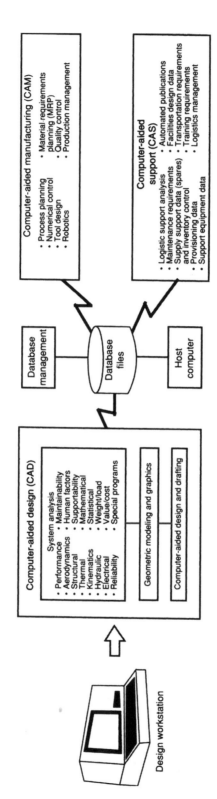

FIGURE 4.5 Major CAD/CAM/CAS interfaces.

for electrical design, mechanical design, and structural design, reliability, maintainability, and human factors, supportability and comparable requirements must be integrated into the design process. In other words, all of the design requirements specified in Figure 3.5 (Chapter 3, Section 3.2) must be appropriately integrated, the design communications network illustrated in Figure 4.2 must be available and utilized, and each of the design workstations assigned to a given project must have the capability to deal with these overall requirements.

In addition to providing a capability that will promote a completely integrated design approach, a provision must be incorporated to allow for the smooth transition of information from the design process to the manufacturing process and to the system support structure. Perhaps, in the not too distant future, the design engineer (with the assistance of a team of discipline-oriented experts) will be able to provide the appropriate design input at the workstation that will, in turn, automatically flow into both the CAM and CAS processes, through the application of effective data communications methods. This provision, of course, will require complete compatibility (language, data format, etc.) among CAD, CAM, and CAS, at both the producer (contractor) and the supplier levels.

Relative to the current status of CAD tools and their applications, many industrial firms have developed and installed CAD capabilities, integrated design workstations, supporting software, and the like. These capabilities vary significantly from one installation to the next. However, in most instances (and particularly for large-scale systems), the use of graphics technology, analytical methods, and database management capabilities has been effectively employed in accomplishing some of the more traditional design activities; for example, electrical circuit design, mechanical packaging design, and structural design. However, very little has been accomplished relative to the integration of reliability, maintainability, human factors, and supportability into this process; that is, the design disciplines discussed throughout Chapter 3.

Although computer technology has been successfully applied in the development of reliability models, maintainability models, logistic support models, and so on, these tools have been utilized on a "stand-alone" basis. In general, the design activities associated with these disciplines have been accomplished independently, not properly integrated into the basic engineering design process, and the tools that have been developed in these areas have not been integrated with each other or into the more traditional design workstations.[12]

In response to the need to become more effective, from the standpoint of contributing to the design process in a timely manner, a great deal of effort has been expended during the past few years in developing reliability, maintainability, and supportability models. If one were to survey the literature in this area, more than 350 models of varying types could be identified, and this number is growing at a rapid rate. A brief description of seven of these models, which are being utilized in support

---

[12]The term *islands of automation* is often used in referring to analytical tools that are not integrated and are employed on an independent basis. There are many areas where this condition exists today.

of different design applications today, follows, to provide some insight as to the type of tools that are available:

1. *Computer-aided system engineering* (CORE). CORE supports both static and dynamic analysis of system requirements, functional behavior, and system architecture, with automatic generation of customizable reports via a simple scripting language. The model enables the user to develop a system description in a well-defined, structured manner, provides a full database framework and flexible schema to support requirements traceability, and can be easily implemented using a PC or Mac. (VITECH Corp., 2020 Kraft Dr., Blacksburg, VA 24060).

2. *Cost analysis strategy assessment* (CASA). CASA is utilized to develop life-cycle cost (LCC) estimates for a wide variety of systems and equipment. It incorporates various analysis tools into one functioning unit and allows the analyst to generate data files, perform life-cycle costing, sensitivity analysis, risk analysis, cost summaries, and the evaluation of alternatives. (Defense Acquisition University, DSS Directorate (DR1-S), Fort Belvoir, VA 22060.)

3. *Equipment designer's cost analysis system* (EDCAS). EDCAS is a design tool that can be used in the accomplishment of a level-of-repair analysis (LORA). It includes a capability for the evaluation of repair versus discard-at-failure decisions, and it can handle up to 3500 unique items concurrently (i.e., 1,500 line replaceable units, 2,000 shop replaceable units). Repair-level analysis can be accomplished at two indenture levels of the system, and the results can be used in determining optimum spare/repair part requirements and in the accomplishment of life-cycle cost analysis (LCCA). (TFD Group, 70 Garden Court, Monterey, CA 93940).

4. *OPUS model*. OPUS is a versatile model used primarily for spare/repair parts and inventory optimization. It considers different operational scenarios and system utilization profiles in determining demand patterns for spares, and it aids in the evaluation of various design packaging schemes. Alternative support policies and structures may be evaluated on a cost-effectiveness basis. (Systecon AB, Linnëgatan 5, Box 5205, S-10245, Stockholm, Sweden.)

5. *Relex PRISM*. This constitutes a reliability software suite, which includes capabilities for the development of reliability block diagrams, reliability predictions, maintainability predictions, failure mode and effects analysis (FMEA)/FMECA, FTA/event trees, Weibull analysis, and life-cycle cost analysis. (Relex Software Corporation, 540 Pellis Road, Greensburg, PA 15601.)

6. *VMETRIC*. This is a spares model that can be used to optimize system availability by determining the appropriate individual availabilities for system components and the stockage requirements for three indenture levels of equipment (e.g., line replaceable units, shop replaceable units, and subassemblies) at all echelons of maintenance. Outputs include optimum stock levels at each echelon of maintenance, Economic Order Quantity (EOQ) quantities, and optimal reorder intervals. (TFD Group, 70 Garden Court, Monterey, CA 93940).

7. *217Plus: The Next Generation Reliability Prediction Models.* This is a series of analytical models applied in predicting system reliability, system-level failure rates, component failure rates, software reliability, approaches from "parts count" to comprehensive reliability analysis, and includes various reliability databases (Reliability Information Analysis Center/RIAC, 6000 Flanagan Road. Suite 3, Utica, NY 13502).

These models are only a representative few in the total spectrum of tools available throughout the industrial and government sectors today. However, as indicated earlier, they have been developed on an independent basis and are being utilized outside the mainstream design effort. These will have two primary objectives in the future:

1. Evaluate and integrate these and other tools, as applicable, so that they can be utilized interactively, as conveyed in the illustration in Figure 2.29. The goal is to develop a set of tools that (a) can be effectively utilized in response to a variety of needs, (b) can address the system as an entity, while being utilized to evaluate different components of the system, and (c) can "talk to each other" in terms of data communications.

2. Incorporate these and other tools, as appropriate, into the workstations for all responsible designers. The design engineer, using the ideal workstation such as shown in Figure 4.5, should have available not only the necessary tools for the accomplishment of a structural analysis, but also the tools needed for the accomplishment of a reliability analysis as well. The same individual may not accomplish both tasks; however, he or she needs to view the results of both (and their impacts on each other) concurrently in order to make intelligent design decisions.

In regard to system engineering, as it applies to CAD applications, the challenges for the future are directly in line with these objectives. The appropriate tools must be available and well integrated into the typical design workstation, in such a manner as to promote the proper consideration of all disciplines in the system design process.

4.6   COMPUTER-AIDED MANUFACTURING (CAM)

Computer-aided manufacturing (CAM) refers to the application of computerized technology to the manufacturing or production process. This application, reflected in the context of the system life cycle presented in Figure 4.3, primarily includes the use of automated methods as they pertain to the following four activities:

1. *Process planning.* Throughout the production process, there may be a series of steps required to fabricate a component, to assemble a group of items, or both. Process planning addresses the entire flow of activities, evolving from the definition of a given design configuration to the finished product delivered

to the consumer, and CAM applications include those activities that can be automated. Whereas the activities associated with process planning have been known and practiced for a long time, the use of computer technology in accomplishing these activities is a relatively recent innovation.[13]

2. *Numerical control* (NC). Within the production process, there may be many instances in which machine tools are required for milling, drilling, cutting, routing, welding, bending, or a combination of these operations. The application of computerized technology for the control of machine tools, with prerecorded coded information, to make a part has been practiced for many years. However, these activities have often been accomplished in isolation from other activities in the production process. NC instructions have been prepared by programmers taking information from engineering drawings, programs have been tested, revised, retested, and so on. As this can be quite expensive, the goal in the future is to be able to generate NC input instructions directly from the design database, developed through the CAD applications discussed in Section 4.5.

3. *Robotics*. At various stages in the production process, there may be applications in which robots can be effectively employed for the purpose of materials handling (i.e., the carrying of parts from one location to the next) or for the positioning of tools and workpieces in preparation for NC applications. In some instances, robots are actually being used to operate drills, welders, and other tools. Computer technology is, of course, employed in the programming of robots for these and related production operations.

4. *Production management*. Throughout the production process, there is an ongoing management activity in which computer applications can be effectively utilized in support of production forecasting, scheduling, cost reporting, material requirements planning (MRP) activities, the generation of management reports, and so on. There is a requirement to develop a management information system (MIS) that enables the review and control of production functions.

The application of CAM methods to the production process will be different for a "flow shop" as compared with a "job shop" operation, or for the production of multiple quantities versus the construction of a "one-of-a-kind." Independent of the function, the entire process must be addressed as a system, possible CAM applications (i.e., combinations of NC, robotics, and data processing methods) must be identified, and a well-integrated and flexible manufacturing capability must be developed. The objective is to design a production capability with the proper mix of people and degrees of automation.

---

[13]In evaluating the entire production process, some activities may be found to be similar in nature, for which standard procedures can be utilized. This is particularly true in the manufacture of components where, by appropriate grouping, a common and standardized process can be applied using CAM methods. This approach is often defined as *group technology*.

In the initial design, and subsequent monitoring and control, of a production capability, the principles covered under concurrent engineering and quality engineering must prevail. The manufacturing tolerances, allowable through the application of NC tools and robotics, must be consistent with the initial requirements for system design. Care must be exercised to ensure that unexpected variances, causing possible product degradation through the production process, do not occur. This is of particular concern relative to meeting the objectives of system engineering.

4.7 COMPUTER-AIDED SUPPORT (CAS)[14]

Computer-aided support (CAS) refers to the application of computerized technology to the entire spectrum of logistics and system support activities described in Section 3.4.8. For years, the consideration of logistics and maintenance support requirements has been relegated to an activity downstream in the system life cycle. Further, in the initial definition of system support requirements, the process that has been implemented includes the generation of an extensive amount of data and documentation, much of which is distributed through a network involving many different locations. As systems are produced and delivered for operational use, the overall logistics and support activity often assumes an additional degree of complexity, involving a great deal of data/documentation, the processing and distribution of many different components, and the requirements for a significant amount of data communications. The support requirements, particularly for large-scale systems, have not always been responsive to the system need, and the results have been costly.

In addressing logistics in the context of the system life cycle, there are design-related activities, analysis activities, technical publications activities, provisioning and procurement activities, fabrication and assembly activities, inventory and warehousing activities, transportation and distribution activities, maintenance and product support activities, and management activities across the spectrum. In Figure 4.4, the applications are broad, and there is a certain degree of overlap with CAD and CAM requirements.

To provide an indication as to the variety of computer technology applications in the logistics field, five examples are noted:

1. *Logistics engineering.* The utilization of reliability, maintainability, and supportability models in the accomplishment of design trade-offs is a major requirement throughout system development (e.g., level-of-repair analysis, spares requirements, maintenance loading, transportation analysis, life-cycle cost models). This activity

---

[14]In the 1980s, the Department of Defense promoted the concept of computer-aided logistic support (CALS), with the objective of improving quality, timeliness, and responsiveness in the acquisition of future system support requirements. A good reference is MIL-HDBK-59, Military Handbook, *Computer-Aided Acquisition and Logistics Support Guide* (Washington, DC: DOD). These objectives, along with the further enhancement of computer-aided and electronic commerce (EC) methods, are still applicable and are used in the collection and processing of *logistics management information* (LMI) for the acquisition of new systems today.

should be integrated into the CAD effort described in Section 4.5, and the use of graphics technology, analytical methods, and database management capabilities is required.

2. *Supportability analysis (SA)*. Through the evaluation of a given design configuration, SA data are developed with the objective of identifying the specific requirements for system support; for example, spare parts, test and support equipment, personnel quantities and skill levels. These data must be generated, processed, stored, retrieved, and fed back in a timely manner if the design for supportability is to be realized. Data processing and database management capabilities are required.

3. *Logistics management information (LMI)*. This category covers the requirements for spares and repair parts provisioning data, support equipment provisioning data, design drawings and change notices, technical procedures, training manuals, and various reports. The development of technical manuals (i.e., system operating procedures and maintenance instructions) through automated processes is included. Computerized technology applications require the use of spreadsheets, word processing, graphics, and database management capabilities.

4. *Distribution, transportation, and warehousing*. The ongoing maintenance and support of the system throughout its planned life cycle requires the distribution, transportation, handling, and warehousing activities pertaining to spares and repair parts. In addition to the data processing and database management requirements associated with inventory control and MRP activities, the application of automated materials handling equipment and robotics can be effectively employed in the performance of warehousing functions. The automatic ordering and picking of components from the warehouse shelves, in response to a defined need, provides an excellent example of the use of computerized technology in the logistics area. Further, the opportunities for the future incorporation of "expert systems," or "artificial intelligence," are apparent in considering the "if-then" type of decisions that are necessary throughout materials handling process.

5. *Maintenance and support*. The customer service activities, in support of the system in the field throughout its planned life cycle, include the accomplishment of scheduled and unscheduled maintenance actions. This, in turn, leads to the consumption of maintenance personnel resources, the utilization of test equipment, the requirements for spare parts, the need for formal maintenance procedures, and so on. Although computerized methods have been utilized in the generation of spares/repair parts provisioning data and technical publications, there are additional applications relative to test equipment. In a few instances, handheld testers, with supporting software, have been used for maintenance diagnostics purposes. This area, supported with some selected applications of artificial intelligence, is a prime candidate for future growth.

As is the case relative to CAD, the CAS objective is being implemented at the detail level (i.e., a bottom-up approach). Initially, there was a great deal of effort expended in the development of requirements and interface specifications. In addition, there are many activities under way that are directed toward the development of technical manuals and procedures using automated processes. The development and processing of supportability analysis data is one area in which much has been accomplished. Although these efforts currently represent "islands of automation," the overall objective is to integrate the CAS requirements with CAD/CAM, as illustrated in Figure 4.5.

4.8 SUMMARY

The successful implementation of the system engineering process is highly influenced by technology and the tools available to the designer. The advent of computer technology, information technology, and the proper use of graphic methods and displays, analytical models, spreadsheets and word processing, database management capabilities, and so on, have enabled the designer to accomplish much more, in a shorter time frame, earlier in the system life cycle, and with less overall risk. As such, the design process has undergone the first phase in the transition from a long series of manually performed tasks to a more efficient integrated and automated process.

Relative to the future, the challenge is to complete the next phase. This involves the integration of the many analytical methods/tools, currently being utilized on an individual stand-alone basis, into the overall design process. The objective is to (1) develop a workstation concept providing the appropriate communications between all responsible design engineering functions (illustrated in Figure 4.2) and (2) develop a capability allowing the smooth and automated flow of information from the CAD capability to CAM and CAS (illustrated in Figure 4.5). Not only must these capabilities be able to "talk to each other" within the context of a given project, but the design information produced must be compatible and transferable between other projects having similar objectives. Thus, care must be exercised relative to the selection of an appropriate language, data format, and data structure. Nevertheless, a great deal of progress and future growth in this area can be anticipated.

QUESTIONS AND PROBLEMS

1. Describe, in your own words, what is meant by IT, EDI, EC, and the Internet. How are they related (if at all)?

2. Why is it important (in the context of system engineering) to become familiar with some of the analytical methods identified in Section 4.2? Provide some specific examples.

3. Describe, in your own words, what is included in each of the following: CAD, CAM, and CAS.

4. Identify and describe some of the technologies that are being applied in the design process. Provide some examples of typical applications.

5. Describe some of the benefits associated with the application of computerized methods in the design process. How do these methods relate to the objectives of system engineering?

6. Identify some of the problems associated with the application of computerized methods in the design process. What cautions must be observed?

7. Relating to the design disciplines identified in Chapter 3, Section 3.4, a listing of typical program tasks is provided. In each of the following disciplines, identify those tasks that can be accomplished using computerized methods. Include some specific examples.

   (a) Reliability engineering tasks

   (b) Maintainability engineering tasks

   (c) Human-factors engineering tasks

   (d) Logistics and supportability tasks

   (e) Quality engineering tasks

   (f) Value/cost engineering tasks

8. Describe what is meant by *artificial intelligence*. How can it be applied?

9. How does the application of computer technology impact concurrent engineering?

10. Select a system of your choice (describing the tasks in design), and develop a flowchart showing the application of CAD as it is being implemented.

11. How does CAM relate to system engineering? Describe some possible impacts.

12. How does CAS relate to system engineering? Describe some possible impacts.

13. Draw a flowchart showing the interface relationships among CAD, CAD, and CAS.

5

DESIGN REVIEW AND EVALUATION

System design is an evolutionary process, progressing from an abstract notion to something that has form and function, is fixed, and can be reproduced in specified quantities to satisfy a designated consumer need. Initially, a requirement (or need) is identified. From this point, design evolves through a series of phases; that is, conceptual design, preliminary system design, and detail design and development, as illustrated in Figure 1.12.

As the design progresses, there are natural degrees of system definition. Requirements are defined, leading to a "functional" baseline. This includes the definition of operational requirements and the maintenance concept, trade-off study reports and the results of the feasibility analysis, the identification of technical performance measures (TPMs), and the system specification (type A). Functional analysis and requirements allocation are accomplished, the results of which are defined through an "allocated" baseline. This baseline may be defined through a combination of development, process, product, and/or material specifications (types B, C, D, and E) as applicable. This configuration is progressively expanded, through numerous iterations, until a "product" baseline is defined, and so on. These natural phases of system definition are reflected by the activities and milestones identified in Figure 1.26.

In viewing the overall design process, the necessary checks and balances must be incorporated to ensure that the system configuration being developed will indeed fulfill the initially specified requirements. These checks and balances, accomplished through the conductance of design reviews, are provided early in the system life cycle when changes can be accomplished with relative ease and usually without great cost. A design review and evaluation function must be integral within the design process. Within the design review function, there must be feedback provisions for corrective action and the incorporation of design changes as necessary. The basic philosophy of design evolution, with the necessary review and feedback provisions, is shown in Figure 1.27. The purpose of this chapter is to explain this concept by describing evaluation methods, informal and formal design reviews, and the associated feedback and corrective-action loop.

5.1   DESIGN REVIEW AND EVALUATION REQUIREMENTS

One of the objectives in establishing a formal mechanism for design review and evaluation is to ensure, on a progressive and continuing basis, that the results of design reflect a configuration that will ultimately meet the stated consumer need.

Design evolves from the initial definition of requirements for a given system, through a series of iterations following a top-down approach, to a firm system configuration ready for production and/or construction. As one progresses through this series of steps, it is important that one initiate the requirements verification process from the beginning, because the earlier that potential problems are detected, the easier it will be to incorporate changes if needed. Thus, an ongoing design review and evaluation effort is required.

In evaluating the various stages of design, illustrated in Figures 1.12 and 3.1, the overall review process can be effectively accomplished through a combination of several approaches. First, there is an informal day-to-day review and evaluation activity that occurs as design decisions are made and data are developed (refer to Sections 2.9 and 2.10). This activity may involve many different design disciplines, making decisions on a relatively independent basis and generating design data based on the results. Second, formal design reviews are conducted at designated stages in the evolution of design, and these serve as a vehicle for communications and the formal approval of design data. These two main areas of activity are reflected in Figure 5.1 and are discussed further in Sections 5.2 and 5.3, respectively.

In response to the "whys" of design review, the objective is to ensure that system requirements are being met. These requirements, which are included in the system specification (refer to Section 3.2), are stated in both quantitative and qualitative terms. The purpose of the design review process is to evaluate the system configuration at different stages in terms of these requirements.

In addressing the aspect of *requirements*, there are program-level requirements, system-level technical requirements, detailed design requirements at the component level, and so on. Not only these requirements are viewed in a hierarchical sense, but the level of emphasis placed on these requirements will shift as we progress from conceptual design to the detail design and development phase. For example, it may be appropriate to establish a hierarchical relationship of system parameters such as that shown in Figure 2.25. Many of these parameters can be expressed in terms of a specific quantitative measure of system performance; that is, the identification of a technical performance measure (refer to Section 2.6). Some of these measures are applicable at the system level, some are more appropriately applied at the subsystem level, and some are directly related to the assembly or component level. In any

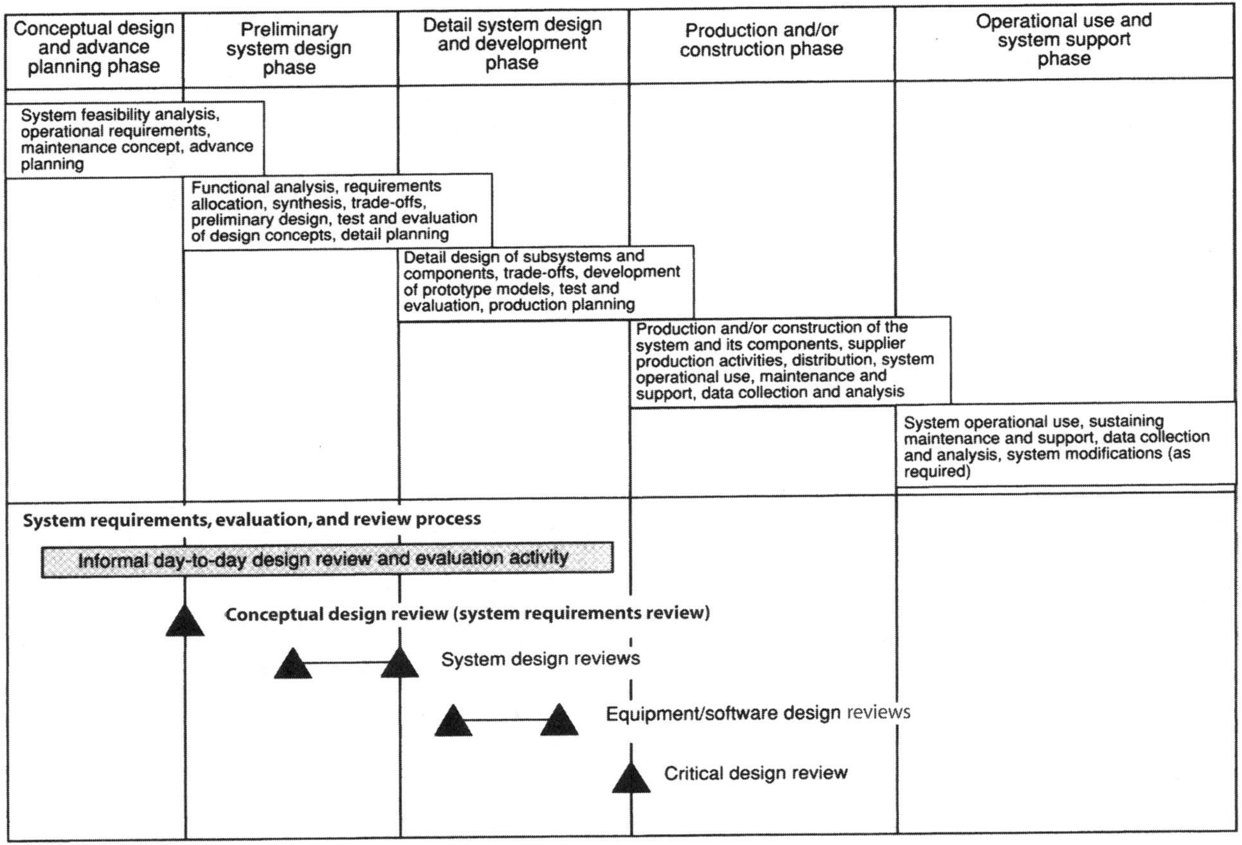

Conceptual design and advance planning phase	Preliminary system design phase	Detail system design and development phase	Production and/or construction phase	Operational use and system support phase
System feasibility analysis, operational requirements, maintenance concept, advance planning	Functional analysis, requirements allocation, synthesis, trade-offs, preliminary design, test and evaluation of design concepts, detail planning	Detail design of subsystems and components, trade-offs, development of prototype models, test and evaluation, production planning	Production and/or construction of the system and its components, supplier production activities, distribution, system operational use, maintenance and support, data collection and analysis	System operational use, sustaining maintenance and support, data collection and analysis, system modifications (as required)

System requirements, evaluation, and review process

Informal day-to-day design review and evaluation activity

Conceptual design review (system requirements review)

System design reviews

Equipment/software design reviews

Critical design review

FIGURE 5.1 Design review and evaluation.

245

event, the system specification (and its supporting specifications) should establish the "order" of evaluation parameters on the basis of priority and importance.

From the desired hierarchical relationship(s) of evaluation parameters, it is now possible to establish some specific criteria against which the results of design are compared. This, of course, leads to the identification of design review requirements for conceptual design, for preliminary system design, and for the detail design and development phase. In conceptual design, the design review process must address top system-level performance measures, functional-level relationships, and so on (as included in the system type A specification). In the detail design and development phase, although the system-level requirements are still important, the emphasis may be on the selection and standardization of parts, the mounting of components in a module design, the accessibility of an item requiring frequent maintenance, and the labeling of panel displays and controls. These factors must be integrated into the overall design review and evaluation process presented in Figure 5.1.

Given the criteria against which the design is to be evaluated, it is important to identify the disciplines that have the greatest impact on the design relative to compliance with a specific requirement. For instance, in meeting an equipment diagnostics requirement in the design of an electronics system, electrical engineering, mechanical engineering, and maintainability engineering, in accomplishing their respective design tasks, may have the greatest impact on the corrective maintenance downtime ($\overline{M}ct$) figure of merit for the system. In assessing the level of participation in the design review process, it is necessary that these design disciplines be adequately represented. In other words, along with identification of the criteria for evaluation, the design "responsibility" must be identified.

Design responsibility (and participation in design reviews) is covered further, from the organizational perspective, in subsequent sections of this text. However, at this point, it is worthwhile to consider some of the requirements for design review participation. As in Figure 2.25, a hierarchy of system evaluation parameters should be established and tailored for each major system being developed. Those parameters considered to be important can be identified, as shown in Figure 5.2. At the same time, a "degree-of-interest" relationship can be established between the various technical performance measures (TPMs) and the applicable disciplines participating in the design process. The level of interest indicated (i.e., high, medium, and low) pertains to the actual, or perceived, impact that the activity of the discipline has on a designated TPM for the system. This, in turn, should lead to establishing the organizational requirements for design review and evaluation as one progresses from conceptual design to the detail design and development phase. Sections 5.2 and 5.3 cover this area more comprehensively.

5.2   INFORMAL DAY-TO-DAY REVIEW AND EVALUATION

As shown in Figure 5.1, the design review and evaluation process includes two basic categories of activity: (1) an informal activity in which the results of design are reviewed and discussed on a day-by-day basis and (2) a structured series of formal

Technical Performance Measures (TPMs) \ Engineering Design Functions	Aeronautical Engineering	Components Engineering	Cost Engineering	Electrical Engineering	Human-factors Engineering	Logistics Engineering	Maintainability Engineering	Manufacturing Engineering	Materials Engineering	Mechanical Engineering	Reliability Engineering	Structural Engineering	Systems Engineering
Availability (90%)	H	L	L	M	M	H	M	L	M	M	M	M	H
Diagnostics (95%)	L	M	L	H	L	M	H	M	M	H	M	L	M
Interchangeability (99%)	M	H	M	H	M	H	H	H	M	H	H	M	M
Life Cycle Cost ($350K / unit)	M	M	H	M	M	H	H	L	M	M	H	M	H
$\overline{M}$ct (30 min.)	L	L	L	M	M	H	H	M	M	M	M	M	M
MDT (24 hrs.)	L	M	M	L	L	H	M	M	L	L	M	L	H
MMH/ OH (15)	L	L	M	L	M	M	H	L	L	L	M	L	H
MTBF (300 hrs.)	L	H	L	M	L	L	M	H	H	M	H	M	M
MTBM (250 hrs.)	L	L	L	L	L	M	H	L	L	L	M	L	H
Personnel Skill Levels	M	L	M	M	H	M	H	L	L	L	L	L	H
Size (150 ft. by 75 ft.)	H	H	M	M	M	M	M	H	H	H	M	.H	M
Speed (450 mph.)	H	L	L	L	L	L	L	L	L	L	L	M	H
System Effectiveness (80%)	M	L	L	M	L	M	M	L	L	M	M	M	H
Weight (150K pounds)	H	H	M	M	M	M	M	H	H	H	L	H	M

H= high interest; M= medium interest; L= low interest

FIGURE 5.2   The relationship between TPMs and responsible design disciplines.

design reviews conducted at specific times in the overall system development process. The output from the day-to-day informal activity leads into the formal design reviews; this relationship is shown in Figure 5.3.

Design is generally initiated by the electrical engineer, the mechanical engineer, the structural engineer, the process engineer, and/or others who are directly

248

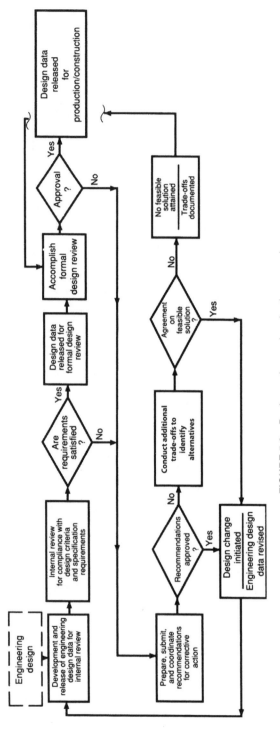

FIGURE 5.3 Design review and evaluation procedure.

responsible for the design of various components of the system. The results, usually produced independently from these different sources, are described through a combination of drawings, parts lists, reports, computerized databases, and supporting design documentation. As this definition process evolves, there are two major objectives:

1. The results of design must be properly communicated in a clear, effective, and timely manner to all members of the design team. Everyone involved in the design process must work from the *same* database.

2. The results of design must be compatible with the initially defined requirements for the system. Although each responsible designer should be familiar with the total spectrum of system requirements (e.g., electrical and reliability requirements), the physical separation of design disciplines and the lack of appreciation for the interfaces often result in discrepancies of one type or another (i.e., conflicts, omissions, incompatibilities between system components). These discrepancies must, of course, be corrected as soon as possible.

The design review activity is intended to satisfy both objectives. This can be accomplished through a series of steps involving the distribution of drawings, parts lists, and data to all affected areas of design, the review and sign-off of data for approval, the generation of recommendations for change in the event of noncompliance with a given requirement or for the purposes of product improvement, review of the change recommendations by the responsible designer, and so on. This is a day-to-day process with design data evolving from many different sources, and the amount of data can be rather extensive, depending on the nature of the system being developed and the size of the project.

In the past, particularly in regard to large projects in which members of the design team are remotely located from one another, the process of data distribution and approval has often been somewhat lengthy in terms of the time required to proceed through the cycle of events. For this reason, combined with the need for the designer to "get on with the design," many organizations have chosen to skip these steps of data distribution, review, and approval in the interest of saving time. In other words, the individual designer makes a decision (often independently), design documentation is prepared and released, component parts are procured and/or fabricated, and so on. Although it is hoped that all design interfaces have been recognized and that system requirements have been met, this has not always been the case. In the rush to complete the design, there have been omissions, conflicts, and/or problems associated with the incompatibility of system components. These problems have become evident later on during a formal design review (when formal design reviews have been conducted) or during system test and evaluation. Further, the implementation of design changes has been more costly than it might have been had these changes been incorporated earlier in the design process.[1]

---

[1]These problems can be partially solved through the implementation of an integrated database, as illustrated in Figure 2.33.

Relative to the future, the implementation of the informal design review and evaluation process shown in Figure 5.3 is, of course, highly desirable. Yet this procedure has to be accomplished both efficiently and in a timely manner. Although the series of data review steps accomplished in the past may have been somewhat time-consuming, the advent of the computerized methods described in Chapter 4 should result in a definite improvement. The utilization of computer-aided design (CAD) technology and the establishment of an effective communications network, will help to ensure an efficient flow of the necessary information. Design data can be distributed to many different locations expeditiously and on a concurrent basis, data review and approval sign-off can be accomplished through the electronics media, and data revisions can be implemented in a relatively short time frame. With these capabilities available, it is hoped that the process illustrated in Figure 5.3 can be implemented in an effective manner.

Concerning the review and evaluation itself, the depth of the review is a function of the complexity of design, whether the item being designed is new (i.e., promoting the state of the art) or is made up of existing off-the-shelf components, and whether the item is being developed by an outside supplier or designed in-house. Items that are complex or include the application of new technology will be investigated to a greater extent than standard components that are available and have been used in other systems.

In evaluating a given design configuration for compliance with a specified set of requirements, the reviewer may wish to develop a series of checklists based on applicable criteria. For example, through the review of selected design standards, component parts data, human-factors anthropometric data, maintainability accessibility factors, safety standards, and so on, the various design review activities can develop criteria that are directly applicable to the system in question. These criteria, summarized in the form of a checklist, are referenced as the evaluation of a given item is being conducted. The checklist serves as an aid in facilitating the review process. Figure 5.4 shows a sample checklist identifying typical topic areas for system-level reviews. Figure 5.5 presents an example of some specific questions that amplify the topics in Figure 5.4. In preparing for the various informal day-to-day design reviews, checklists of this nature can be very helpful.[2]

The results from the day-to-day informal review process, in the context of approved (signed-off) design documentation, are identified as items to be addressed in the formal design review. This includes not only design drawings and parts lists, but also trade-off study reports that support critical design decisions.

---

[2]A sample design review checklist is included in Appendix D. The development of such a checklist, tailored to a particular system, can be very beneficial. The results of a QFD analysis indicating areas of importance in design can lead to the preparation of design-related questions which, when applied, can help to emphasize the nature of the criteria that should be reflected in the ultimate design configuration being reviewed. By asking the right questions, one can provide the proper emphasis where required.

```
┌─────────────────────────────────────────────────────────────────────┐
│                    System Design Review Checklist                      │
├─────────────────────────────────────────────────────────────────────┤
```

General Requirements:

Have the technical and program requirements for the system been adequately defined through

1.	Feasibility analysis	5.	Functional analysis and allocation
2.	Operational requirements	6.	System specification
3.	Maintenance concept	7.	Supplier requirements
4.	Effectiveness factors	8.	System Engineering Management Plan (SEMP)

Design Features:

Does the design reflect adequate consideration of

1.	Accessibility	19.	Panel displays and controls
2.	Adjustments and alignments	20.	Personnel and training
3.	Cables and connectors	21.	Producibility
4.	Calibration	22.	Reconfigurability
5.	Data requirements	23.	Reliability
6.	Disposability	24.	Safety
7.	Ecological requirements	25.	Selection of parts/Materials
8.	Economic feasibility	26.	Servicing and lubrication
9.	Environmental requirements	27.	Societal requirements
10.	Facility requirements	28.	Software
11.	Fasteners	29.	Standardization
12.	Handling	30.	Storage
13.	Human factors	31.	Supportability
14.	Interchangeability	32.	Support equipment requirements
15.	Maintainability	33.	Survivability
16.	Mobility	34.	Testability
17.	Operability	35.	Transportability
18.	Packaging and mounting	36.	Quality

In reviewing the design (layouts, drawings, parts lists, reports), this checklist may be beneficial in covering major program requirements and design features applicable to the system. The items listed are supported with more detailed questions and criteria included in Appendix D. The response to each item listed should be YES.

FIGURE 5.4   Sample design review checklist.

5.3 FORMAL DESIGN REVIEWS

A *formal design review* constitutes a coordinated activity (i.e., a structured meeting or a series of meetings) directed toward the final review and approval of a given design configuration, whether it be the overall system configuration, a subsystem, or an element of the system. Although the informal day-to-day review process discussed in Section 5.2 covers specific aspects of the design; this coverage usually involves a series of independent fragmented efforts representing a variety of engineering disciplines. The purpose of the formal review is to provide a mechanism whereby *all*

Detailed Design Review Checklist

---

18. Packaging and Mounting

    a.   Is the packaging design attractive from the standpoint of consumer appeal (e.g., color, shape, size)?

    b.   Is functional packaging incorporated to the maximum extent possible? Interaction effects between packages should be minimized, and it should be possible to limit maintenance to the removal of one module (the one containing the failed part) when a failure occurs and not require the removal of two, three, or four modules in order to solve the problem.

    c.   Are equipment modules and/or components that perform similar operations electrically, functionally, and physically interchangeable?

    d.   Is the packaging design compatible with the level-of-repair analysis decisions? Repairable items are designed to include maintenance provisions such as test points, accessibility, and plug-in components. Items classified as "discard-at-failure" should be encapsulated and maintenance provisions are not required.

    e.   Are disposable modules incorporated to the maximum extent practical? It is highly desirable to reduce overall product support through a "no-maintenance" design concept as long as the items involved are high in reliability and relatively low in cost.

    f.   Are plug-in modules and components utilized to the maximum extent possible (unless the use of plug-in components significantly degrades the equipment reliability)?

    g.   Are the accesses between modules adequate to allow for hand grasping (refer to Design Handbook "X" for recommended accessibility provisions)?

    h.   Are modules and components mounted such that the removal of any single item for maintenance will not require the removal of other items? Component stacking should be avoided where possible.

    i.   In areas where component stacking is necessary because of limited space, are the modules mounted in such a way that access priority has been assigned in accordance with the predicted removal and replacement frequency? Items that require frequent maintenance should be more accessible.

    j.   Are modules and components, not of the plug-in variety, mounted with four fasteners or less? Modules should be securely mounted, but the number of fasteners should be held to a minimum.

    k.   Are shock-mounting provisions incorporated where shock and vibration requirements are excessive?

    l.   Are provisions incorporated to preclude the installation of the wrong module?

    m. Are plug-in modules and components removable without the use of tools? If tools are required, they should be of the standard variety.

    n.   Are guides (slides or pins) provided to facilitate module installation?

    o.   Are modules and components properly labeled?

FIGURE 5.5   Partial listing of design review questions.

interested and responsible members of the design team can meet in a coordinated manner, communicate with each other, and agree on a recommended approach. The formal design review process usually includes the following steps:

1. A newly designed item, designated as being complete by the responsible design engineer, is selected for formal review and evaluation. The item may be the overall system configuration as an entity or a major element of the system, depending on the program phase and the category of review conducted.

2. A location, date, and time for the formal design review meeting are specified.

3. An agenda for the review is prepared, defining the scope and anticipated objectives of the review.

4. A design review board (DRB) representing the organizational elements and the disciplines *affected* by the review is established. Representation from electrical engineering, mechanical engineering, structural engineering, reliability engineering, logistics engineering, manufacturing or production, component suppliers, management, and other appropriate organizations is included as applicable. This representation, of course, will vary from one review to the next. A well-qualified and unbiased chairperson is selected to conduct the review.

5. The applicable specifications, drawings, parts lists, predictions and analysis results, trade-off study reports, and other data supporting the item being evaluated must be identified prior to the formal design review meeting and made available during the meeting for reference purposes as required. It is hoped that each of the selected design review board members will be familiar with the data prior to the meeting.

6. Selected items of equipment (breadboards, service test models, prototypes), mock-ups, and/or software may be utilized to facilitate the review process. These items, of course, must be identified early.

7. Reporting requirements and the procedures for accomplishing the necessary follow-up action(s) stemming from design review recommendations must be defined. Responsibilities and action-item time limitations must be established.

8. Funding sources for the necessary preparations, for conducting the formal design review meetings, and for the subsequent processing of outstanding recommendations must be identified.

The formal design review meeting generally includes a presentation (or a series of presentations) on the item being evaluated, by the responsible design engineer, to the selected design review board members. This presentation should cover the proposed design configuration, along with the results of trade-off studies and analyses that support the design approach. The objective is to summarize what has been established earlier through the informal day-to-day design activity. If the design review board members are adequately prepared, this process can be accomplished in an efficient manner.

The formal design review must be well organized and firmly controlled by the design review board chairperson. Design review meetings should be brief and to the point, objective in terms of allowing for *positive* contributions, and must not be allowed to drift away from the topics on the agenda. Attendance should be limited to those who have a direct interest in and can contribute to the subject matter being presented. Design specialists who participate should be authorized to speak and make decisions concerning their areas of specialty. Finally, the design review activity must make provisions for the identification, recording, scheduling, and monitoring of corrective actions. Specific responsibility for follow-up action must be designated by the design review board chairperson.

With the conductance of formal design review meetings, five purposes are served:

1. The formal design review meeting provides a forum for communications across the board. The necessary coordination and integration are not adequately accomplished through the informal day-to-day review process, even with the availability of computerized technology. "Person-to-person" contact is required.

2. It provides for the definition of a common configuration baseline for all project personnel; that is, everyone involved in the design process must work from the *same* baseline. The responsible design engineer is given an opportunity to explain the proposed design configuration, and representatives from the various supporting disciplines are provided an opportunity to learn of the designer's problems. This, in turn, creates a better understanding between design and support personnel.

3. It provides a means for solving outstanding interface problems, and it promotes the assurance that all elements of the system are compatible. Those conflicts that were not resolved through the informal day-to-day review are addressed. Moreover, those disciplines not properly represented through earlier activity are provided an opportunity to be heard.

4. It provides a formalized check (i.e., audit) of the proposed system/product design configuration with respect to specification and contractual requirements. Areas of noncompliance are noted, and corrective action is initiated as appropriate.

5. It provides a formal report of major design decisions that have been made and the reasons for making them. Design documentation, analyses, predictions, and trade-off study reports that support these decisions are properly recorded.

The conductance of formal design review meetings tends to increase the probability of mature design, as well as the incorporation of the latest design techniques where appropriate. Group reviews may lead to the identification of new ideas, the application of simplier processes, and the realization of cost savings. A good "productive" formal design review activity can be very beneficial. Not only can it cause a reduction in the producer's risk relative to meeting specification and contractual requirements, but the results often lead to an improvement in the producer's methods of operation.

As stated earlier, formal design review meetings are generally scheduled prior to each major evolutionary step in the design process; for example, after the definition of a functional baseline, but prior to the establishment of an allocated baseline. Although the quantity and type of design reviews scheduled may vary from program to program, four basic types are easily identifiable and common to most programs. They are the conceptual design review, the system design review, the equipment or software design review, and the critical design review. The relative time phasing of these reviews is illustrated in Figure 5.1.

5.3.1 Conceptual Design Review

The *conceptual design review* (or system requirements review) is usually scheduled toward the end of the conceptual design and prior to entering the preliminary system design phase of the program (preferably not longer than one to two months after program start). The objective is to review and evaluate the functional baseline for the system, and the material to be covered through this review should include the following:[3]

1. Feasibility analysis (the results of technology assessments and early trade-off studies justifying the system design approach being proposed)
2. System operational requirements
3. System maintenance concept
4. Functional analysis (top-level block diagrams)
5. Significant design criteria for the system (e.g., reliability factors, maintainability factors, and logistics factors)
6. Applicable effectiveness figures of merit (FOMs) and technical performance measures (TPMs)
7. System specification (type A; refer to Section 3.2 and Figure 3.2)
8. System Engineering Management Plan (SEMP)
9. Test and Evaluation Master Plan (TEMP)
10. System design documentation (layout drawings, sketches, parts lists, selected supplier components data)—description of system architecture.

The conceptual design review deals primarily with top *system-level requirements,* and the results constitute the basis for follow-on preliminary system design and development activity. Participation in this formal review should include selected representation from both the consumer and producer organizations. Consumer representation should involve not only those personnel who are responsible for the acquisition of the system (i.e., contracting and procurement), but also those who will ultimately be responsible for the operation and support of the system in the field. Individuals with experience in operations and maintenance should participate in the system requirements review. On the producer side of the spectrum, those lead engineers responsible for *system* design should participate, along with representation from various design disciplines and production (as necessary). It is important to ensure that the disciplines identified in Chapter 3 are adequately represented in the formal design review process from the beginning.

---

[3]It is recognized that some of these requirements may not be adequately defined during the conceptual design phase, and that the review of such may have to be accomplished later. However, in promoting the desired generic approach described herein (and particularly with regard to system engineering), maximum effort should be made to complete these requirements early, even though changes may be necessary as system design progresses. The object is to encourage (or "force") early system definition, even if the "baseline" changes later.

In summary, the conceptual design review is extremely important for all concerned, as it represents the first opportunity for formal communication relative to system requirements from the top down. It can provide an excellent baseline for all subsequent design effort. Unfortunately, for many projects in the past, the conductance of a conceptual design review has not been accomplished. Further, if such a review were conducted, the results were not always made available to responsible design engineering personnel assigned to the project. This, in turn, has resulted in a series of efforts conducted in somewhat of a vacuum and not well coordinated or integrated. Thus, with the objectives of system engineering in mind, it is essential that a good functional baseline for the system be defined and properly evaluated through the conductance of an effective conceptual design review.

5.3.2 System Design Reviews

System design reviews are generally scheduled during the preliminary design phase when functional requirements and allocations are defined, preliminary design layouts and detailed specifications are prepared, system-level trade-off studies are conducted, and so on (refer to Figure 5.1). These reviews are oriented to the overall system configuration, rather than individual equipment items, software, and other components of the system. As the design evolves, it is important to ensure that the requirements described in the system specification are maintained. There may be one or more formal reviews scheduled, depending on the size of the system and the complexity of design. System design reviews cover a variety of topics, such as the following:

- Functional analysis and the allocation of requirements (beyond what is covered in the conceptual design review).
- Development, process, product, and material specifications as applicable (types B, C, D, and E).
- Design data defining the overall system (layouts, drawings, parts/materials lists, supplier data).
- Analyses, reports, predictions, trade-off studies, and related design documentation. This includes material that has been prepared in support of the proposed design configuration, and analyses/predictions that provide an assessment of what is being proposed. Reliability and maintainability predictions, logistic support analysis data, and so on, are included.
- Assessment of the proposed system design configuration in terms of applicable technical performance measures (TPMs).
- Individual program/design plans (e.g., reliability and maintainability program plans, human-factors program plan, and logistics plan).

Participation in system design reviews should include representation from both the consumer and producer organizations, as well as from major suppliers involved in the early phases of the system life cycle.

5.3.3 Equipment/Software Design Reviews

Formal design reviews covering equipment, software, and other components of the system are scheduled during the detail design and development phase of the life cycle. These reviews, usually oriented to a particular item, include coverage of the following six aspects:

1. Process, product, and material specifications (types C, D, and E—beyond what is covered in the system design reviews).

2. Design data defining major subsystems, equipment, software, and other elements of the system as applicable (assembly drawings, specification control drawings, construction drawings, installation drawings, logic diagrams, schematic diagrams, materials and detailed parts lists, and so on).

3. Analyses, reports, predictions, trade-off studies, and other related design documentation as required in support of the proposed design configuration and/or for assessment purposes. Reliability and maintainability predictions, human-factors task analysis, supportability analysis data, and so on, are included.

4. Assessment of the proposed system design configuration in terms of the applicable technical performance measures (TPMs). An ongoing review and evaluation are required to ensure that these system-level requirements are maintained throughout the various stages of detail design and development.

5. Engineering breadboards, laboratory models, service test models, mock-ups, and prototype models used to support the specific design configuration being evaluated.

6. Supplier data covering specific components of the system as applicable (drawings, materials and parts lists, analysis and prediction reports, and so on).

Participation in these formal reviews should include representation from the consumer (i.e., customer), producer (i.e., contractor), and applicable supplier organizations.

5.3.4 Critical Design Review

The *critical design review* is generally scheduled after the completion of detail design, but prior to the release of firm design data for production or construction. Design is essentially "frozen" at this point, and the proposed configuration is evaluated in terms of adequacy and producibility. The critical design review may address the following:

1. A complete set of final design documentation covering the system and its components (manufacturing drawings, materials and parts lists, supplier component parts data, drawing change notices, and so on).

2. Analyses, predictions, trade-off studies, test and evaluation results, and related design documentation (final reliability and maintainability predictions, human-factors and safety analyses, logistic support analysis records, test reports, and so on).

3. Assessment of the final system design configuration (i.e., the product baseline) in terms of applicable technical performance measures (TPMs).

4. A detailed production/construction plan (description of proposed manufacturing methods, fabrication processes, quality control provisions, supplier requirements, material flow and distribution requirements, schedules, and so on).

5. A final logistics and maintenance support plan covering the proposed life-cycle maintenance and support of the system throughout the consumer utilization phase.

The results of the critical design review describe the final system/product configuration baseline prior to entering into production and/or construction. This review constitutes the last in a series of progressive evaluation efforts, reflecting design and development from a historical perspective and showing growth and maturity in design as the engineering project evolved. It is important to view the design review process in *total* and to provide an overall evaluation of certain designated system attributes as the project progresses, particularly because close continuity is required between the various reviews. An example of designated system attributes that should be assessed on a continuing basis is presented in Figure 5.6.

5.4 THE DESIGN CHANGE AND SYSTEM MODIFICATION PROCESS

The objective thus far has been to develop a system on a progressive basis and to establish a firm configuration baseline through the formal review and evaluation process. In essence, the results from the conceptual design review lead to the definition of system-level requirements, the results from the system design reviews constitute a more in-depth description of the system packaging concepts, and so on. As we progress through the series of design reviews described in Section 5.3, the system definition becomes more refined, and the configuration baseline (updated from one review to the next) is established. This baseline, which constitutes a single point of reference for all individuals who are involved in the design process, is critical from the standpoint of meeting the system engineering objectives described earlier.

Once a configuration baseline has been established, it is equally important that any variations, or changes, with respect to that baseline be tightly controlled. It is certainly not anticipated that a given baseline will remain as such forever, particularly during the early stages of system development. However, in evolving from one design configuration to the next, it is important that all changes be carefully recorded and documented in terms of their possible impact on the initially specified system requirements. The process of configuration identification, the control of changes, and

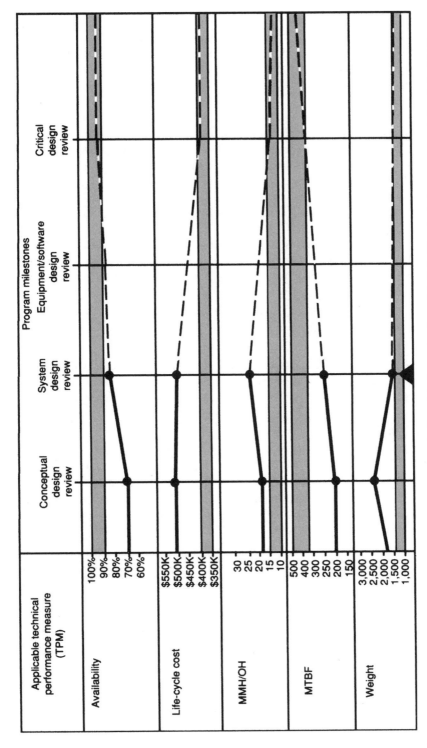

FIGURE 5.6 System parameter measurement and evaluation at design review (sample). *Note:* The shaded area represents the desired goal.

maintaining the integrity and continuity of design are accomplished through configuration management (CM).[4]

In the defense sector, *configuration management* is often related to the concept of *baseline management*. As shown in Figure 1.12, *functional, allocated*, and *product* baselines are established as the system development process evolves. These baselines are described through a family of specifications (types A, B, C, D, and/or E), drawings and parts lists, reports, and related documentation. The formal design review process provides the necessary authentication of these baseline configurations, and the configuration identification (CI) function is accomplished. CI relates to a particular baseline, and the configuration status accounting (CSA) function is a management information system that provides traceability of configuration baselines and changes thereto, and facilitates the effective implementation of changes. CSA includes the documentation in evolving from one configuration baseline to the next.

Proposed design changes, or proposed changes to a given baseline (i.e., a CI design), may be initiated from any one of a number of sources during any phase in the overall system life cycle. Such changes, prepared in the form of an engineering change proposal (ECP), may be classified as follows:

1. *Class 1 changes.* Design changes will affect form, fit, and/or function (e.g., changes that will impact system performance, reliability, safety, supportability, life-cycle cost, and/or any other system specification requirement).

2. *Class 2 changes.* Design changes are relatively minor in nature and will not affect system specification requirements (e.g., changes covering material substitutions, documentation clarifications, drawing nomenclature, producer deficiencies).

Changes may be categorized as "emergency," "urgent," or "routine," depending on priority and on the criticality of the change.

A simplified version of the system control procedure is illustrated in Figure 5.7. Proposed changes to a given baseline may be initiated during any phase of system development, production, and/or operational use. Each proposed change is presented in the form of an engineering change proposal (ECP) submitted for review, evaluation, and approval. In general, each ECP should include the following:[5]

- A statement of the problem and a description of the proposed change
- A brief description of alternatives that have been considered in responding to the need

---

[4]Configuration management (CM) is the process that identifies the functional and physical characteristics of an item during its life cycle, controls changes to those characteristics, and records and reports change processing and implementation status.

[5]In many organizations, the procedures related to configuration management and change control are a little more complex than those presented here. The procedure may involve engineering change requests (ECRs), design revision notices (DRNs), interface control documents (ICDs), and so on. The objective here is to present a *simplified* approach, providing a basic understanding of the importance of change control as part of the system engineering process

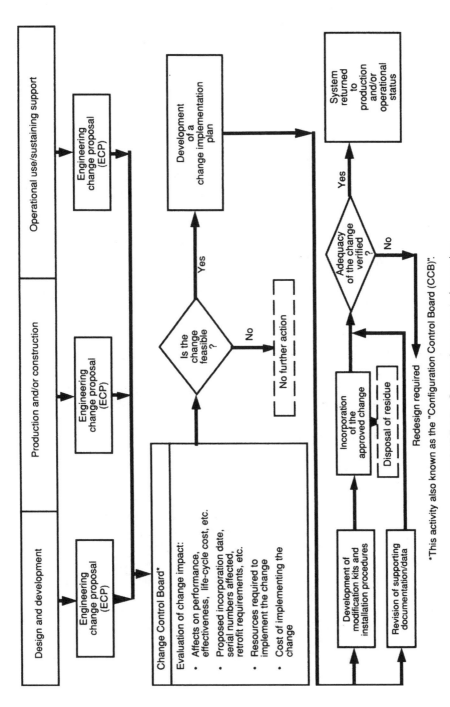

FIGURE 5.7   System change control procedure.

*This activity also known as the "Configuration Control Board (CCB)".

- An analysis showing how the change will solve the problem
- An analysis showing how the change will impact system performance, effectiveness factors, packaging concepts, safety, elements of logistic support, life-cycle cost, and so on. What are the impacts (if any) on system specification requirements? What is the effect on life-cycle cost?
- An analysis to ensure that the proposed solution will not cause the introduction of new problems
- A preliminary plan for incorporating the change; that is, proposed date of incorporation, serial numbers affected, retrofit requirements, and verification test approach (as applicable)
- A description of the resources required to implement the change
- An estimate of the costs associated with implementing the change
- A statement covering the impact on the system if the proposed change is *not* implemented; that is, an identification of the possible risks associated with a "do-nothing" decision

As shown in Figure 5.7, ECPs are processed through the change control board (sometimes known as the configuration control board, or the CCB) for review and evaluation. The CCB should function in a manner similar to the design review board (DRB) discussed in Section 5.3. Board representation should cover those design disciplines impacted by the change, including customer and supplier representation as necessary. Not only is it necessary to review and evaluate the original design, but it is important to ensure that all proposed design changes are handled in a similar manner. On occasion, when project schedules are "tight," the designer will generate data just to have something available for the record, and the *real* design configuration will be reflected through the "change process." Although this is not a preferred practice, it does occur in a number of instances when the objective is to save time. In any event, the review of design changes must be treated with the same degree of importance as is specified for the formal design review.

On completion of the formal design change review by the CCB, approved ECPs will be supported with the development of a plan for incorporating the change(s) in the system. This plan should include coverage of not only the modifications required for the prime equipment, but also the modifications associated with test and support equipment, spares and repair parts, facilities, software, and technical documentation. All elements of the system must be addressed on an integrated basis.

The actual incorporation of changes to the system is accomplished using a variety of approaches, depending on when the change is to be implemented. The time of implementation is a function of priority and/or criticality. Emergency or urgent changes may require immediate action, whereas routine changes may be grouped and incorporated at some convenient later point in time. Approved changes initiated during system design and development, prior to the availability of any hardware, software, or other physical components, may be incorporated through the preparation of design change notices (DCNs), or equivalent, attached to the applicable drawings/documentation covering those areas of design affected by the change. As the project progresses, these "paper" (or database) changes will be reflected in the new design configuration.

In the event that changes are initiated during the production/construction phase when multiple quantities of identical items are being produced, a designated serial-numbered item needs to be identified to indicate effectivity; that is, the change will be incorporated on the production line in serial number "X" and on later models. This should ensure that all applicable items scheduled to be produced in the future will automatically reflect the updated configuration.

For those system components that are already in use, changes may be incorporated through the installation of a modification kit in the field at the consumer's operational site. Such kits are installed, and the system is tested to verify the adequacy of the change. At the same time, the system support capability (e.g., test equipment, spares, and technical data) needs to be upgraded for compatibility with the prime mission-oriented segments of the system. Optimally, the installation process should take place at a time when the system is not in demand or being utilized in the performance of a mission.

This overall process is illustrated in Figure 5.7. With the incorporation of validated changes, the system configuration is updated and a new baseline is established. In situations in which the adequacy of the change is not verified, some additional redesign may be required.

5.5 SUMMARY

This chapter primarily addresses the basic review, evaluation, and feedback process illustrated in Figure 1.27. This process, which is critical in regard to the objectives of system engineering, must be tailored to the specific system development effort and must be properly controlled. An ongoing measurement and evaluation activity is essential and must be initiated from the beginning. Performing a one-time review and evaluation after the system has been produced and is in operational use may be costly in terms of possible modifications for corrective action. In addition, the incorporation of design changes on a continuing basis without the proper controls may be costly from the standpoint of system support. In essence, there must be a well-planned program approach, with the proper controls, in order to ensure a total integrated system configuration in the end.

QUESTIONS AND PROBLEMS

1. Describe the checks and balances in the design process (as you see them).

2. How is design review and evaluation accomplished? Why is it important relative to meeting system engineering objectives?

3. What is included in the establishment of a "functional" baseline? "allocated" baseline? "product" baseline? Why is baseline management important?

4. Select a system of your choice, and construct a sequential flow diagram of the overall system development process. Identify the major tasks in system development, and develop a plan/schedule of formal design reviews. Briefly describe what is covered in each.

5. Identify some of the benefits derived through formal design review. Describe some of the concerns.

6. In developing an agenda in preparing for a formal design review, what considerations must be addressed in the selection of items to be covered in the review process? How are review and evaluation criteria identified? Describe the steps and resources required in preparing for the design review.

7. How are technical performance measures (TPMs) considered in the design review process?

8. In the event that a deficiency is identified during design review, what steps are required for corrective action?

9. How are design changes initiated? How are priorities established?

10. How are design changes implemented? Identify the steps involved in system modification.

11. Describe the functions of the CCB.

12. What is *configuration management* (CM)? Define *configuration identification* (CI) and *configuration status accounting* (CSA).

13. How does configuration management (CM) relate to system engineering? Why is it important? What is likely to occur if configuration management practices (or equivalent) are not followed?

6

SYSTEM ENGINEERING
PROGRAM PLANNING

The first five chapters of this text have dealt with the system engineering process, the major steps in the process, and some of the available technologies and the tools that can be applied throughout. Given the definition of the process as a baseline, the remaining challenge lies with its *implementation*. The objective in Chapters 6, 7, and 8 is to respond to the requirements described throughout Chapters 1 through 5. In other words, the presumption in this text is that there is a firm *system* requirement first and then a *planning, organization*, and *implementation* requirement in response (i.e., bringing the required system into being).[1]

To meet the objectives described herein requires the application of a combination of both *technology* and *organization and management* skills. Although it is essential that one completely understand the technical process and available tools, this will not guarantee success unless the proper organizational "environment" is created where the applicable management skills can be effectively implemented in fulfilling the stated goals. As shown in Figure 6.1 (and referring to Section 1.5), there are technology-related activities and there are planning and organization activities that must be jointly applied throughout the top-down life-cycle process that has been the thrust in the earlier chapters.

The key to the successful implementation of any program is *early planning*. Planning for system engineering activities commences at program inception. As the

---

[1]In many situations, there is a tendency deal first with a organizational entity and then attempt to justify a specific system requirement around that organization, whether there is a need for that system or not. In this instance, it is important to establish the proper thought process that first justifies the requirement for a system, and then develop an approach for responding to that requirement through good planning, the implementation of an effective organizational approach, and so on.

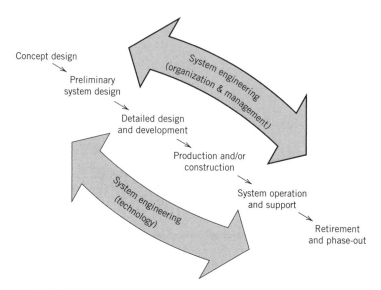

FIGURE 6.1 Management and technology applied to the system engineering process.

"need" for a system is identified and feasibility studies are conducted in selecting a technical design approach, requirements are established to define a program structure that can be implemented to bring the system into being. Planning is initiated with the definition of program requirements and the subsequent development of a *Program Management Plan* (*PMP*), shown in Figure 1.26. This, in turn, leads to the identification of system engineering requirements and the preparation of a detailed *System Engineering Management Plan* (*SEMP*), or *System Engineering Plan* (*SEP*).[2]

The SEMP is developed during the conceptual design and advanced planning phase, as illustrated in Figure 6.2, and includes a description of the system engineering functions and tasks to be accomplished, work packages and a work breakdown structure (WBS), task schedules and cost projections, an organizational structure and its interfaces, key policies and procedures, documentation and reporting requirements, and so on. The SEMP constitutes the overall planning document, which includes the necessary directives and guidance material for the successful implementation of the requirements described throughout the first five chapters of this text.[3]

---

[2]Usually, there is one overall planning document for every program/project, which covers all program requirements at a high level and leads to a variety of lower-level plans that address specific areas of activity. Although the specific nomenclature may vary from one program to the next, the title *Program Management Plan* (*PMP*) was selected in this instance to represent this top-level plan (refer to Figure 3.2, Chapter 3).

[3]In the preparation and implementation of the SEMP, it should be noted that system engineering activities may be implemented by the customer (consumer), by the prime producer (contractor), and/or by a major supplier. In some instances (particularly for large programs), an initial SEMP for the overall *system* may be prepared by the customer, with a lower-level SEMP prepared by the producer. Obviously, the second must evolve from and support the first. In any event, the intent herein is to describe what material might be included in the SEMP overall, and not attempt to differentiate between who does what. It is assumed that the proper "tailoring" will be accomplished, depending on the individual program requirements.

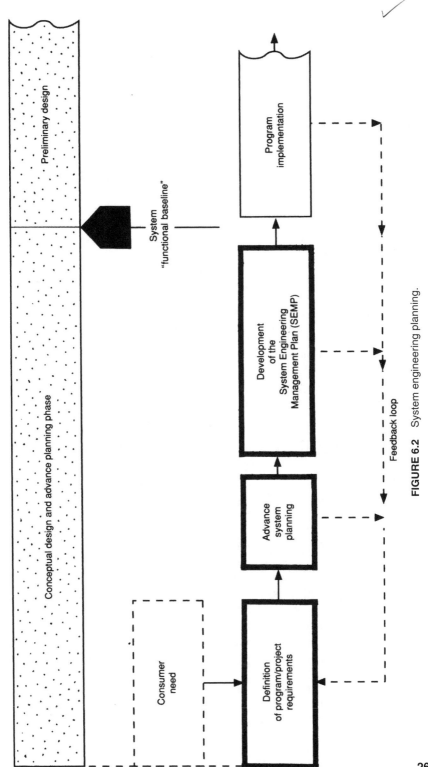

FIGURE 6.2 System engineering planning.

This chapter covers system engineering program planning, the first step in system management, illustrated in Figure 6.2. The material presented leads into the discussion of the organization for system engineering (Chapter 7) and system engineering program evaluation (Chapter 8). Implementing the requirements described in the first five chapters is highly dependent on the thoroughness of planning from the beginning and in the follow-on organization, management, and control later on.

6.1 SYSTEM ENGINEERING PROGRAM REQUIREMENTS

As shown in Figure 6.2, the first step in the planning process involves the definition of program (or project) requirements. Although this may appear to be rather basic, every program is different and it is essential that system engineering requirements be tailored accordingly. The concepts and methods described throughout this text, however, are applicable to all programs. Only the nature and depth of application may vary from one program to the next.

6.1.1 The Need for Early System Planning

The successful implementation of the concepts and methods of system engineering is highly dependent on (1) employing a top-down approach in the development of a system, (2) integrating early the design and related supporting activities, (3) viewing requirements in terms of the entire system life cycle, and (4) preparing complete *requirements documentation* from the beginning (i.e., applicable specifications and plans). As early system concepts are generated and feasibility studies are conducted to determine possible alternative technical solutions in response to a given design problem, the appropriate level of planning must be initiated to ensure that the ideas generated through analysis are properly consumated and integrated into a final product configuration in a cost-effective manner.

System engineering planning commences at program inception during the needs analysis (described in Section 2.2). Liaison activities with the customer (consumer/ user) are required to ensure that the defined need is interpreted correctly and described accurately, and that the translation from the stated need to the definition of system requirements is responsive. This is a critical step in the early definition of system requirements, as there is often a communications gap and the *real* need is not thoroughly understood. This, of course, may result in the development of planning information for a design configuration that will not perform the functions intended.

Given the identification and description of the need, the next step involves the accomplishment of feasibility studies (refer to Section 2.3). Future technological opportunities are identified and alternative approaches are investigated for possible design application. The feasibility of each of the alternatives being considered is a function not only of meeting the necessary performance requirements, but also of being responsive to the following questions:

1. Have the resource requirements associated with each alternative been defined (i.e., human, material, equipment, software, and data needs)? Have the sources

of supply been identified? Will the necessary resources be available when required?

2. Does the alternative being considered for selection reflect a cost-effective approach based on a life-cycle analysis?

3. Has an *impact analysis* been accomplished to determine whether there are possible secondary and/or tertiary effects as a result of selecting a given alternative? It is hoped that the selection of a specific alternative will not have a detrimental impact relative to the environment; that is, in regard to social, political, or ecological concerns as presented in Figure 3.36 (Section 3.4.11). In addition, interactions with other systems should be minimized.

4. Have all major interfaces, in a *system-of-systems (SOS)* context, been defined (as applicable)? It is hoped that other systems within the same overall hierarchy will not have any detrimental effects on the new system being introduced.

It is at this stage in the life cycle when early system engineering planning is important. The analysis effort is directed toward the *system* level, potential suppliers are identified, the overall system integration process commences, the interaction effects (both internal and external) are assessed, and potential areas of risk are identified. As system definition continues through the development of operational requirements, the maintenance concept, and the prioritization of technical performance measures (TPMs), the planning process evolves through another series of iterations. The requirements for system integration are greater in terms of both the *technical* integration of the various elements of the system and the integration of the many and varied organizational entities participating in the system development effort.

System planning is continuous, commencing with the definition of a need and extending through the development of the SEMP. As system-level requirements are defined, the planning process leads to the identification of those activities that must be accomplished in order to provide a system configuration that will fulfill these requirements. Design and management decisions at this stage in the system life cycle have a great impact on program activities later on. Thus, it is imperative that a complete and well-integrated planning effort be implemented from the beginning.

6.1.2 Determination of Program Requirements

Although the concepts, methods, and processes describing system engineering are generally applicable to all categories of systems, they must be tailored for each individual application. Further, the applications are numerous and varied:

- Large-scale systems with many different components such as a space system, an urban transportation system, a hydroelectric power generating system, a health-care system, an educational system, a recreational system, and so on.
- Small systems with relatively few components, such as a local communications system, a computer system, a hydraulic system, a mechanical braking system, and so on.
- Systems in which there is a great deal of new design and development effort required; that is, the application of new technologies.

- Systems in which the design is based primarily on the utilization of existing standard off-the-shelf components.
- Systems that are highly equipment-intensive, software-intensive, facilities-intensive, or human-intensive; for example, a production system versus a ground command and control system, versus a data distribution system, versus a maintenance capability.
- Systems in which there are a large number of suppliers involved in the design and development process, at both the national and international levels.
- Systems in which there are a number of different organizations involved in the design and development process.
- Systems being designed and developed for utilization in the government sector, the private sector, and so on.

Although this text basically addresses only a few of the major categories of systems described in Chapter 1 (i.e., the manmade, open-loop, dynamic system), there are still a wide variety of applications, as illustrated in Figure 6.3.

In each individual situation, the system engineering process described in Chapter 2 is applicable. Although the extent and depth of effort will vary, the steps required for bringing a system into being are basically the same. A needs analysis and feasibility analysis are accomplished, operational requirements and the system maintenance concept are defined, and functional analysis and requirements allocations are completed. Even though one may be dealing with a relatively simple case, such as a small system made up of standard off-the-shelf components, there is still a

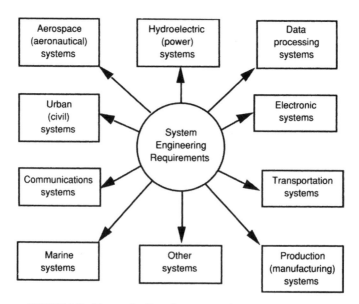

FIGURE 6.3   The application of system engineering requirements.

need to accomplish a top-down requirements analysis, a functional analysis and allocation, and so on. In other words, there is a *system* design requirement, even though new design may not be required at the subsystem or component level.

Following the general steps reflected in Figure 1.12 (Chapter 1) is a good overall approach to the design and development of any new system. As one progresses from the needs analysis through the accomplishment of feasibility studies and the definition of system-level requirements, the process evolves from the identification of the "whats" to the "hows"; that is, *what* is required in terms of a system functional capability, and *how* is this to be accomplished from the standpoint of a technical design approach. Evaluation of the early responses to the proposed technical design approach leads to the identification of specific *program* (or *project*) requirements.

Program requirements, in this context, refer to the management approach and the steps to be followed in the procurement and/or acquisition of the system in response to a stated need, along with the identification of the resources required to fulfill this objective. A program structure should be established that will enable the design and development, production and/or construction, and delivery of the system to the consumer in a cost-effective manner. This includes the identification of program functions and detailed tasks, the development of an organizational structure, the development of a work breakdown structure (WBS), the preparation of program schedules and cost projections, the implementation of a program evaluation and control capability, and so on. This information, presented in the form of a program plan, provides the necessary day-to-day management guidance required in the realization of any technical objective.

In fulfilling the system engineering objectives described in the earlier chapters of this text, a SEMP is developed as part of the early planning requirements for each program (refer to Figure 6.2). Although the detailed content may vary from one instance to the next, some of the major features of the SEMP are noted in Section 6.2.

6.2 SYSTEM ENGINEERING MANAGEMENT PLAN (SEMP)[4]

As shown in Figure 3.2, the SEMP is developed based on the Program Management Plan (PMP) and covers all management functions associated with the performance of system engineering activities for a given program. The SEMP constitutes the chief engineer's plan for identifying and integrating all major engineering activities; i.e., the top *technical* plan that allows the integration of the many more subordinate plans, such as the mechanical engineering design plan, the software engineering plan, the reliability and maintainability plan(s), the human-factors and safety program plan(s), and so on.

---

[4]Three additional sources that include coverage of the SEMP are (1) A. P. Sage, and J. E. Armstrong, *Introduction to Systems Engineering* (New York: John Wiley & Sons, Inc., 2000); (2) EIA/IS-632, *Processes for Engineering a System* (Washington, DC: Electronic Industries Association); and (3) IEEE-1220, *Standard for Application and Management of the Systems Engineering Process* (New York: Institute of Electrical and Electronics Engineers, IEEE).

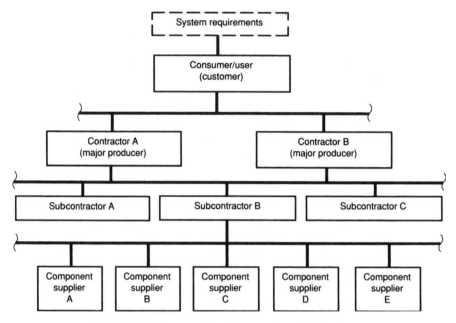

FIGURE 6.4   Consumer, producer, and supplier interfaces.

Preparation of the SEMP is the responsibility of the "system manager" and may be accomplished by the customer (consumer/user) or by a major contractor (producer), depending on the program. The relationships between the customer, prime contractors or major producers, subcontractors, suppliers, and so on, particularly for large-scale systems, may take the form illustrated in Figure 6.4. In such instances, the customer/user is the system manager and is responsible for the SEMP, but may delegate the overall system integration and management responsibility to a prime contractor (i.e., Contractor A or Contractor B in the figure).

In the event that the customer prepares the SEMP, then Contractor A and Contractor B must each prepare a SEMP covering their respective system engineering activities, each being responsive to the higher-level SEMP. On the other hand, if the system integration and management responsibility is delegated to Contractor A (for example), then the responsibility for preparation of the SEMP, and for implementing the tasks defined therein, will be at this level.[5]

The process discussed here may appear to be self-evident. However, it is important to be clear that if the SEMP is to be meaningful and accomplish its objectives, it must be developed directly from the top-level PMP. Further, the responsibility for the SEMP, and for the accomplishment of the functions described within, must be clearly defined and supported by the program manager (or program director). When system management responsibility is delegated to Contractor A (in Figure 6.4), then Contractor A must be given both the *responsibility* and the *authority* to perform

---

[5]In such instances, the SEMP is usually prepared and included as part of the contractor's proposal to the customer.

all system-level functions (described in the SEMP) on behalf of Contractor B, as well as for all subcontractors and applicable suppliers.[6] Finally, the SEMP must be appropriately identified as the key top-level design engineering plan in the overall program documentation tree structure.

In terms of material content, the SEMP must be *tailored* to the system requirements, the program size and complexity, and the nature of the procurement and acquisition process. To indicate the nature of the information that may be included in a SEMP, a proposed outline is presented in Figure 6.5. Although this outline is certainly not "fixed" (as there are likely to be all kinds of variations in use today), it will serve as a guide for further discussion of some of the content. Not every topic in the detailed SEMP outline is discussed here, but a few selected areas deserve additional coverage.

6.2.1 Statement of Work

The *statement of work* (SOW) is a narrative description of the work required for a given project. In regard to the SEMP, it must be developed from the overall project SOW described in the PMP and should include the following:

- *A summary statement of the tasks to be accomplished.* An identification of the major system engineering tasks is presented in Section 6.2.2. These, in turn, must be supported by elements of work included in the work breakdown structure (WBS) discussed in Section 6.2.4.
- *An identification of the input requirements from other tasks.* These may include the results from other tasks accomplished within the project, tasks completed by the customer, and/or tasks accomplished by a supplier.
- *References to applicable specifications (including the system A specification), standards, procedures, and related documentation as necessary for the completion of the defined scope of work.* These references should be identified as key requirements in the documentation tree described in Section 6.2.5.
- *A description of the specific results to be achieved.* This may include deliverable equipment, software, design data, reports, and/or related documentation, along with the proposed schedule of delivery as presented in Section 6.2.7.

In preparing the SOW, the following five general guidelines are considered appropriate:

1. The SOW should be relatively short and to the point (not to exceed two to three pages), and must be written in a clear and precise manner.
2. Every effort must be made to avoid ambiguity and the possibility of misinterpretation by the reader.

---

[6]Quite often, contractors are held responsible for the development and implementation of a system engineering program (and the SEMP), but are not delegated the authority to successfully complete the task. Because the requirements for system engineering must start at the top, it is essential that the proper level of authority be delegated along with the responsibility.

System Engineering Management Plan (SEMP)
1.0 Overview
2.0 Applicable Documents
3.0 General Description of System Architecture
4.0 System Engineering Process
4.1 System Operational Requirements
4.2 Maintenance Concept
4.3 Technical Performance Measures (TPMs)
4.4 Functional Analysis (System Level)
4.5 Allocation of Requirements
4.6 System Synthesis, Analysis, and Design Optimization
4.7 System Test and Evaluation
4.8 Construction/Production Requirements
4.9 System Utilization and Sustaining Support
4.10 System Retirement and Material Recycling/Disposal
5.0 Technical Program Planning, Implementation, and Control
5.1 Program Requirements/Statement of Work
5.2 Organization (Customer/Producer/Supplier Structure and Interrelationships)
5.2.1 Producer/Contractor Organization (Project/Functional/Matrix)
5.2.2 System Engineering Organization
5.2.3 Program Tasks
5.2.4 Supplier Requirements
5.3 Key Organizational Interfaces
5.4 Work Breakdown Structure (WBS)
5.5 Project Schedule and Milestone Charts
5.6 Technical Performance Measurement (TPM) "Tracking"
5.7 Program Cost (Projections/Reporting)
5.8 Technical Communications (Program Reports/Documentation)
5.9 Program Monitoring and Control
6.0 Engineering Specialty Integration (Identification of Key Engineering Specialties, How They Relate to System Engineering, and Their Interrelationships with Each Other)
6.1 "Functional" Engineering (e.g., Electrical, Mechanical, Structural, Industrial, etc.)
6.2 Software Engineering
6.3 Reliability Engineering
6.4 Maintainability Engineering
6.5 Human Factors Engineering
6.6 Safety Engineering
6.7 Security Engineering
6.8 Manufacturing and Production Engineering
6.9 Logistics and Supportability Engineering
6.10 Disposability Engineering
6.11 Quality Engineering
6.12 Environmental Engineering
6.13 Value/Cost Engineering
6.14 Other Engineering Disciplines (as appropriate)
7.0 Configuration Management (CM)
8.0 Data Management (DM)
9.0 Program Technology Requirements (Computer-Aided methods, EC/EDI/IT Applications etc.)
10.0 Special International Requirements
11.0 Risk Management
12.0 References (Specifications, Standards, Plans, Procedures, Pertinent Documentation)

FIGURE 6.5 System Engineering Management Plan (SEMP) outline.

3. Describe the requirements in sufficient detail to ensure clarity, considering both practical applications and possible legal interpretations. Do not *underspecify* or *overspecify*!

4. Avoid unnecessary repetition and the incorporation of extraneous material and requirements. This can result in unnecessary cost.

5. Do not repeat detailed specifications and requirements that are already covered in the applicable referenced documentation.

The SOW will be read by many different individuals with a variety of backgrounds (e.g., engineers, accountants, contract managers, schedulers, lawyers), and there must be no unanswered questions as to the scope of work desired. This statement forms a basis for the definition and costing of detailed tasks, for the establishment of subcontractor and supplier requirements, and so on.

6.2.2 Definition of System Engineering Functions and Tasks

System engineering, as defined throughout this text, covers a broad spectrum of activity. It may even appear that the "systems engineer," or the system engineering organization, does everything! Although this is not practical, the fulfillment of system engineering objectives does require some involvement, either directly or indirectly, in almost every facet of program activity. The challenge is to identify those functions (or tasks) that deal with the overall *system* as an entity and, when successfully completed, will have a positive impact on the many related and subordinate tasks that must be accomplished.

For the purpose of identifying a select number of *key* tasks for system engineering, the process described in Chapter 2 can be considered a framework for further discussion. As a start, a review of some of the overall basic goals is appropriate. There are six objectives of system engineering:

1. Ensure that the requirements for system design and development, test and evaluation, production, operation, and support are developed in a timely manner through a top-down, iterative requirements analysis.

2. Ensure that system design alternatives are properly evaluated against meaningful, quantifiable criteria that relate to all of the desired characteristics; for example, performance factors, effectiveness factors, reliability and maintainability characteristics, human factors and safety factors, supportability characteristics, and life-cycle cost.

3. Ensure that all applicable design disciplines and related specialty areas are appropriately integrated into the *total engineering effort* in a timely and effective manner.

4. Ensure that the overall system development effort progresses in a logical manner with established configuration baselines, formal design reviews, the proper documentation supporting design decisions, and the necessary provisions for corrective action as required.

5. Ensure that the various elements (or components) of the system are compatible with each other and are combined to provide an entity that will perform its required functions in an effective and efficient manner.

6. Ensure that the interfaces with other systems within the same system-of-systems (SOS) configuration are properly identified, and that these external systems are compatible with the new system being introduced.

A review of these general goals leads to the question, *what detailed program (or project tasks) should be performed in order to successfully meet the objectives of system engineering?* Although each individual program is different and the associated activities must be tailored accordingly, the tasks presented in Figure 6.6 are considered applicable in most instances. For better clarification, an abbreviated description of each follows:

1. Perform a needs analysis and conduct feasibility studies (refer to Sections 2.1, 2.2, and 2.3). These activities should be the responsibility of the system engineering organization, because they deal with the system as an entity and are fundamental in the initial interpretation and subsequent definition of system requirements.

2. Define system operational requirements, the system maintenance concept, and the technical performance measures (TPMs) (refer to Sections 2.4, 2.5, and 2.6). The results of these activities are included in the overall definition of system-level requirements and are the basis for top-down system design.

3. Prepare the system type A specification (refer to Section 3.2). This is the top technical document for system design, and fulfilling the objectives of system engineering is dependent on the completeness and comprehensiveness of this specification. The B, C, D, and E specifications are based on the requirements of the A specification.

4. Prepare the Test and Evaluation Master Plan (refer to Section 2.11). This document reflects the approach, methods, and procedures that are to be followed in the overall evaluation of the system in terms of compliance with the initially specified requirements. Although there are many different and relatively small facets of testing, it is the compilation of these that provides an overall evaluation of the system as an entity.

5. Prepare the System Engineering Management Plan (refer to Sections 1.5 and 6.2). This, of course, is the top management document for all system engineering program activities.

6. Accomplish functional analysis and the allocation of requirements (refer to Sections 2.7 and 2.8). Functional analysis, which is the process of translating system-level requirements into detailed design criteria, provides the foundation for the development of many different individual design disciplinary tasks (refer to Section 2.7.4). The allocation process defines the specific design requirements for different components of the system, whether developed through supplier activities or procured off the shelf. In any event, the responsibility for this effort is appropriate

	System Engineering Tasks	Task Input Requirements	Task Output Requirements
1.	Perform needs analysis and conduct feasibility studies.	Consumer/customer requirements documentation; technical information reports covering technology applications; selected research reports; trade-off study reports supporting design approach.	Feasibility study reports; trade-off study reports justifying system-level design decisions.
2.	Define operational requirements and the system maintenance concept.	Consumer/customer requirements documentation; customer specifications and standards; feasibility study reports; trade-off study reports supporting design approach.	System requirements documentation (operational requirements and maintenance concept); trade-off study reports justifying system-level design decisions; list of prioritized TPMs; functional analysis (system-level).
3.	Prepare the system type A specification.	Technical information reports covering technology applications; feasibility study reports; system requirements documentation (operational requirements and maintenance concept); trade-off study reports justifying system-level design decisions; list of prioritized TPMs; functional analysis (system-level).	System type A specification.
4.	Prepare the Test and Evaluation Master Plan (TEMP).	System type A specification; customer test specification and standard; test requirements sheets (individual discipline test requirements).	Test and Evaluation Master Plan (TEMP).
5.	Prepare the System Engineering Management Plan (SEMP).	Consumer/customer requirements documentation; customer program specifications and standards; system requirements documentation (operational requirements and maintenance concept); system type A specification; Test and Evaluation Master Plan (TEMP); advance system planning information ; Program Management Plan (PMP).	System Engineering Management Plan (SEMP).
6.	Accomplish functional analysis and the allocation of requirements.	System requirements documentation (operational requirements and maintenance concept); system specification; trade-off study reports justifying system-level design decisions.	Functional analysis reports—functional flow diagrams (operational and maintenance functions); timeline analysis sheets, requirements allocation sheets (RASs), trade-off study reports, test requirements sheets, design criteria sheets.
7.	Accomplish system analysis, synthesis, and design integration.	Consumer/customer requirements documentation; customer specifications and standards; functional analysis reports; system type A specification; System Engineering Management Plan (SEMP); Test and Evaluation Master Plan (TEMP); individual design discipline program planning requirements.	Selected design data: system integration reports; supplier data and reports; trade-off study reports justifying design decisions; selected design discipline reports (predictions and analyses).
8.	Plan, coordinate, and conduct formal design review meetings.	Program Management Plan (PMP); System Engineering Management Plan (SEMP); applicable design data (drawings, parts and material lists, reports, databases); trade-off study reports justifying design decisions;individual design discipline reports (predictions, analyses, etc.).	Design review meeting minutes; action-item lists with designated responsibilities; approved/ released design data and supporting documentation.
9.	Monitor and review system test and evaluation activities.	Test and Evaluation Master Plan (TEMP); System Engineering Management Plan (SEMP); individual test data and reports.	System test and evaluation report(s).
10.	Plan, coordinate, implement, and control design changes.	Configuration management data and reports (description of design baseline); proposed engineering change proposals; change control requirements and actions.	Change implementation plans, change verification data/reports.
11.	Initiate and maintain production/ construction liaison; supplier liaison; and customer service activities.	System design data; production/construction requirements; approved design changes; system operating and maintenance procedures; consumer/customer operations and system utilization requirements; field data and failure reports.	Field data and failure reports; customer service reports on field operations.

FIGURE 6.6 System engineering tasks.

277

because it facilitates the necessary design integration effort by providing a common baseline definition of the system in functional terms.

7. Accomplish system analysis, synthesis, and design integration functions on a continuing basis throughout the overall design and development process (refer to Sections 2.9 and 2.10 and Chapter 3). System integration is iterative by nature and includes both the technical considerations dealing with the physical and functional interfaces of equipment, software, personnel, facilities, and so on, and the management considerations pertaining to organizational interfaces. From the management perspective, the system engineering organization is responsible for ensuring that (a) all program design-related functions/tasks are initially defined, (b) appropriate responsibilities and working relationships are established, (c) organization and communication channels are identified, and (d) program requirements are completed in a satisfactory manner. The system engineering organization is responsible for ensuring that the proper level of communications, coordination, and integration exists between the various design disciplines as applicable. Of particular interest are the task requirements for reliability engineering (Figure 3.15), maintainability engineering (Figure 3.21), human-factors engineering (Figure 3.27), safety engineering (Figure 3.30), logistics (Figure 3.34), software engineering (Section 3.4.1), manufacturing and production engineering (Section 3.4.7), quality engineering (Section 3.4.10), value/cost engineering (Section 3.4.12), and environmental engineering (Section 3.4.11).

8. Plan, coordinate, and conduct formal design review meetings; for example, conceptual design review, system design reviews, equipment/software design reviews, and critical design review (refer to Chapter 5). The system engineering organization is responsible for ensuring that an ongoing design evaluation is performed. This is partially accomplished through the scheduling of periodic design review meetings. The conductance of these meetings must be accomplished by a unbiased individual, and the overall results must be supportive of *system-level* design objectives.

9. Monitor and review system test and evaluation activities (refer to Section 2.11). It is essential that the system engineering organization be involved from the standpoint of interpreting and integrating individual test results into the evaluation of the system as a whole.

10. Plan, coordinate, implement, and control design changes as they evolve from engineering change proposals (ECPs) initiated from either the informal day-to-day review activity or as a result of formal design reviews (refer to Section 5.4). The system engineering organization is responsible for establishing and maintaining system "baselines" through the design and development process; for example, "functional" baseline, "allocated" baseline, and the "product" baseline in Figure 1.12. System engineering is essentially responsible for configuration management as the system evolves through its planned life cycle.

11. Initiate and maintain production/construction liaison, supplier liaison, and customer service activities. As the system configuration progresses from the design and development phase into production and/or construction, and subsequently into operational use, there is a requirement for a specified level of engineering support.

The purpose is to provide some engineering assistance relative to training and the understanding of system design, the incorporation of approved engineering changes into the system, and the acquisition of data from production activities and consumer operations in the field. The system engineering organization must be able to track the system throughout its planned life cycle.

The 11 basic program tasks just described constitute an example of what might be appropriate for a typical program, although the specific requirements may vary from one program to the next. The goal is to identify tasks that are oriented to the *system*, and that are *critical* relative to meeting the five major system engineering objectives stated earlier. More specifically, it is essential that an overall *system's* approach be followed from the initial establishment of requirements. As design progresses, it is essential that the system configuration being developed includes the desired characteristics. Finally, it is essential that the product output be validated in terms of meeting the initially established requirements.

In accomplishing this, there are requirements definition tasks, there are design review and approval tasks, there are configuration control tasks, and there are final test and evaluation tasks. These activities are undertaken through the combination of providing key documentation (specifications, plans, and reports), conducting carefully scheduled design reviews with the appropriate feedback provisions, and providing the necessary ongoing coordination and integration efforts. These activities must address all *system* functions accomplished throughout the various levels depicted in Figure 6.4.

To provide a more in-depth understanding of the 11 tasks, Figure 6.6 presents a summary, listing these tasks and showing typical input and output requirements. Although the majority of the input-output requirements are self-explanatory, through a review of the appropriate sections of this text, some additional discussion is necessary in support of the output requirements of task 6, dealing with functional analysis and allocation.

Functional analysis encompasses the process of translating system-level requirements into detailed design criteria and results in the complete definition of the system configuration in functional terms (refer to Section 2.7). The accomplishment of functional analysis is facilitated through the development of functional flow block diagrams, described in Section 2.7.1. Based on these diagrams, the system engineer may wish to evaluate the various functions further from the standpoint of series-parallel relationships, time durations, and, ultimately, the identification of major resource requirements. In addition, specific functional requirements need to be communicated to program/project personnel through time line analysis sheets, requirements allocation sheets (RASs), trade-off study reports, test requirements sheets, design criteria sheets, and the like. Time line analysis sheets and requirements allocation sheets are briefly discussed in the following paragraphs.

Although the functional block diagrams convey general series-parallel relationships, these requirements may be developed further through the use of *time line analysis sheets*. Time line analysis adds considerable detail in defining the durations of various functions. Concurrency, overlap, and the sequential relationships of functions/tasks can be projected. Moreover, time-critical functions can be readily identified; that is, those functions that directly affect system availability, operating time, and maintenance downtime. An example of a time line analysis sheet format is presented in Figure 6.7.

A *requirements allocation sheet (RAS)* is often used as the primary document for the identification of specific design requirements based on the functional analysis. The RAS is developed for each block in the functional flow block diagram. Performance requirements are described, which include (1) the purpose of the function, (2) the detailed performance characteristics that the function must accomplish, (3) the criticality of the function, and (4) applicable design constraints. Performance requirements must address design characteristics such as size, weight, volume, output, throughput, reliability maintainability, human factors, safety, supportability, economic factors, and so on. Both qualitative and quantitative performance requirements resulting from an analysis of the function are identified by the RAS. These requirements are expanded in sufficient detail to allow for the synthesizing and evaluation of alternative concepts for satisfying each functional need, employing a combination of resources in terms of equipment, personnel, software, and facilities. An initial definition of these resource requirements is included in the RAS. Figure 6.8, which is an extension of Figure 2.21, presents an example of a requirements allocation sheet (RAS) format.

The specific output from task 6 (Figure 6.6) will vary in structure and format, depending on the type of system and the stage of design and development. For large-scale systems involving many different interfaces, the relationships illustrated in Figure 6.9 may exist. Referring to the figure, the prime objective is to promote (i.e., force) the documentation process, or formal traceability, down to the development of specific *design criteria* for the various components of the system, as applicable. On the other hand, for smaller systems in which the design is relatively simple, the utilization of all of these data outputs may not be feasible.

6.2.3 System Engineering Organization

One of the most important sections of the SEMP describes the organizational structure that is being proposed for implementation of the objectives and the tasks defined in Section 6.2.2. A specific and individual system engineering department or group, by itself, will not be able to complete all of the work required for the 11 tasks presented in Figure 6.6. However, it is not the intent here to justify a large organization for working out the details. Yet the system engineering organization, through its system-level technical expertise and its leadership abilities, must take the lead and ensure that these task requirements are completed in an effective, efficient, and timely manner. In other words, the system engineering organization must be able to work with, influence, and inspire many other groups within (and external to) the project if the specified tasks are to be successfully completed. The system engineering organization must have the respect and cooperation of the other required functions in order for the proper integration to occur.

Figure 6.10 presents an abbreviated illustration to show how a system engineering department/group might fit within and relate to other major functions within the

System:		Subsystem:	Description of requirement:													
Source (functional flow block diagram)	Function number	Location		Elapsed time (hours)												Total time
Task number	Task description	Personnel	0.5	1.0	1.5	2.0	2.5	3.0	3.5	4.0	4.5	5.0	5.5	6.0	

FIGURE 6.7 Time line analysis sheet.

281

System:	Subsystem:	Description of requirement:							
Source (functional flow block diagram)	Function number	Location							
Functional performance and design requirements	Personnel requirements				Equipment requirements		Software/data requirements		Facility requirements	
	Tasks	Task time	Performance requirements	Training	Nomenclature	Specification	Nomenclature	Specification	Nomenclature	Specification

FIGURE 6.8 Requirements allocation sheet.

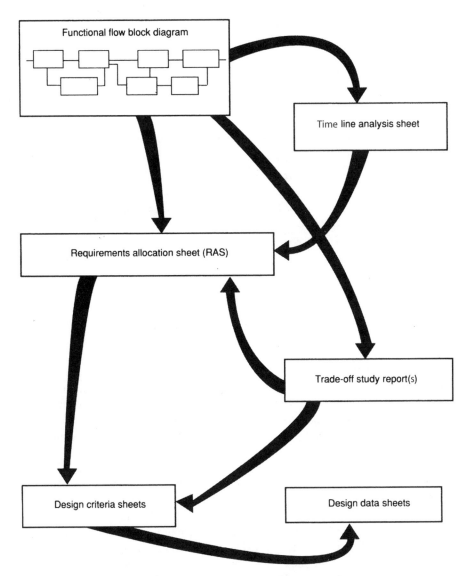

FIGURE 6.9   System engineering documentation.

overall organization for a large contractor. In delving further into the subject of "organization," it is clear that there are many different structures and approaches that may apply. For example, the primary system engineering organization may be contained within the customer's organization, with various responding subgroups within the contractor's organization. In a contractor's organization, the basic structure may constitute a *functional* approach, a *project/product line* approach, a *matrix* approach, or various combinations thereof. There are advantages and disadvantages associated with each of these approaches, which are essential to recognize if the

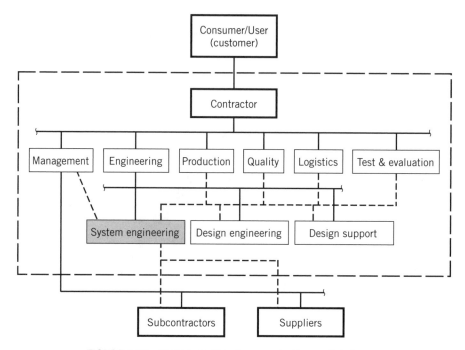

FIGURE 6.10   System engineering organization and interfaces.

system engineering organization is to work effectively within the structure provided. Further, there are the external interactions involving subcontractors and suppliers, which, in turn, may be critical for the accomplishment of the work required.

The subject of organization—the development of organizational structures, the staffing of an organization, and the "organization for system engineering"—is covered in detail in Chapter 7. However, at this point, it should be emphasized that a complete and thorough discussion of the organizational approach to be implemented for the project in question must be included in the SEMP for the system being developed or modified. Of particular interest are the numerous interfaces that must exist, across the board, if the objectives described throughout this text are to be met. Effective communication links, represented by the dotted lines in Figure 6.10, must be in place and functioning from the beginning. Although the organizational "makeup" within the system engineering block (in the figure) may look great on paper, it will not work unless the many noted interfaces are operational on a day-to-day basis.

6.2.4   Development of a Work Breakdown Structure (WBS)[7]

One of the initial steps in program planning, after the generation of the statement of work (SOW) and identification of the organizational structure, is the development of

---

[7]WBS and work packages are covered in most texts dealing with project management. A good reference is H. Kerzner, *Project Management: A Systems Approach to Planning, Scheduling, and Controlling*, 9th ed. (Hoboken, NJ: John Wiley & Sons, 2005).

the work breakdown structure (WBS). The WBS is a product-oriented tree that leads to the identification of the activities, functions, tasks, subtasks, work packages, and so on, that must be performed for the completion of a given program. It displays and defines the system (or product) to be developed and portrays all of the elements of work to be accomplished. The WBS is *not* an organizational chart in terms of project personnel assigned and responsibilities, but does represent an organization of work packages prepared for the purposes of program planning, budgeting, contracting, and reporting.[8]

Figure 6.11 illustrates an approach to the development of the WBS. During the early stages of system planning, a summary work breakdown structure (SWBS) is usually prepared by the customer and included in a request for proposal (RFP) or an invitation for bid (IFB). This structure, developed from the top down, primarily for budgetary and reporting purposes, covers *all* programs functions and generally includes three levels of activity:

1. *Level 1*: Identifies the total program scope of work, or the system to be developed, produced, and delivered to the customer. Level 1 is the basis for the authorization and "go-ahead" (or release) for all program work.

2. *Level 2*: Identifies the various projects, or categories of activity, that must be completed in response to program requirements. It may also include major elements of the system and/or significant project activities; for example, subsystems, equipment, software, elements of support, program management, and system test and evaluation. Program budgets are usually prepared at this level.

3. *Level 3*: Identifies the activities, functions, major tasks, and/or components of the system that are directly subordinate to the Level 2 items. Program schedules are generally prepared at this level.

As program planning progresses and individual contract negotiations are consumated, the SWBS is developed further and adapted to a particular contract or procurement action, resulting in a contract work breakdown structure (CWBS). Referring to Figure 6.4, for example, the customer may develop the SWBS with the objective of initiating program work activity. This structure will usually reflect the integrated efforts of all organizational entities assigned to the project and should not be related to any single department, group, or section. The SWBS, included in the customer's RFP, is the basis for the definition of all internal and contracted work to be performed on a given program. Through the subsequent preparation of proposals, contract negotiations, and related processes, Contractor A is selected to accomplish all work associated with the preliminary system design phase, and Contractor B is selected to complete all work associated with the detail design and development phase. From the definition of individual statements of work, a CWBS is developed to identify the elements of work for each program phase.

---

[8]Although the 11 tasks described in Section 6.2.2 reflect a *generic* approach for a system engineering organization in general, the development of the WBS will lead to the identification of the *specific* task requirements for the system being covered by the SEMP. A *tailored* approach is required.

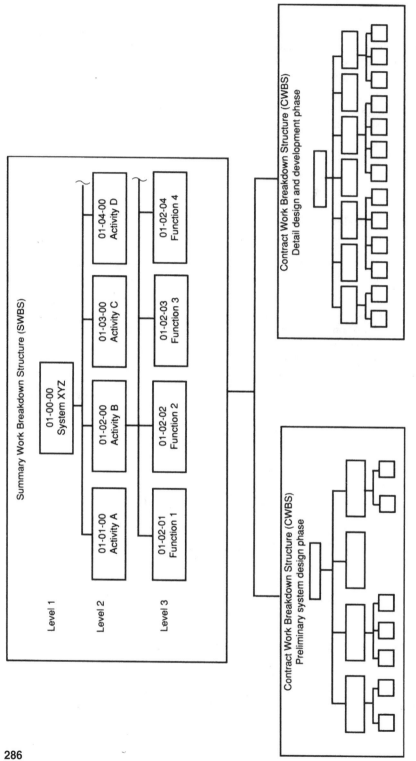

FIGURE 6.11 Partial work breakdown structure development.

The CWBS is tailored to a specific contract (or procurement action) and may be applicable to prime contractors, subcontractors, and/or suppliers, as shown in Figure 6.4.

The WBS constitutes a top-down hierarchical breakout of project activities that can be further divided into functions, functions into tasks, tasks into subtasks, subtasks into levels of effort, and so on. Conversely, detailed tasks (with defined starting and ending dates) can be combined into work packages, and work packages can be integrated into functions and activities, with the accumulation of all work being reflected at the top program or system level.

In developing a WBS, care must be exercised to ensure that (1) a continuous flow of work-related information is provided from the top down, (2) all applicable work is represented, (3) enough levels are provided to allow the identification of well-defined work packages for cost/schedule control purposes, and (4) the duplication of work effort is eliminated. If the WBS does not contain enough levels, then management visibility and the integration of work packages may prove to be difficult. However, if too many levels exist, too much time may be wasted in performing program review and control actions.

Figure 6.12 presents an example of a summary work breakdown structure (SWBS) covering the development of a large system. As program requirements are defined through a contractual (or procurement) arrangement, the SWBS can be readily converted into a CWBS to reflect the actual work required under the contract. The CWBS, as it appears in a contractual document, is also presented at three levels in order to provide a good baseline for planning purposes while allowing for some flexibility within the contractor's organization. An expansion of the CWBS can be accomplished as necessary to provide for internal cost/schedule controls.

Figure 6.13 shows an expansion of the system engineering activities to the fifth level; that is, those work packages under 3B1100 in Figure 6.12. The purposes are to recognize the major system engineering tasks presented in Figure 6.6 and to provide a breakout of these tasks in a CWBS format to the extent necessary for proper cost/schedule visibility. Note that Figure 6.13 includes two different CWBSs, one covering the work to be performed during the conceptual design and advance planning phase and the other directed toward the work required during the preliminary system design phase. Each individual CWBS is derived from the SWBS and the overall program CWBS. Further, there must be a close tie between the two, as the CWBS for preliminary design must reflect the activities that evolve directly from the earlier phase.

The elements of the WBS may include an identifiable item of equipment or software, a deliverable data package, an element of logistic support, a human service, or a combination thereof. WBS elements should be selected to permit the initial structuring of budgets and the subsequent tracking of technical performance measures (TPMs) against cost. Thus, in expanding the WBS to successively lower levels, the requirements for day-to-day task management must be balanced against the overall reporting requirements for the program. In essence, program activities are broken down to the lowest level that can be associated with both an organization and a cost account, as illustrated in Figure 6.14. From this, schedules are developed,

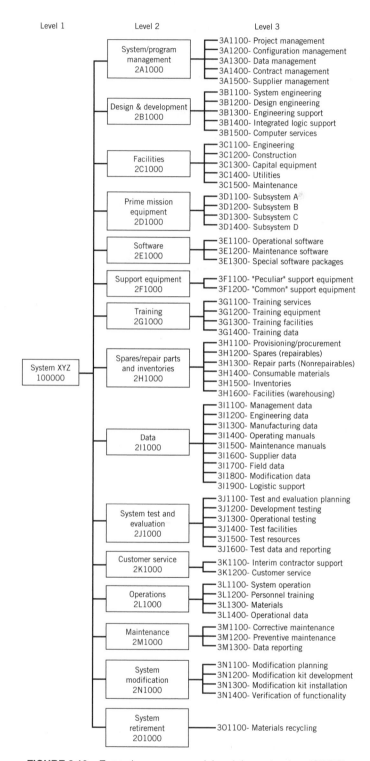

Level 1	Level 2	Level 3

System XYZ
100000

System/program management
2A1000
- 3A1100- Project management
- 3A1200- Configuration management
- 3A1300- Data management
- 3A1400- Contract management
- 3A1500- Supplier management

Design & development
2B1000
- 3B1100- System engineering
- 3B1200- Design engineering
- 3B1300- Engineering support
- 3B1400- Integrated logic support
- 3B1500- Computer services

Facilities
2C1000
- 3C1100- Engineering
- 3C1200- Construction
- 3C1300- Capital equipment
- 3C1400- Utilities
- 3C1500- Maintenance

Prime mission equipment
2D1000
- 3D1100- Subsystem A
- 3D1200- Subsystem B
- 3D1300- Subsystem C
- 3D1400- Subsystem D

Software
2E1000
- 3E1100- Operational software
- 3E1200- Maintenance software
- 3E1300- Special software packages

Support equipment
2F1000
- 3F1100- "Peculiar" support equipment
- 3F1200- "Common" support equipment

Training
2G1000
- 3G1100- Training services
- 3G1200- Training equipment
- 3G1300- Training facilities
- 3G1400- Training data

Spares/repair parts and inventories
2H1000
- 3H1100- Provisioning/procurement
- 3H1200- Spares (repairables)
- 3H1300- Repair parts (Nonrepairables)
- 3H1400- Consumable materials
- 3H1500- Inventories
- 3H1600- Facilities (warehousing)

Data
2I1000
- 3I1100- Management data
- 3I1200- Engineering data
- 3I1300- Manufacturing data
- 3I1400- Operating manuals
- 3I1500- Maintenance manuals
- 3I1600- Supplier data
- 3I1700- Field data
- 3I1800- Modification data
- 3I1900- Logistic support

System test and evaluation
2J1000
- 3J1100- Test and evaluation planning
- 3J1200- Development testing
- 3J1300- Operational testing
- 3J1400- Test facilities
- 3J1500- Test resources
- 3J1600- Test data and reporting

Customer service
2K1000
- 3K1100- Interim contractor support
- 3K1200- Customer service

Operations
2L1000
- 3L1100- System operation
- 3L1200- Personnel training
- 3L1300- Materials
- 3L1400- Operational data

Maintenance
2M1000
- 3M1100- Corrective maintenance
- 3M1200- Preventive maintenance
- 3M1300- Data reporting

System modification
2N1000
- 3N1100- Modification planning
- 3N1200- Modification kit development
- 3N1300- Modification kit installation
- 3N1400- Verification of functionality

System retirement
2O1000
- 3O1100- Materials recycling

FIGURE 6.12 Example summary work breakdown structure (SWBS).

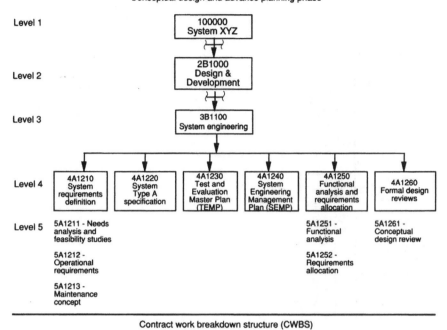

Contract work breakdown structure (CWBS)
Conceptual design and advance planning phase

Level 1 — 100000 System XYZ

Level 2 — 2B1000 Design & Development

Level 3 — 3B1100 System engineering

Level 4:
- 4A1210 System requirements definition
- 4A1220 System Type A specification
- 4A1230 Test and Evaluation Master Plan (TEMP)
- 4A1240 System Engineering Management Plan (SEMP)
- 4A1250 Functional analysis and requirements allocation
- 4A1260 Formal design reviews

Level 5:
- 5A1211 - Needs analysis and feasibility studies
- 5A1212 - Operational requirements
- 5A1213 - Maintenance concept
- 5A1251 - Functional analysis
- 5A1252 - Requirements allocation
- 5A1261 - Conceptual design review

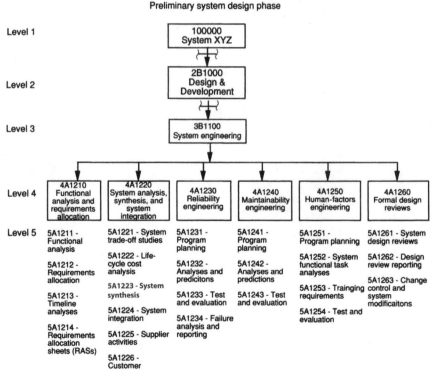

Contract work breakdown structure (CWBS)
Preliminary system design phase

Level 1 — 100000 System XYZ

Level 2 — 2B1000 Design & Development

Level 3 — 3B1100 System engineering

Level 4:
- 4A1210 Functional analysis and requirements allocation
- 4A1220 System analysis, synthesis, and system integration
- 4A1230 Reliability engineering
- 4A1240 Maintainability engineering
- 4A1250 Human-factors engineering
- 4A1260 Formal design reviews

Level 5:
- 5A1211 - Functional analysis
- 5A1212 - Requirements allocation
- 5A1213 - Timeline analyses
- 5A1214 - Requirements allocation sheets (RASs)

- 5A1221 - System trade-off studies
- 5A1222 - Life-cycle cost analysis
- 5A1223 - System synthesis
- 5A1224 - System integration
- 5A1225 - Supplier activities
- 5A1226 - Customer liaison

- 5A1231 - Program planning
- 5A1232 - Analyses and predicitons
- 5A1233 - Test and evaluation
- 5A1234 - Failure analysis and reporting

- 5A1241 - Program planning
- 5A1242 - Analyses and predictions
- 5A1243 - Test and evaluation

- 5A1251 - Program planning
- 5A1252 - System functional task analyses
- 5A1253 - Trainging requirements
- 5A1254 - Test and evaluation

- 5A1261 - System design reviews
- 5A1262 - Design review reporting
- 5A1263 - Change control and system modificaitons

FIGURE 6.13 CWBS expansion showing system engineering activities.

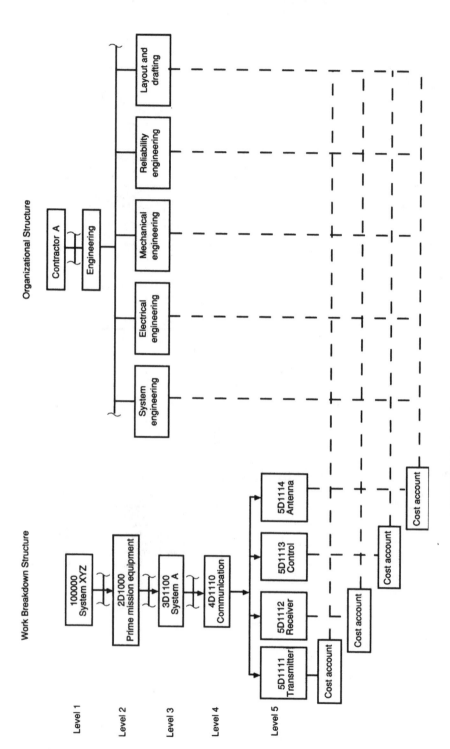

FIGURE 6.14 Organizational integration with CWBS.

cost estimates are generated, accounts are established, and program activities are monitored for purposes of schedule/cost control.

In developing the WBS, it is essential that a good comprehensive "WBS dictionary" be prepared. This is a document containing the terminology and definition of each element of the WBS. Traceability must be maintained from the top down, and all applicable work must be included. This is facilitated by assigning a number to each work package in the WBS. In Figure 6.11, the total program is represented by 01-00-00, and the numbers are broken down for activities, functions, tasks, subtasks, and so on. In Figure 6.12, a slightly different numbering system is used. Although the numbering systems will vary for different programs (and with different contractors), it is important to ensure that both activities and budgets/costs can be traced, both upward and downward. In the initial generation of a CWBS by a contractor during the preparation of a proposal, budgets may be allocated downward to specific tasks. After contract award, as tasks are being accomplished, costs are being incurred and charged to the appropriate cost account. These costs are then collected upward for reporting purposes. The WBS provides the vehicle for measuring work package progress in terms of schedule and cost.

In summary, the work breakdown structure (WBS) provides many benefits:

- The total program, or system, can be easily described through the logical break-out of its elements into nicely definable work packages.
- The discipline associated with the development of the WBS provides a greater probability that every program activity will be accounted for.
- The WBS is an excellent vehicle for linking program objectives and activities with available resources.
- The WBS facilitates the initial allocation of budgets and the subsequent collection and reporting of costs.
- The WBS provides an excellent matrix for the assignment of tasks and work packages to various organizational departments, groups, and/or sections. Responsibility assignments can be readily identified.
- The WBS is an excellent vehicle for the reporting of system technical performance measures (TPMs) against schedule and cost.

Finally, the WBS is an excellent tool for the promotion of program communications at various levels. As such, it must be updated to reflect program/system changes, consistent with configuration management actions. Maintaining currency is essential in meeting system engineering objectives.

6.2.5 Specification/Documentation Tree

In Section 3.1 (Figures 3.2 and 3.3), *specifications* are utilized primarily for acquiring items and/or for some element of work; that is, contracting for the design and development of a new item, the procurement of a commercial off-the-shelf (COTS) item, the testing and verification of a product, the construction of a new facility, the production of x quantity of items, and so on. Specifications may be applied on contracts and imposed on major contractors, subcontractors, and the suppliers of goods and services. They are *requirements*-oriented and *performance*-oriented, and they should state the "whats" (i.e., *what* to do versus *how* to do it) and must be written in a clear and concise manner. Vague, redundant, nebulous, and ambiguous language should be eliminated. Requirements should be quantifiable and verifiable, and the need to use judgment for interpretation should be avoided; that is, the use of phrases such as "best design practices" and "good workmanship" should be avoided. Specifications establish requirements relative to both design and performance characteristics. Management information, statements of work, procedural data, schedules and cost projections, and so on, should not be included.

Relative to applications, there are (1) general specifications, (2) program-related specifications, (3) military specifications and standards, (4) industrial standards, (5) specific company standards, and (6) international specifications and standards.[9] The different categories of specifications described in Section 3.1 refer primarily to program-unique specifications, or those that are directed toward a particular program requirement and/or a specific system component. In addition, there are specifications and standards that cover components, materials, and processes across the board, independent of application. In any event, there may be a wide variety of specifications and standards applied on a given program.

In applying specifications, extreme care must be exercised to ensure that they are prepared to the proper depth of detail and applied at the appropriate level in the system hierarchy. Specification documents must be detailed to the extent required to impact design in terms of component selection, the utilization of materials, and the identification of processes. By contrast, applying specifications with too much detail and at a level too low in the system hierarchy can be extremely detrimental. This may not only tend to inhibit innovation and creativity by not allowing for possible trade-offs, but "overspecification" can be quite costly. Applying a detailed specification to a small commercially available off-the-shelf component may result in an overdesign situation, which, in turn, can significantly increase the cost of that component.[10]

Another concern pertaining to the application of specifications involves possible areas of conflict. Experience shows that conflicts (i.e., contradictions in direction) are sometimes introduced with the application of general specifications and standards across the board. These documents are prepared by different individuals, at different times with different applications in mind, and are not necessarily consistent in terms of detailed requirements. Often in the development of program requirements, there is a tendency to follow the most expeditious and easiest approach by attaching a long

---

[9]Industrial standards can vary significantly and are developed by organizations such as the American National Standards Institute (ANSI), International Standards Organization (ISO), American Society for Testing and Materials (ASTM), Electronic Industries Association (EIA), Institute of Electrical and Electronics Engineers (IEEE), and the National Standards Association (NSA).

[10]For example, in a number of instances in the defense sector, military specifications and standards have been imposed on the procurement of small components, hand tools, and so on. The "blind" imposition of specifications on commercially available off-the-shelf items can turn out to be quite costly. Not only will an overdesign situation result, but the number of available suppliers will be reduced.

list of these specifications and standards to a SOW, with an accompanying statement: "The contractor must comply with the attached list of specifications and standards in fulfilling program requirements." This blind application can result in conflicts pertaining to component part selection, manufacturing process variations, test and evaluation parameters, and so on. In such instances, there is a question as to which specification takes precedence. What are the priorities, in order of importance?

With the objective of promoting clarification and eliminating the areas of possible conflict, the preparation of a *specification tree* (or documentation tree) is recommended. This is a family tree of specifications and documents that supports the system hierarchy, establishes order of precedence in the event of conflicts, and relates to the elements of work in the work breakdown structure (WBS). Figure 6.15 illustrates a simplified specification tree.

The tree in Figure 6.15 is developed from the top down, commencing with the preparation of the system specification (refer to Section 3.2). Subsequently, additional specifications are applied, following the system hierarchy illustrated in Figure 2.23. As one progresses, the application of specifications must be consistent with the work requirements described by the WBS in Figure 6.11. Further, this application must be adapted to the contracting structure between major contractors, subcontractors, and suppliers (refer to Figure 6.4).

The critical task here is the tailoring of specifications to the particular system application. Even though the design requirements may dictate the use of an available off-the-shelf item, the application of that item in this system may be quite different from comparable applications in other systems. Thus, the major components of the system should be described through a series of program-related specifications, as shown in Figure 6.15; for example, development specifications, product specifications, and process specifications. Below this level, it may be appropriate to apply general specifications, as long as they support the overall requirements in system design. When there are a number of different specifications and/or standards applied to the same system component, they must be complementary and mutually supportive. In the event of conflicts in direction or concerns relative to priorities of importance, the specification tree must provide an indication as to which document takes precedence.

The development of design requirements from the top down is critical in meeting the system engineering objectives stated herein. Thus, extreme care must be exercised in the initial identification and application of specifications and standards. Although this function is sometimes viewed as being relatively minor, the results can be rather costly if the proper level of attention is not directed to this area from the beginning. Conflicts, changes in specification requirements resulting in contractual modifications, and so on, can be extremely detrimental to a program. The inclusion of a complete specification tree in the SEMP may assist in avoiding potential problems later.

6.2.6 Technical Performance Measures (TPM)

In Section 2.6 (Chapter 2), technical performance measures (TPMs) for the system are identified through the development of operational requirements and the

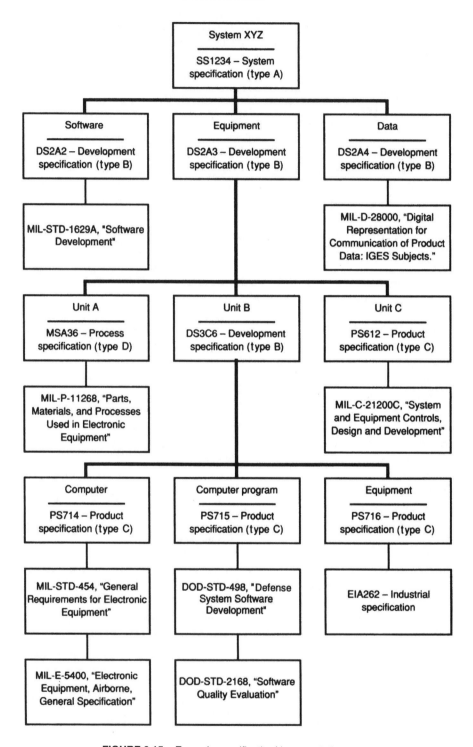

FIGURE 6.15 Example specification/documentation tree.

maintenance concept, and are prioritized using quality function deployment (QFD) (or equivalent) methods. Figure 2.11 provides an example of the results. As indicated in the figure, *velocity, availability,* and *size* are the top three in priority. Employing the QFD approach will help in developing the criteria and characteristics in design that must be built-in in order to ensure that the velocity, availability, and size requirements are ultimately met.

As the system development effort progresses, periodic design reviews will be conducted, as described in Chapter 5. The known design configuration at that time will be evaluated with the high-priority TPMs in mind. Checklists may be utilized to aid in the evaluation process, identifying those characteristics that have been incorporated and that relate to and directly support the TPM objectives (refer to Figure 5.5). Design parameters and the applicable TPMs will be measured and tracked, as shown in Figure 5.6. This must be accomplished on a continuing basis, and the results should be included as an inherent part of the regular program management review process. Those high-priority TPMs, assumed to be critical, should receive the most attention in the review and evaluation process.

See Figure 5.6; if there is a deviation from the specified TPM value (either upward or downward), the "causes" for such must be identified and the appropriate corrective action must be initiated accordingly. Inherent within the program management information system structure is the requirement to plot future trends, predict potential deviations, and identify the possible consequences and associated risks in the event that no corrective action is initiated. In essence, technical performance measurement must be built into the regular program management and control process, and the prioritization of TPMs is a necessary input for the risk management plan (refer to 11.0 in the SEMP outline shown in Figure 6.5).

6.2.7   Development of Program Schedules

In line with the statement of work (SOW) and the work breakdown structure (WBS), individual program tasks are presented in terms of a time line; that is, a beginning time and an ending time. Schedules are developed to reflect the work requirements throughout all phases of a program.

Schedule planning commences with the identification of major program milestones at the top level and proceeds downward through successively lower levels of detail. A system engineering master schedule (SEMS) is initially prepared, laying out the major program activities on the basis of elapsed time. This serves as the frame of reference for a family of subordinate schedules, developed to cover subdivisions of work as represented by the WBS. Progress against a given schedule is measured at the bottom level, and task status information is related to the appropriate cost account identified by the WBS element and the responsible organization (refer to Figure 6.14).

Program task scheduling may be accomplished using one or a combination of techniques. Seven of the more common methods are briefly described in the following paragraphs.

1. *Bar chart.* A simple bar chart presents program activities in terms of sequences and the time span of efforts. Specific milestones and the assignment of resources are not covered. Figure 6.16 illustrates a partial bar chart.

2. *Milestone chart.* A presentation of specific program events (i.e., identifiable outputs) and required start and completion times by calendar date is included. Deliverable items required under contract are noted. Figure 6.17 shows a sample milestone chart.

3. *Combined milestone/bar chart.* The combining of activities and milestones into an overall project schedule is a common approach for many programs. Figure 6.18 presents the primary system engineering tasks, included in Figure 6.6, in a program time line format. This, of course, serves as the basis for the assignment of resources and the development of cost projections.

4. *Program networks.* Network scheduling methods include the Program Evaluation and Review Technique (PERT), the Critical Path Method (CPM), and various combinations of these. PERT and CPM are ideally suited for early planning where precise task time data are not readily available, and the aspects of probability are introduced to help define risk leading to improved decision making. These techniques provide visibility and enable management to control one-of-a-kind projects as opposed to repetitive functions. Further, the network approach is effective in showing the interrelationships of combined activities.[11] Figure 6.19 shows an example of a network diagram consisting of 17 "events" and 29 major "activities." Events are usually designated by circles and are considered as checkpoints showing specific milestones; that is, dates for starting a task, completing a task, and delivering an item under contract. Activities are represented by the lines between the circles, indicating the work that needs to be accomplished to complete an event. Work can start on the next activity only after the preceding event has been completed. The numbers on the activity lines indicate the time required in days, weeks, or months. The first number reflects an optimistic time estimate, the second number indicates the expected time, and the third number indicates a pessimistic time estimate.[12]

In applying PERT/CPM to a project, one must identify all interdependent events and activities for each phase of the project. Events are related to program milestone dates that are based on management objectives. Figure 6.20 describes the major activities that are reflected by the lines in Figure 6.19. Managers and programmers work with engineering organizations to define these objectives and identify tasks and

---

[11]Two good references covering project management scheduling methods are (1) H. Kerzner. *Project Management: A Systems Approach to Planning, Scheduling, and Controlling*, 9th ed. (New York: John Wiley & Sons, Inc., 2005); and (2) D. J. Cleland, *Project Management; Strategic Design and Implementation*, 3rd ed. (New York: McGraw-Hill, 1998).

[12]The level of detail and depth of network development (i.e., the number of activities and events included) are based on the critically of tasks and the extent to which program evaluation and control are desired. Milestones that are critical in meeting the objectives of the program should be included, along with activities that require extensive interaction for successful completion. The author has had experience dealing with PERT/CPM networks including 10 to 700 events. The number of events/activities, of course, will vary with the project.

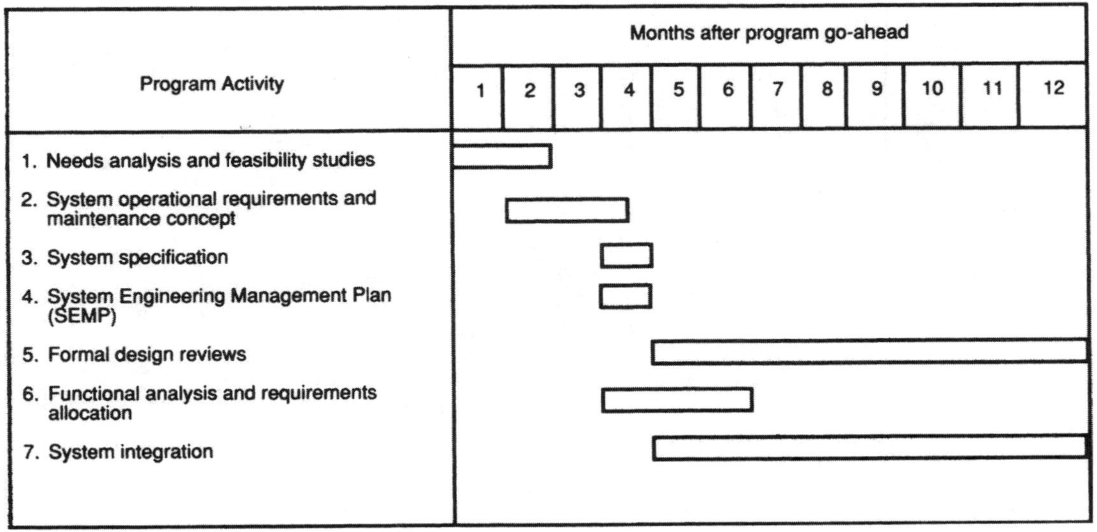

FIGURE 6.16 Partial bar chart.

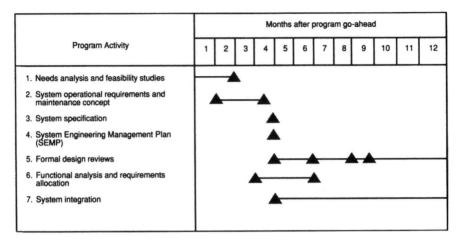

FIGURE 6.17   Sample milestone chart.

subtasks. When this is accomplished to the necessary level of detail, networks are developed, starting with a summary network and working down to detailed networks covering specific segments of a program. The development of networks is a team approach.

When actually constructing networks, one starts with an end objective (i.e., event 17 in Figure 6.19) and works backward in developing the network until event 1 is identified. Each event is labeled, coded, and checked in terms of program time frame. Activities are then identified and checked to ensure that they are properly sequenced. Some activities can be performed concurrently, and others must be accomplished in series. For each completed network, there is one *beginning event* and one *ending event*, and all activities must lead to the ending event.

The next step in developing a network is to estimate activity times and to relate these times in terms of probability of occurrence. An example of the calculations that support a typical PERT/CPM network is presented in Figure 6.21 and described in the following list.

(a) Column 1
List each event, starting from the last event and working backward to the beginning (i.e., from event 17 to event 1 in Figure 6.19).

(b) Column 2
List all previous events that lead into, or are shown as being prior to, the event listed in column 1 (e.g., events 15 and 16 lead into event 17).

(c) Columns 3 to 5
Determine the optimistic time (t_a), the most likely time (t_b), and the pessimistic time (t_c) in weeks or months for each activity. Optimistic time means that there is very little chance that the activity can be completed before this time, whereas pessimistic time means that there is little likelihood that the activity will take longer. The most likely time (t_b) is located at the highest probability point or

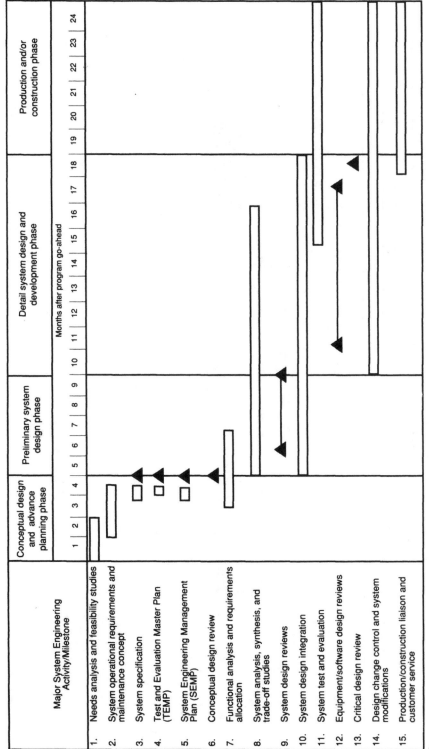

FIGURE 6.18 Major system engineering activities and milestones.

300

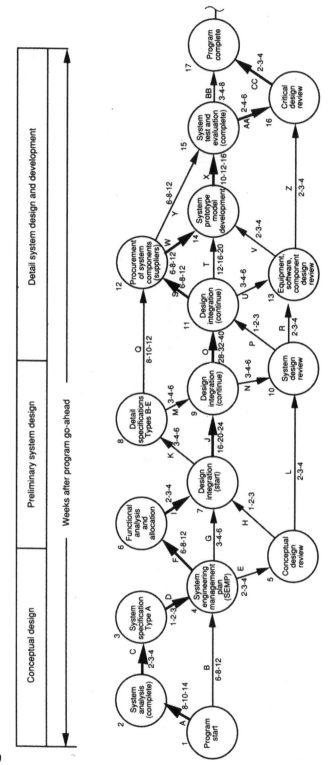

Conceptual design | Preliminary system design | Detail system design and development

Weeks after program go-ahead

Critical Path: 1-2-3-4-6-7-9-11-12-14-15-16-17

FIGURE 6.19 Partial program summary network.

Activity	Description of Program Activity
A	Perform needs analysis, conduct feasibility studies, and and accomplish systems analysis (operational requirements, maintenance concept, and functional definition of the system).
B	Conduct advance planning, perform initial management functions, and complete the System Engineering Management Plan (SEMP).
C	Prepare the system specification (type A).
D	Develop system-level technical requirements for inclusion in the System Engineering Management Plan (SEMP).
E	Prepare system-level design data and supporting materials for the conceptual design review.
F	Accomplish functional analysis and the allocation of overall system requirements to the sub-system level and below (as required).
G	Develop the necessary organizational and related infra-structure in preparation for the accomplishment of the required program design integration tasks.
H	Translate the results from the conceptual design review to the appropriate design activities (i.e., approved design data, recommendations for improvement/corrective action).
I	Translate the results from the functional analysis and allocation activity into specific design criteria required as an input for the design integration process.
J	Accomplish preliminary design and related design integration activities.
K	Translate the results from system-level design into specific requirements at the subsystem level and below. Prepare development, process, product, and/or material specifications as required.
L	Conduct the necessary planning and prepare for the system design review.
M	Translate the requirements contained within the various applicable specifications into specific design criteria required as an input for the design integration process.
N	Prepare design data and supporting materials (as a result of preliminary design) for the system design review:
O	Accomplish detail design and related design integration activities.
P	Translate the results from the system design review. to the appropriate design activities (i.e., approved design data, recommendations for improvement/corrective action).

Activity	Description of Program Activity
Q	Identify the appropriate system component suppliers, impose the necessary specification requirements through contracts, and monitor supplier activities.
R	Conduct the necessary planning and prepare for the equipment/software/component design reviews (there may be a series of individual design reviews covering different system components).
S	Provide detail design data (as necessary) to support supplier operations.
T	Develop a prototype model, with associated support, in preparation for system test and evaluation.
U	Prepare design data and supporting materials (as a result of detail design) for the equipment/software/component design reviews.
V	Translate the results from the equipment/software/component design reviews for incorporation into the prototype model(s) as applicable. The prototype model that is to be utilized in test and evaluation must reflect the latest design configuration.
W	Provide supplier components, with supporting data, for the development of the system prototype to be utilized in test and evaluation activities.
X	Prepare for and conduct system test and evaluation (implement the requirements of the Test and Evaluation Master Plan).
Y	Provide test data and logistic support, from the various suppliers, throughout the system test and evaluation phase. Test data are required to cover individual tests conducted at supplier facilities, and logistic support (i.e., spare/repair parts, test equipment, etc.) is necessary to support system testing activities.
Z	Conduct the necessary planning and prepare for the critical design review.
AA	Test results, in the form of either design verification or recommendations for improvement/corrective action, are provided as an input the critical design review.
BB	Prepare system test and evaluation report.
CC	Translate the results from the critical design review. for incorporation into the final system configuration prior to entering the production and/or construction phase of the program.

FIGURE 6.20 List of activities in the program network.

1 Event number	2 Previous event number	3 t_a	4 t_b	5 t_c	6 t_e	7 s^2	8 TE	9 TL	10 TS	11 TC	12 Probability (%)
17	16	2	3	4	3.0	0.111	115.2	115.2	0	110	6.4
	15	3	4	8	4.5	0.694	112.1	115.2	3.1	115	47.9
16	15	2	4	6	4.0	0.444	112.1	112.2	0	120	91.9
	13	2	3	4	3.0	0.111	86.5	112.2	25.7		
15	14	10	12	16	12.3	1.000	108.2	108.2	0		
	12	6	8	12	8.3	1.000	95.9	108.2	12.3		
14	13	2	3	4	3.0	0.111	86.5	95.9	9.4		
	12	6	8	12	8.3	1.000	95.9	95.9	0		
	11	12	16	20	16.0	1.778	95.3	95.9	0.6		
13	11	3	4	6	4.2	0.250	83.5		13.6		
	10	2	3	4	3.0	0.111	53.8		42.1		
12	11	6	8	12	8.3	1.000	87.6	87.6	0		
	8	8	10	12	10.0	0.444	60.8	87.6	26.8		
11	10	1	2	3	2.0	0.111	52.8	79.3	26.5		
	9	28	32	40	32.7	4.000	79.3	79.3	0		
10	9	3	4	6	4.2	0.250	50.8		30.7		
	5	2	3	4	3.0	0.111	21.3		59.0		
9	8	3	4	6	4.2	0.250	35.0	46.6	11.6		
	7	16	20	24	20.0	1.778	46.6	46.6	0		
8	7	3	4	6	4.2	0.250	30.8		15.8		
7	6	2	3	4	3.0	0.111	26.6	26.6	0		
	5	1	2	3	2.0	0.111	20.3	26.6	6.3		
	4	3	4	6	4.2	0.250	19.5	26.6	7.1		
6	4	6	8	12	8.3	1.000	23.6	23.6	0		
5	4	2	3	4	3.0	0.111	18.3		9.3		
4	3	1	2	3	2.0	0.111	15.3	15.3	0		
	1	6	8	12	8.3	1.000	8.3	15.3	7.0		
3	2	2	3	4	3.0	0.111	13.3	13.0	0		
2	1	8	10	14	10.3	1.000	10.3	10.3	0		

FIGURE 6.21 Example of program network calculations.

the peak of the distribution curve. These times may be predicted by someone who is experienced in estimating. The time estimates may follow different distribution curves, where P represents the probability factor (see Figure 6.22). The three time estimates are also included in Figure 6.19 for each activity (A, B, C, etc.).

(d) Column 6
Calculate the expected or mean time, t_e, from

$$t_e = \frac{t_a + 4t_b + t_c}{6} \tag{6.1}$$

(e) Column 7
In any statistical distribution, one may wish to determine the various probability factors for different activity times. Thus, it is necessary to compute the variance (σ^2) associated with each mean value. The square root of the variance, or the standard deviation, is a measure of the dispersion of values within a distribution and is useful in determining the percentage of the total population sample that falls within a specified band of values. The variance is calculated from equation (6.2):

$$\sigma^2 = \left(\frac{t_c - t_a}{6}\right)^2 \tag{6.2}$$

(f) Column 8
The earliest expected time for the project, TE, is the sum of all times, t_e, for each activity, along a given network path, or the cumulative total of the expected times through the preceding event remaining on the same path throughout the network. When several activities lead to an event, the highest time value (t_e) will be used. For instance, in Figure 6.19, path 1–4–7–9–11–14–15–17 totals 98; path 1–2–3–4–7–9–11–14–15–17 totals 105; and path 1–2–3–4–6–7–9–11–12–14–15–16–17 totals 115.2. The highest

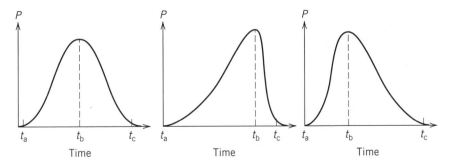

FIGURE 6.22 Sample distribution curves.

value for *TE* (if one were to check all network paths) is 115.2 weeks, and this is the value selected for event 17. The *TE* values for events 16, 15, and so on, are calculated in a similar manner, working backward to event 1.

(g) Column 9

The latest allowable time for an event, *TL*, is the latest time for completion of the activities that immediately precede the event. *TL* is calculated by starting with the latest time for the last event (i.e., where *TE* equals 115.2 in Figure 6.21) and working backward, subtracting the expected time (t_e) for each activity, remaining on the same path. The *TL* values for events 16, 15, and so on, are calculated in a similar manner.

(h) Column 10

The slack time, *TS*, is the difference between the latest allowable time (*TL*) and the earliest expected time (*TE*):

$$TS = TL - TE \tag{6.3}$$

(i) Columns 11 and 12

TC refers to the required scheduled time for the network based on the actual need. Assume that management specifies that the project reflected in Figure 6.19 must be completed in 110 weeks. It is now necessary to determine the likelihood, or probability (*P*), that this will occur. This probability factor is determined as follows:

$$Z = \frac{TC - TE}{\sqrt{\Sigma \text{ path variances}}} \tag{6.4}$$

where *Z* is related to the area under the normal distribution curve, which equates to the probability factor. The "path variance" is the sum of the individual variances along the longest path, or the critical path, in Figure 6.19 (i.e., path 1–2–3–4–6–7–9–11–12–14–15–16–17).

$$Z = \frac{110 - 115.2}{\sqrt{11.666}} = -1.522$$

From the normal distribution tables, the calculated value of -1.522 represents an area of approximately 0.064; that is, the probability of meeting the scheduled time of 110 weeks is 6.4%. If the management requirement is 115 weeks, then the probability of success would be approximately 47.9%; or if 120 weeks were specified, the probability of success would be around 91.9%.

When evaluating the resultant probability value (Column 12 of Figure 6.21), management must decide on the range of factors allowable in terms of risk. On the one hand, if the probability factor is too low, additional resources may be applied to the project in order to reduce the activity schedule times and improve the probability of success. On the other hand, if the probability factor is too high (i.e., there is practically no risk involved), this may indicate that excess resources are being applied, some of which may be diverted elsewhere. Management must assess the situation and establish a goal.

In Figure 6.19, the critical path, which is reflected by the heavy arrows (i.e., path 1–2–3–4–6–7–9–11–12–14–15–16–17), includes the series of activities requiring the greatest amount of time for completion. These are *critical* activities where slack times are zero, and a slippage of schedule in any one of these activities will cause a schedule delay in the overall program. Thus, these activities must be closely monitored and controlled throughout the program.

The network paths representing other program activities shown in Figure 6.19 include slack time (*TS*), which constitutes a measure of program scheduling flexibility. The slack time is the interval of time in which an activity could actually be delayed beyond its earliest scheduled start without necessarily delaying the overall program completion time. The availability of slack time will allow for a possible reallocation of resources. Program scheduling improvements may be possible by shifting resources from activities with slack time to activities along the critical path.

An additional point relative to program schedules is that a hierarchy of individual networks may be developed following a pattern similar to the WBS development approach illustrated in Figure 6.11. To provide the proper monitoring and control actions, scheduling may be accomplished at different levels. Figure 6.23 shows a breakdown of the program network (illustrated in Figure 6.19) into a lower-level network covering reliability program requirements. A similar network may be developed for maintainability, another network for electrical design, and additional detailed networks as appropriate. These lower-level networks must, of course, directly support the overall program network.

The utilization of the PERT/CPM scheduling technique offers four advantages:

(a) It is readily adaptable to advanced planning and essentially forces the detailed definition of tasks, task sequences, and task interrelationships. All levels of management and engineering are required to think through and evaluate the entire project carefully.

(b) With the identification of task interrelationships, it tends to force the initial definition and subsequent management and control of the interfaces between customer and contractor, organizations within the contractor's structure, and between the contractor and various suppliers. Management and engineering gain a greater appreciation of the project in terms of total resource requirements.

(c) It enables management and engineering to predict with some degree of certainty the probable time that it will take to achieve an objective. Areas of program risk/uncertainty can be readily identified.

(d) It enables the rapid assessment of progress and allows for the early detection of possible delays and problems.

The implementation of PERT/CPM in a comprehensive and timely manner is possible because the technique is particularly adaptable to computer methods. In fact, there are a number of computer models and associated software that are available for network scheduling.

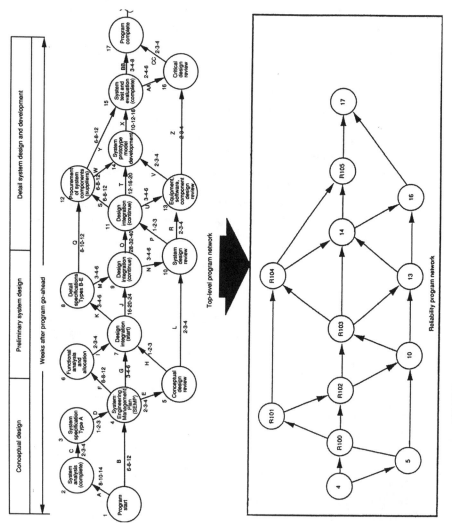

FIGURE 6.23 Top-level network breakdown by program element.

5. *Network/cost*. PERT/CPM networks may be extended to include cost by superimposing a cost structure on the time schedule. When implementing this technique, there is always the time–cost option, which enables management to evaluate alternatives relative to the allocation of resources for activity accomplishment. In many instances, time can be saved by applying more resources. Conversely, cost may be reduced by extending the time to complete an activity.

The time–cost option can be attained through the following general steps:

(a) For each activity in the network, determine possible alternative time and cost estimates (and cost slope) and select the lowest cost alternative.

(b) Calculate the critical path for the network. Select the lowest cost option for each network activity, and check to ensure that the total of the incremental activity times does not exceed the allowable overall program completion time. If the calculated value exceeds the program time permitted, review the activities along the critical path and select the alternative with the lowest cost slope. Reduce the time value to be compatible with the program requirement.

(c) After the critical path has been established in terms of the lowest cost option, review all network paths with slack time, and shift activities to extend the times and reduce costs wherever possible. Activities with the steepest time–cost slopes should be addressed first.

PERT/CPM–COST has proven to be a very useful technique in the planning of program events and activities, and it allows for the necessary program schedule–cost status monitoring and control requirements accomplished throughout system development.

6. *Gantt chart*. This technique is used primarily in production and/or construction planning to show activity or job requirements, facility loading, and work status on a day-to-day basis. It was designed for and is most successfully utilized to support highly repetitive operations. An example of one basic form of a Gantt chart is shown in Figure 6.24. Gantt charts, used for both long-range planning and short-range scheduling on a day-to-day basis, may take the form of machine-loading control charts, labor-loading control charts, and/or job progress control charts.

7. *Line of balance (LOB)*. This technique is similar to the Gantt chart relative to determining production/construction status. Although the Gantt technique primarily relates information on the effective and efficient utilization of resources expended (e.g., labor loading, machine loading), LOB is more product-oriented. LOB is not directly concerned with the resources expended, but is utilized in determining production progress in terms of percentage of task completion. Major bottlenecks in the production process are emphasized.

Application of the scheduling methods described herein will vary from project to project and from one organization to the next. In addition, the technique used may be different for each phase of the system life cycle. For instance, the use of PERT/CPM may be readily adaptable to a research and development program, whereas Gantt charts are more appropriate for a production program.

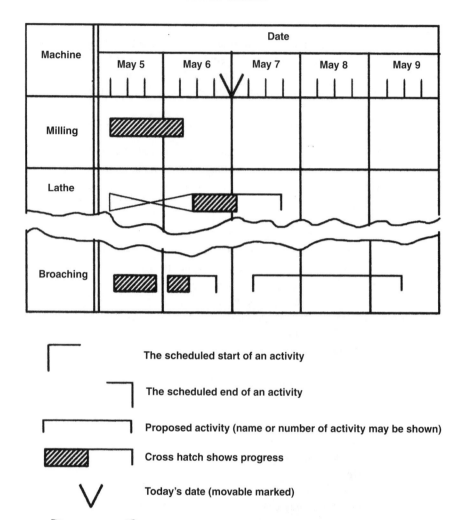

FIGURE 6.24   Gantt chart for a machine used in production.

In considering the objectives of system engineering, the use of PERT/CPM (or an equivalent network approach), as compared with bar charts or milestone charts, seems appropriate. There are many different one-of-a-kind tasks accomplished relatively early in the system life cycle, and the organizational task interfaces are numerous. There is a need for a high degree of visibility across the project, and it is important that potential problems be detected as early as possible. The use of the network scheduling technique should help in maintaining the necessary communications and in providing the appropriate monitoring and control functions.

6.2.8 Preparation of Cost Projections[13]

Good cost control is important to all organizations, regardless of size. This is particularly true in our current environment where resources are limited and competition is high.

Cost control starts with the initial development of cost estimates for a given program and continues with the functions of cost monitoring and the collection of data, the analysis of such data, and the initiation of corrective action before it is too late. Cost control implies good overall cost management, which includes cost estimating, cost accounting, cost monitoring, cost analysis, reporting, and the necessary control functions. More specifically, five activities are applicable.

1. *Define elements of work.* Develop a statement of work (SOW) in accordance with the requirements described in Section 6.2.1. Detailed project tasks are identified in Section 6.2.2 (refer to Figure 6.6).

2. *Integrate tasks into the work breakdown structure* (WBS). Combine project tasks into work packages, and integrate these elements of work into the work breakdown structure (WBS). Work packages are identified with each block in the WBS. These packages and WBS blocks are then related to organizational groups, branches, departments, suppliers, and so on. The WBS is structured and coded in such a manner that project costs can be initially allocated (or targeted) and then collected for each block. Costs may be accumulated both vertically and horizontally to provide summary figures for various categories of work. WBS objectives and requirements are described in Section 6.2.4 (refer to Figures 6.12 and 6.13 for system engineering functions in the WBS).

3. *Develop cost estimates for each project task.* Prepare a cost projection for each project task, develop the appropriate cost accounts, and relate the results to elements of the WBS.

4. *Develop a cost data collection and reporting capability.* Develop a method for cost accounting (i.e., the collection and presentation of project costs), data analysis, and the reporting of cost data for management information purposes. Major areas of concern are highlighted; that is, current or potential cost overruns and high-cost "drivers."

5. *Develop a procedure for evaluation and corrective action.* Inherent within the overall requirement for cost control is the provision for feedback and corrective action. As deficiencies are noted, or potential areas of risk are identified, project management must initiate the necessary corrective action in an expeditious manner.

---

[13]The various aspects of cost estimating, cost/schedule control, cost analysis, cost performance measurement, cost variance reporting, and related areas, are covered in most texts on program/project management. A good reference is H. Kerzner, *Project Management: A Systems Approach to Planning, Scheduling, and Controlling,* 9th ed. (New York: John Wiley & Sons. Inc., 2005). It should be noted that the emphasis in this section is primarily on the costing of internal projects versus the application of life-cycle cost analysis methods described in Appendix B, although the results here constitute an integral part of an overall life-cycle cost analysis projection.

An initial step in developing a good cost control capability is cost estimating and the preparation of cost projections. Each task is broken down into subtasks and other detailed elements of work, and personnel projections are developed on a month-to-month basis. Figure 6.25 identifies selected activities for a project involving the design and development of a relatively large-scale system. In this instance, a 12-month design period is assumed, and projections are made in terms of the number of individuals by job classification required to complete the task, scheduled on a month-to-month basis. For instance, under system engineering there is a need for the assignment of four individuals with the grade of "senior engineer" during Month 3 of the project. Although not completely shown in the figure, *all* major program activities should be covered through an appropriate breakout of job classification requirements; that is, principal engineer, senior engineer, engineer, junior engineer, engineering technician, analyst, draftsperson, data specialist, and shop mechanic. These resource requirements are projected for each project task and are related to the WBS (e.g., 3B1 100 in Figures 6.12 and 6.13).

Given a projection, presented in terms of labor requirements by grade, the next step is to convert these into cost factors on a month-to-month basis. Most organizations have established job classifications with computed salary pay scales. These factors are used in estimating the direct labor costs for a designated activity extended into the future. In addition, material costs are determined for each month, and the appropriate inflationary factors are added to both labor and material. The net results include a projection of *direct labor costs* and *direct material costs,* inflated as necessary to cover future economic contingencies. These projections must, of course, support all program tasks identified in Sections 6.2.1 and 6.2.2 and should be compatible with the related task schedules described in Section 6.2.7.

As individual project activities are being further defined through the preparation of cost estimates, not only must these activities be tied to a particular block in the WBS, but the results must be assigned to a specific cost account (refer to Figure 6.14). A partial breakdown structure for a project is presented in Figure 6.26. The objective is to show the various project cost accounts in a hierarchical manner, indicating the structure that will be used for subsequent cost accounting and reporting purposes.

Relative to application, cost estimating may be accomplished at any time or during any phase of the system life cycle. Sometimes during the early phases of conceptual and/or preliminary system design, when the availability of engineering data are limited, estimates may take the form of "rough orders of magnitude;" that is, approximations within plus or minus 30% of reality. The use of regression analysis, linear and nonlinear estimating relationships, learning curves, parametric analysis, or a combination of these, aids in the development of cost figures of merit (FOMs). Later, as engineering experience is acquired, estimating methods are more precise. Plans, specifications, design data, supplier cost proposals, updated project *cost-to-complete* reports, and so on, are available. Cost estimates, using actual engineering data and/or the development of data through analogous methods, are prepared with an expected accuracy in the order of plus or minus 5%.

On completion of the cost projections for individual tasks, one can then combine these into an overall cost projection for the project as a whole, as shown in

Program	WBS no.	Cost account	Projection (months)												Total
			1	2	3	4	5	6	7	8	9	10	11	12	
Project management	2A1000	2000	1	1	2	3	4	4	4	4	4	4	3	3	37
System engineering	3B1100	3000													
Principal engineer			-	1	1	1	1	1	1	1	1	1	1	1	12
Senior engineer			2	3	4	4	4	3	2	2	2	2	2	2	32
Design engineering	3B1200	7000													-
Principal engineer			2	3	3	3	3	2	2	2	2	2	2	2	28
Senior engineer			3	6	8	8	8	7	6	5	5	4	4	3	67
Engineer			5	7	10	15	17	20	20	20	20	17	16	9	176
Junior engineer			3	4	5	6	10	10	10	10	15	15	20	20	128
Engineering technician			1	1	1	5	7	7	8	9	10	10	12	15	86
Design data	2I1200	4000	2	2	3	5	5	10	15	15	20	25	25	25	152
System software	2E1000	8000	1	1	2	2	3	3	5	7	7	10	12	15	68
Design support	3B1300	5000	2	2	5	5	5	10	10	25	30	30	30	25	179
Intergrated logistic support	3B1400	6000	2	3	3	3	5	6	6	10	10	15	15	15	93
System test & evaluation	2J1000	9000	1	1	1	1	1	1	5	5	10	15	15	20	76
Total			26	35	48	61	73	84	94	115	136	150	157	155	1134

FIGURE 6.25 Project labor projection (man-months).

311

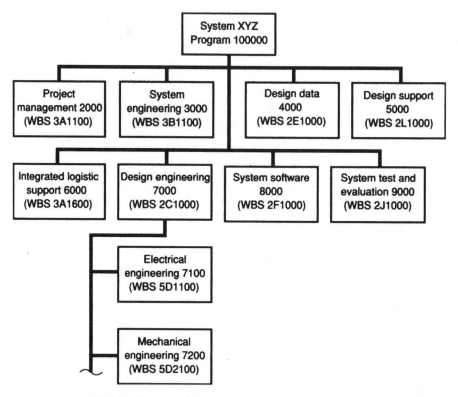

FIGURE 6.26   Partial cost account code breakdown structure.

Figure 6.27. Initially, an estimate for all direct labor is developed, with an organizational overhead factor applied on top. Direct material costs are then determined, and a second burden rate (i.e., a general-and-administration factor) is applied to cover some additional indirect costs associated with both labor- and material-related activities. The net result is an overall cost projection for the project, including both direct and indirect costs.

6.2.9   Program Technical Reviews and Audits

Technical reviews are an integral part of the system engineering process. These reviews can vary from the very formal design reviews described in Chapter 5 to the informal reviews concerned with specific project activities or task elements of the work breakdown structure (WBS). All such reviews share the common objective of determining the technical adequacy of the existing system design configuration and whether or not its meets the initially specified requirements. Further, as the design and development effort evolves, the reviews become more detailed and definitive.

The type, number, and basic objectives of the formal design reviews conducted for a given program will vary with the nature and complexity of the system being

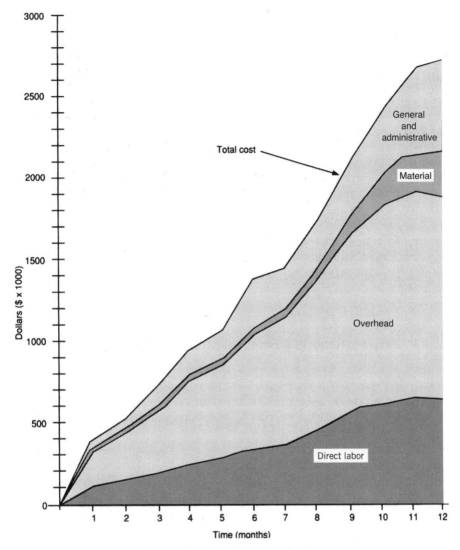

FIGURE 6.27 Project cost projection.

developed, the organizational structure and type of contracting mechanism in place, and so on. In Chapter 5, formal design reviews include four basic categories of reviews; *conceptual, system, equipment/software,* and *critical.* These are considered to be basic and representative for most programs. However, for many large-scale defense programs, there may be many more reviews, including system requirements review (SRR), system functional review (SFR), system design reviews (SDRs), preliminary design reviews (PDRs), software specification reviews (SPRs), system verification review (SVR), critical design review (CDR), test readiness review (TRR), production readiness review (PRR), and so on.[14] Although the scheduling of design reviews has many benefits, as conveyed in Chapter 5, it is essential that care be taken so as not to schedule so many that they become meaningless. The conductance of such reviews may be quite costly, considering the personnel time and resources required.

In addition to the formal design reviews conducted on many projects, there may be a number of formal *program management reviews* scheduled as well. Sometimes the design reviews are perceived as being oriented to only *engineering* and involving responsible engineers representing the appropriate *engineering specialties.* Key levels of program management are not involved, even though many of the design decisions discussed may have significant implications from an overall program management perspective. On the other hand, during the periodic management-oriented reviews, where the emphasis is often directed to current status in terms of *performance, cost,* and *schedule,* there are decisions made that can have a direct impact on design. Under certain conditions, the two categories of reviews can be counterproductive unless care is taken to ensure that the technical and management reviews are mutually supportive. A system engineering goal is to facilitate the communications process and the scheduling of both categories of reviews so that the results are complementary in meeting the overall program/project objectives.

6.2.10   Program Reporting Requirements

Inherent within the planning process is the establishment of both *technical* and *management* requirements at program inception. In addition, one needs to review progress against these requirements on a periodic basis as system design and development evolves. A procedure must established for the initiation of corrective action, as necessary, in the event of problems.

In response, a management information system (MIS) should be developed to provide ongoing visibility and the reporting of progress against designated cost, schedule, and performance measures. Schedule and cost information is derived in accordance with the procedures described in Sections 6.2.7 and 6.2.8. Periodic reports are necessary for purposes of assessing *current* status against *planned* status. The frequency of reporting is a function of the overall project schedule and the risks associated with various design activities. The comparison process should address such questions as these: *Is the project on schedule? Are the program costs within the established budget limitations? Assuming that the current personnel loading continues as is, what tasks are likely to be in a cost overrun position six months from now?* These and similar questions will have to be answered on many occasions throughout the program.

Figure 6.28 presents an extract from a report covering schedule and cost data. The schedule (or time status) information reflects the output from a typical PERT/CPM

---

[14]Refer to (1) *Systems Engineering Fundamentals* (Fort Belvoir, VA: Defense Acquisition University, December 2000); and (2) E1A/IS-632, *Processes for Engineering a System* (Washington, DC; Electronic Industries Alliance, EIA).

Network/Cost Status Report

	Project: System XYZ			Contract Number: 6BSB-1002					Report Date: 8/15/02			
	Item/Identification			Time Status					Cost Status			
WBS. No.	Cost Account	Beginning Event	Ending Event	Exp. Elap. Time (te) (weeks)	Earliest Completion Date(D_E)	Latest Completion Date (D_L)	Slack D_L-D_E (Weeks)	Actual Date Completed	Cost Est. ($)	Actual Cost to Date ($)	Latest Revised Est. ($)	Overrun (Underrun) ($)
4A1210	3310	8	9	4.2	3/4/02	4/11/02	11.6	4/4/02	2500	2250	2250	(250)
4A1230	3762	R100	R102	3.0	5/15/02	4/28/02	-3.3		4500	4650	5000	500
5A1224	3521	7	9	20.0	6/20/02	8/3/02	0		6750	5150	6750	0

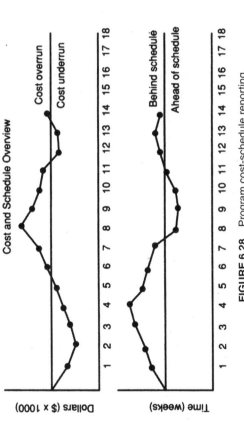

Cost and Schedule Overview

FIGURE 6.28 Program cost-schedule reporting.

network. Relative to performance, the technical performance measures (TPMs) identified in the system specification, and selected as being critical from a periodic review and control standpoint, must be included within the program reporting structure. These TPMs may include factors such as range, accuracy, weight, size, reliability (mean time between failure/MTBF and mean time between maintenance/MTBM), maintainability (mean corrective maintenance time/$\overline{M}$ct and maintenance labor hours per system operating hour/MLHOH), downtime (mean maintenance downtime, MDT), availability, cost, power output, process time, and other parameters that relate directly to the mission of the system being developed. Figure 5.6 (Chapter 5) illustrates the TPM evaluation process as it is tied in with formal design reviews. The measurement, evaluation, and control of these parameters must also be covered through periodic program reporting.

The management information system (MIS) should readily point out existing problems, as well as potential areas in which problems are likely to occur if program operations continue as originally planned. To deal with such contingencies, planning should be initiated to establish a corrective-action procedure that includes the following steps:

1. Identify problems (or potential problem areas) and rank these in order of importance. Ranking should consider the criticality of the system function.

2. Evaluate each problem on the basis of ranking, addressing the most critical problems first. Alternative possibilities for corrective action are considered in terms of (a) effects on program schedule and cost, (b) impact on performance and effectiveness of the system, and (c) the risks associated with the decision as to whether to take corrective action. The most feasible alternative is identified.

3. Given the decision to take corrective action, planning is accomplished to initiate the steps required to resolve the problem. This may be in the form of a system configuration change, a change in management policy, a contractual change, and/or an organizational change.

4. After corrective action has been implemented, some follow-up activity is required to (a) ensure that the incorporated change(s) actually has resolved the problem and (b) assess other aspects of the program to ensure that additional problems have not been created as a result of the change.

Relative to the ranking of problems (and their priorities) that need to be addressed, a Pareto analysis approach might be beneficial in creating visibility pertaining to degrees of importance. See Figure 6.29; the highest-ranked items need the most management attention. The implementation of any changes, of course, must be compatible with the procedures described in Section 5.4.

6.3   DETERMINATION OF OUTSOURCING REQUIREMENTS

The current demands for the delivery of more products, in shorter time frames and at least cost, and in a highly competitive international marketplace environment, has put greater emphasis on the practice of *outsourcing* and on the utilization of many

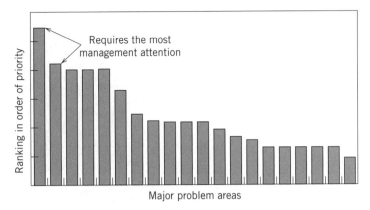

FIGURE 6.29   Pareto diagram identifying problem areas.

different suppliers in fulfilling the requirements for developing, producing, and/or modifying systems. The term *outsourcing* refers to the identification, selection, and contracting with one or more outside (external) suppliers for the procurement and acquisition of materials and services for a given system. The term *supplier* refers to a broad class of external organizations that provide products, components, materials, and/or services to a producer (or prime contractor). This may range from the delivery of a major subsystem or configuration item down to a small component part. More specifically, suppliers may provide services, including (1) the design, development, and manufacture of a major element of a system, (2) the production and distribution of items already designed (providing a manufacturing source), (3) the distribution of commercial and standard component parts from an established inventory (serving as a warehouse and providing parts from various sources of supply), and/or (4) the implementation of a process in response to some functional requirement.

For many systems today, suppliers provide a large number of their elements (e.g., more than 75% of the components in some instances), as well as the spares and repair parts that are required to support maintenance activities. Given the trends toward increased globalization and greater international competition, the suppliers associated with any relatively large-scale program are likely to be geographically located throughout the world, thus creating a worldwide "working" environment, as shown in Figure 6.30. Further, when major suppliers are selected, particularly for the design and development of large system elements, there are likely to be a number of suppliers selected for the production and delivery of some of the smaller components that make up the various subsystems and items of an equivalent level and complexity. Thus, we sometimes find that we may be dealing with a *layering* of suppliers, as illustrated in Figure 6.31.

With the involvement of many different suppliers in the design, development, manufacture, and support of systems, there is an ever-increasing need for the implementation of good system engineering practices and methods/techniques. Major suppliers, as key participants in the design process, must be involved from the

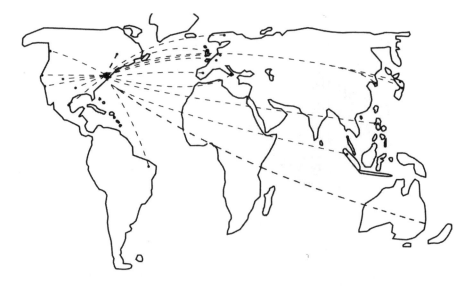

FIGURE 6.30 Potential suppliers for system XYZ.

beginning. The System Engineering Management Plan (SEMP) must include coverage of supplier functions and activities. The system specification (type "A") must provide a good functional baseline from which the various lower-level specifications can be developed. Of further significance, the *functional interfaces* (described in Section 2.7) must be well defined in the applicable specification. In essence, major suppliers must be brought into the design process early, must participate as members

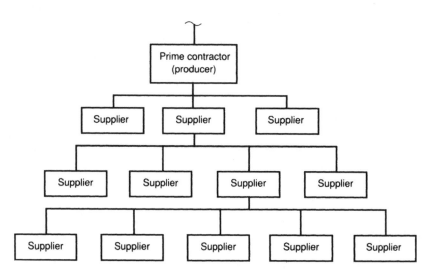

FIGURE 6.31 Typical structure involving the layering of suppliers.

of the design team, and must be committed to the implementation of the system engineering process.

In regard to these requirements, the following sections discuss the identification of potential suppliers for a given program, the development of a *request for proposal* (RFP) soliciting supplier response, the review and evaluation of supplier proposals, the ultimate selection of suppliers, and the subsequent contracting for a defined level of activity. Supplier functions, organizational relationships, and responsibilities are covered further in Chapter 7.

6.3.1 Identification of Potential Suppliers

A review of the system engineering process described in Chapter 2 will illustrate a number of steps, commencing with the identification of a consumer need and extending through the definition of operational requirements, the maintenance concept, the identification of technical performance measures (TPMs), functional analysis and the allocation of requirements, and the preparation of the system specification (type A). These steps are represented by the first two blocks in Figure 6.32.

As indicated, the system is described in *functional* terms identifying the "whats," and each functional entity is evaluated and trade-off studies are conducted with the objective of determining "how" the function(s) can best be accomplished (refer to Section 2.7, Chapter 2). The basic question in each instance is this: should the function be accomplished through the application of equipment, software, facilities, data/information; the utilization of human resources; or a combination of these? The results of these trade-off studies are presented in the form of specific resource requirements.

The next step is to identify the possible sources of supply. Should the design and/or manufacture of an item of equipment, the development of a software package, or the completion of a process be accomplished in house by the producer or prime contractor, or should an external source of supply be selected? The objective is to establish the "where" in determining the *source* in responding to resource requirements.

In many industrial organizations, a "make-or-buy" committee, or an equivalent activity within the producer's organization, is established, with representation from program management, engineering, logistics, manufacturing, purchasing, quality assurance, and other supporting organizational activities as required. Participating engineering should include the system engineering organization and the appropriate design disciplines. Decisions are based on the evaluation of a combination of factors, such as the criticality of need (when is the item required?), item complexity, the availability of internal technical capabilities and required resources versus the use of potential outside suppliers, related social and political factors, and cost.[15]

---

[15]On certain occasions, decisions may be based on social, economic, and/or political considerations, such as the identification of a need to improve the local economy by selecting a supplier in a given geographical area, the desire to increase the amount of subcontracting, the need to establish a manufacturing and/or support capability in a designated foreign nation, the need to respond to an existing unemployment crisis, the desire to support a given political position, and so on.

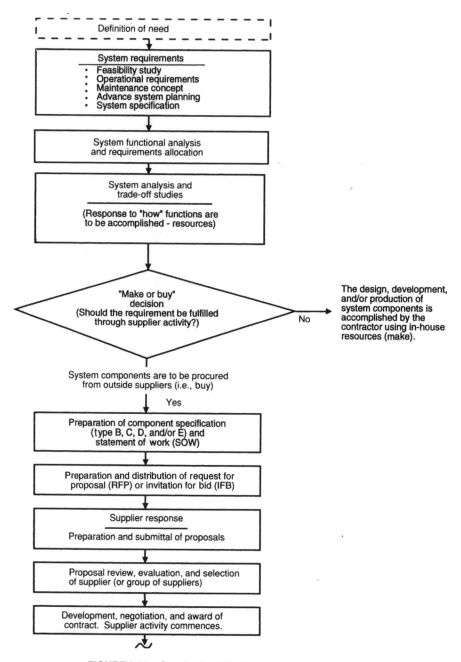

FIGURE 6.32   Supplier identification and procurrent process.

From a system engineering perspective, items that are relatively complex, involving the application of new technologies, and are critical to the overall system development effort, should be handled internally if at all possible. These activities will, in all likelihood, require frequent monitoring and the application of tight controls (both management and technical), which may be difficult to accomplish should a remotely located supplier be selected for the task.

As shown in Figure 6.32, the results from of the deliberations of the "make-or-buy" committee will lead to specific recommendations as to the potential sources of supply for the fulfillment of various functional requirements in a given system development effort. Potential external "candidate" suppliers are identified, which, in turn, will lead to the next step: the development of a formal request for proposal (RFP), request for quotation (RFQ), invitation for bid (IFB), or the equivalent.

6.3.2 Development of a Request for Proposal (RFP)

Having evaluated the alternatives and come to the ultimate decision to "buy," the contractor (in this instance) must develop the necessary materials for incorporation into a *request for proposal* (RFP). The objective is to develop a data package that can be distributed to potential suppliers for the purposes of soliciting a proposal.

In general, the RFP is a formal mechanism by which the contractor specifies the requirements for a product, or for a service, in response to a designated need. The need for a system component has been identified, a decision has been made to procure the item from an outside source, and the contractor must translate the requirements for this item in a detailed and precise manner. These requirements are described in a data package, attached to a letter of invitation to bid, and sent to prospective suppliers interested in responding to the RFP. More specifically, the content of the data package should include the following:

1. A technical specification describing the product, its performance and effectiveness characteristics, physical features, logistics and quality provisions, and so on. This document, tailored to the application, may constitute a type B, C, D, or E specification, depending on the particular requirement (refer to Figures 1.12, 3.2, and 6.15).

2. An abbreviated management plan describing overall program objectives, contractor organizational responsibilities and interfaces, the WBS, program tasks, task schedules, applicable policies and procedures, and so on. This information primarily relates to contractor activities; however, individual suppliers must understand their respective roles in the context of the overall program.

3. A statement of work (SOW) describing detailed tasks, task schedules, deliverable items, supporting data, and reports that are to be provided by the supplier. This information, derived from a combination of the specification and the management plan, constitutes a summary of the work to be performed and serves as the basis for the supplier's proposal.

Meeting the objectives of system engineering is highly dependent on initial supplier selection, applicable follow-on activities, and the ongoing evaluation and control efforts imposed by the contractor. As an input to this process, the technical specification (i.e., the type B, C, D, or E specification, as applicable) must be *comprehensive* in covering *all* of the system-level requirements as they are allocated (or apportioned), down to the element of the system being procured. A top-down approach is an important aspect of system engineering, and the technical specification must support system requirements to the extent applicable.

The degree of influence of the system specification (type A) on the lower-tier specifications is, of course, dependent on the item being procured from the supplier. A large developmental effort will require a very comprehensive type B specification, whereas a standard commercial off-the-shelf component may be covered by a relatively short and simple type C specification. It is important to ensure that the appropriate "traceability" is maintained as one progresses down through the applicable specification tree (refer to Figure 6.15).

Although the top-down technical requirements are maintained through the *specification track*, the appropriate management-oriented requirements must be imposed on the supplier through the management plan and the statement of work (SOW). Organizational continuity must be ensured from the top down, tasks specified for the supplier must directly support those tasks being accomplished by the contractor, schedules must be compatible, the WBS must show the relationships between the supplier and contractor activities, and so on. In other words, a close continuity must be ensured in the transition of work from the contractor to the supplier.

The RFP data package, prepared by the contractor to cover planned supplier activity, is extremely important in maintaining the necessary continuity from the top system-level requirements down to the lowest-level component of the system. One of the prime tasks in system engineering is that of *system integration*, and it is an objective in developing the RFP that the appropriate level of system integration be recognized and addressed. So often a document such as this is compiled in a "hurry-up" manner, proposals are generated, contracts are negotiated, and the necessary system integration requirements are put off until the end. This, of course, can be a costly practice. The RFP data package must be considered an extension of the system type A specification and the SEMP.

6.3.3   Review and Evaluation of Supplier Proposals

After the RFP data package has been developed and distributed to interested and qualified suppliers, each recipient must make a "bid/no-bid" decision. Those suppliers deciding to respond will establish a proposal team and will proceed with the preparation of a proposal. The results, of course, must be responsive to the instructions included in the RFP.

The nature of the supplier's proposal activity will depend on the type and scope of the effort described in the RFP. When the acquisition process is directed toward large elements of the system, involving some design and development (e.g., major subsystems), the supplier proposal activity can be rather extensive. A formal project-type organization may be established, specific project tasks are identified, and the level of effort may be somewhat similar in approach to the project configuration(s) described earlier.

In situations in which large proposals require extensive effort, there is usually a requirement for some design and development activity. If the RFP (through a type B development specification) dictates the need for the design of a major system element, the supplier will often attempt to design and construct a prototype model of the item as part of the proposal effort. A mini-project is organized, design and development tasks are completed expeditiously, and a physical model is delivered to the contractor along with the written proposal. Design decisions are consummated early, with the objective of impressing the contractor (i.e., the customer in this instance) relative to both design approach and the capabilities of the supplier. Should the supplier be successful and be selected in this case, the constructed prototype may well be considered as the baseline configuration leading into follow-on detailed design.

In the preceding scenario, subsystem requirements were specified as part of the RFP, design and development activities were completed during the proposal phase, a formal design review occurred through the contractor's review and evaluation of the supplier's proposal, and the resultant configuration became somewhat fixed relative to the possibility of incorporating any design changes. This scenario can be related to the development process described in Chapter 2, except that the time element is compressed significantly. Because of this type of scenario, the preparation of the RFP assumes a great degree of importance from the system engineering viewpoint (as indicated in Section 6.3.2). Further, the ongoing design activity accomplished during the proposal phase must consider the necessary design characteristics supportive of system engineering objectives (e.g., reliability characteristics and maintainability characteristics). Finally, the formal evaluation of supplier proposals must serve as a final check for compliance with system engineering requirements as they apply to the item, or the service, being procured.

On receipt of all proposals (solicited and unsolicited) from prospective suppliers, the contractor proceeds with the review and evaluation process. When competitive bidding occurs, the contractor generally establishes an evaluation procedure directed toward selecting the best proposed approach. Initially, each supplier proposal is reviewed in terms of *compliance* with the requirements specified in the RFP. Noncompliance may result in automatic disqualification, or the contractor may approach the potential supplier and recommend a proposal revision and/or addition.

When two or more suppliers meet the basic RFP requirements, an evaluation of each proposal is then completed, employing certain preestablished criteria. One may commence with the preparation of a supplier checklist, such as presented in Figure 6.33. The items identified cover some general criteria, design characteristics of the subsystem or product being considered for procurement, the supplier's proposed maintenance and support infrastructure for the subsystem/product, and the qualifications of the supplier. The items in Figure 6.33 are supported by the questions presented in Appendix E and are weighted relative to degrees of importance based on the requirements for the system overall.[16]

---

[16]The questions in Appendix E are similar to the design review questions in Appendix D, except that a "supplier orientation" has been provided. However, a checklist tailored to the system and supplier requirements is preferred when possible.

Supplier Evaluation Checklist

Refer to Appendix E for supporting questions.

E.1 General criteria

E.2 Product design characteristics
    E.2.1 Technical performance parameters
    E.2.2 Technical applications
    E.2.3 Physical characteristics
    E.2.4 Effectiveness factors
        1. Reliability
        2. Maintainability
        3. Human factors
        4. Safety factors
        5. Supportability/ serviceability
        6. Quality factors
    E.2.5 Producibility factors
    E.2.6 Disposability factors
    E.2.7 Environmental factors
    E.2.8 Economic factors

E.3 Product maintenance and support infrastructure
    E.3.1 Maintenance and support requirements
    E.3.2 Data/documentation
    E.3.3 Warranty/guarantee provisions
    E.3.4 Customer service
    E.3.5 Economic factors

E.4 Supplier qualifications
    E.4.1 Planning/procedures
    E.4.2 Organizational factors
    E.4.3 Available personnel and resources
    E.4.4 Design approach
    E.4.5 Manufacturing capability
    E.4.6 Test and evaluation approach
    E.4.7 Management controls
    E.4 8 Experience factors
    E.4.9 Past performance
    E.4.10 Maturity
    E.4.11 Economic factors
    E.4.12 Security control
    E.4.13 Cultural factors

FIGURE 6.33 Supplier evaluation checklist.

The contractor develops a list of topic areas considered to be relevant in the evaluation and assigns weighting factors as shown in Figure 6.34. Note that *supplier qualifications, product design characteristics, product maintenance and support infrastructure*, and *general criteria* have been identified in order of precedence.

Through a review of each supplier proposal, using the questions in Appendix E as a guide, the analyst can assess the degree to which the supplier's proposal responds

Evaluation Criteria	Weighting Factor (%)	Proposal A		Proposal B		Proposal C	
		Rating	Score	Rating	Score	Rating	Score
A. General criteria	10	7	70	5	50	6	60
B. Product design characteristics	30						
1. Performance factors	6	3	18	5	30	4	24
2. Technology applications	3	7	21	8	24	6	18
3. Physical characteristics	2	3	6	4	8	5	10
4. Effectiveness factors	7	7	49	7	49	8	56
5. Producibility factors	2	4	8	6	12	5	10
6. Disposability factors	3	5	15	4	12	8	24
7. Environmental factors	2	2	4	3	6	5	10
8. Economic factors	5	4	20	4	20	6	30
C. Product Maintenance and Support Infrastructure	20						
1. Maintenance and support requirements	7	6	42	8	56	7	49
2. Data/documentation	3	3	9	4	12	6	18
3. Warranties/guarantees	3	2	6	3	9	5	15
4. Customer service	5	5	25	8	40	30	30
5. Economic factors	2	4	8	3	6	6	6
D. Supplier Qualifications	34						
1. Planning/procedures	3	5	15	4	12	5	15
2. Organizational factors	2	6	12	5	10	5	10
3. Personnel and resources	2	4	8	3	6	2	4
4. Design approach	4	6	24	4	16	3	12
5. Manufacturing capability	3	7	21	5	15	4	12
6. Test and evaluation	2	6	12	5	10	4	8
7. Management controls	6	7	42	6	36	4	24
8. Experience factors	4	6	24	4	16	4	16
9. Past performance	5	6	30	5	25	7	35
10. Maturity	3	7	21	7	21	6	18
E. Life-Cycle Cost	12	5	60	7	84	4	48
Grand Total	100		570		595		562

FIGURE 6.34 Proposal evaluation results.

to the desired features conveyed through the questions. From Figure 6.34, the topics listed under *evaluation criteria*, taken from Figure 6.33, are weighted on the basis of level of importance, an assessment is made, and a *rating* is given in each area. A more detailed checklist for each topic may be developed to support the designated rating factor. Figure 6.35 shows an example covering item E in Figure 6.34.

In Figure 6.34, the assigned ratings are multiplied by the weighting factors to provide a score for each item. The individual scores are then added, and the highest

Refer to Figure 6.34, Item E

Rating (Points)	Evaluation Criteria—Life-cycle Cost
10–12	The supplier has justified its design on the basis of life-cycle cost, and has included a complete life-cycle cost analysis in its proposal (i.e., cost breakdown structure, cost profile, etc.).
8–9	The supplier has justified its design on the basis of life-cycle cost, but did not include a complete life-cycle cost analysis in the proposal.
6–7	The supplier's design has not been based on life-cycle cost; however, the supplier plans to accomplish a complete life-cycle cost analysis and has described the approach, model, etc., that it proposes to use in the analysis process.
3–5	The supplier's design has not been based on life-cycle cost, but the supplier intends to accomplish a life-cycle cost analysis in the future. No description of approach, model, etc., was included in his proposal.
0–2	The subject of life-cycle cost (and its application) was not addressed at all in the supplier's proposal.

FIGURE 6.35 Sample checklist of evaluation criteria for supplier proposals.

score indicates the supplier with the best overall approach. In this instance, supplier B appears to be the preferred alternative.[17]

In the evaluation of supplier proposals from the system engineering perspective, the following general questions, as they apply to the subsystem or product being procured, are appropriate:

1. Is the supplier's proposal responsive to the contractor's needs as specified in the RFP?
2. Is the supplier's proposal directly supportive of the system requirements specified in the system type A specification and the SEMP?
3. Have the performance characteristics been adequately specified for the item(s) proposed? Are they meaningful, measurable, and traceable according to system-level requirements?
4. Have effectiveness factors been specified (e.g., reliability, maintainability, supportability, and availability)? Are they meaningful, measurable, and traceable according to system-level requirements?
5. In the event that new design is required, has the design process within the supplier's organization been adequately defined? Does the process incorporate the utilization of computer-aided design (CAD)/computer-aided manufacturing (CAM)/computer-aided support (CAS) technologies where appropriate? Have reliability, maintainability, human factors, supportability,

[17]Refer to Case Study C.6. Appendix C, for the results of a similar evaluation.

life-cycle cost, and related characteristics been properly integrated into the design where appropriate? Have design change procedures been developed, and are changes properly controlled through good configuration management practices?

6. Is the design adequately defined through good documentation; that is, drawings, parts lists, reports, software, tapes, disks, and databases? Are the required data available? Have the data rights been specified?

7. Has the supplier addressed the requirement for the test and evaluation of the proposed system element or component? If testing has been accomplished in the past, are the test results documented and available? Have the plans for future testing been properly integrated into the system TEMP?

8. Have the life-cycle support requirements been identified for the item being proposed; that is, maintenance resource requirements, spare/repair parts, test and support equipment, personnel quantities and skill levels, training, facilities, data, maintenance software, and so on? Have these requirements been minimized to the extent possible through good design?

9. Does the design configuration reflect good growth potential? Reconfigurability?

10. Has the supplier developed a comprehensive production/construction plan? Are key manufacturing processes identified, along with their characteristics?

11. Does the supplier have a good quality assurance program? Are statistical quality control methods utilized where appropriate? Does the supplier have a good rework plan to handle rejected items as necessary?

12. Does the supplier's proposal include a good comprehensive management plan? Does the plan cover program tasks, organization structure and responsibilities, a WBS, task schedules, program monitoring and control procedures, and so on? Has the responsibility for system engineering tasks (as applicable) been defined?

13. Does the supplier's proposal address all aspects of *total cost;* that is, acquisition cost, operation and support cost, and life-cycle cost?

14. Does the supplier have previous experience in the design, development, and production of system elements/components that are similar in nature to the item proposed? Was that experience favorable in terms of delivering high-quality products in a timely manner and within cost?

Although these questions may be helpful in the evaluation of a supplier's proposal, there are three additional factors that must be considered before recommending a specific procurement approach:

1. Should a single supplier be selected (i.e., sole source), or should two or more suppliers be selected to fulfill the requirements as stated in the RFP? On the one hand, if the level of effort specified covers a relatively large element of the system and involves some design and development activity, the selection of two (or more)

suppliers to perform the same tasks may be rather costly. On the other hand, for smaller standard off-the-shelf components, it may be appropriate to establish several sources of supply. The objective is to ensure a source of supply that will meet the need as long as required, with a minimum of risk associated with the possibility of the supplier "going out of business."

2. Will the supplier be able to provide the necessary support for the proposed item, both during and after production, throughout the planned life cycle of that item? Of particular interest is the source for spare/repair parts to support sustaining maintenance requirements after the initial production has been completed and the capability for producing additional spares no longer exists—that is, postproduction support. If such support will not be available, then the procurement policy may dictate that enough spare/repair parts be purchased initially to support maintenance operations for the entire life cycle.

3. Should a supplier be selected on the basis of political, social, cultural, and/or economic factors? In this era of international involvement (or globalization), there may be certain political pressures encouraging the procurement of components, or services, from a particular foreign source. However, it may be feasible to select a prospective supplier on the basis of geographic location and economic need. On occasion, it may be specified that at least $x\%$ of the total volume of system development effort must be subcontracted. In any event, supplier selection is sometimes influenced by political, social, and/or economic factors.

The evaluation of supplier proposals may be accomplished using the approach conveyed in Figure 6.34, modified to take into consideration these additional factors; that is, single versus multiple suppliers, postproduction support requirements, and the influence of political and economic factors on supplier selection. This evaluation activity usually includes not only a review of the written proposal itself, but one or more on-site, inspection-type visits to the supplier facility. A recommendation is made, and contract negotiations between the contractor and the supplier are initiated.

As the results of the supplier evaluation and selection process have a significant impact on program success and meeting the objectives of system engineering, it is important that the system engineering organization be represented throughout this process. The proper coordination and integration of supplier activity into the total engineering design and development effort are essential.

6.3.4   Selection of Suppliers and Contract Negotiation

Having identified prospective suppliers through the evaluation and selection process, it is now incumbent on the contractor to develop a formal contractual arrangement with the supplier. The RFP was initiated, proposals from potential suppliers were generated and evaluated, and now a contractual structure (in some form) needs to be established. The type of contractual agreement negotiated can have a significant impact on supplier performance, particularly in the procurement of large system components involving design and development activity.

The objective of contract negotiation is to achieve the most advantageous contractual agreement from the standpoint of technical requirements, deliverables, pricing, the type of contract imposed, and payment schedule. Obviously, the contractor and the prospective supplier each views this objective relative to his or her own individual position in terms of the risks associated with the numerous options that are available. At one extreme in contracting is the firm-fixed-price (FFP) contract in which the program risks are primarily assumed by the supplier. At the other end, there is a cost-plus-fixed-fee (CPFF) structure in which the contractor assumes most of the risk. Between these two extremes, there are a number of relatively flexible options.

The type of contract negotiated is important, because the results may well impact supplier performance, which, in turn, may influence the contractor's ability to develop and produce a system that will meet the specified requirements and in a timely manner. Supplier performance, particularly in the acquisition of large subsystems, is critical to the successful accomplishment of system engineering objectives. Further, the risk factors associated with the type of contractual structure negotiated should be considered in the development of the risk management plan included as part of the SEMP—refer to Section 6.2.

Thus, it is important that the system engineer have some understanding of contracts, because he or she is not only affected by the type of contract negotiated, but is often directly involved in the negotiation process itself. Fixed-price contracts are tightly controlled, with the supplier assuming most of the risk (in this instance). Engineering design should be fairly well defined, as changes subsequent to contract negotiation may be quite costly. On the other hand, the cost-reimbursement type of contracts (i.e., cost-plus-fixed-fee, cost-plus-incentive-fee) are more flexible in terms of making changes after initial contract negotiation, and the bulk of the risk is assumed by the contractor. In any event, the system engineer should have some feel relative to the extent of design definition required and what can and cannot be done by virtue of various contractual arrangements.

Moreover, the system engineer often participates, from both a technical and a cost-estimating standpoint, in the initial preparation of the request for proposal (including the preparation of the development specification, the management plan, and the statement of work) that leads to contract negotiations. When the negotiations actually take place, the system engineer often participates once again relative to the interpretation of specifications and the technical aspects of task accomplishment. Throughout the negotiation process, the intended scope of work may change, and such changes must be evaluated for their impact on other system design and development activities.

To provide some additional understanding, the eight major categories of contracts are briefly described in the following paragraphs:

1. *Firm-fixed-price (FFP) contract.* An FFP is legal agreement to pay a specified amount of money when the items called for by the contract have been delivered and are accepted. No price adjustments are allowed for the contracted work after award, regardless of the actual costs experienced by the supplier. At a specified price, the supplier assumes all financial risks for performance, and the supplier's profits depend on his ability to initially predict cost, to negotiate, and to subsequently control costs.

Concerning application of this type of contract, the component design should be fairly well established through appropriate specifications.

2. *Fixed-price-with-escalation contract.* This is similar to the FFP contract, except that an escalation clause may be added to cover uncontrollable price increases or decreases. Escalation can be applied to both labor and material. Because there are many uncertainties relative to predicting the magnitude of escalation, an escalation ceiling is often established, with the supplier and the contractor sharing the risks up to that point. Unexpected costs above the established ceiling are assumed by the supplier.

3. *Fixed-price-incentive contract.* Applied in situations in which some cost uncertainties exist, there is an excellent possibility that cost reduction can be attained through good supplier management and by providing the supplier with some profit incentive. A target cost, a minimum cost, and a ceiling price are negotiated, along with a profit-adjustment formula. Profit adjustment, from the initial targeted profit, can be made based on total cost performance.

4. *Cost-plus-fixed-fee (CPFF) contract.* A CPFF is cost-reimbursement contract whereby the supplier is reimbursed for all allowable costs associated with the project. A negotiated fixed fee (e.g., 10% of the estimated cost) is paid to the supplier on completion of work. Although this fee is fixed in terms of a percentage of the total cost, fee increases or decreases may occur as changes occur in the scope of work and the contract. This is particularly applicable when the contractor is willing to accept supplier-generated engineering change proposals to perform work beyond the scope of the initial contract.

5. *Cost-plus-incentive-fee (CPIF) contract.* The CPIF is intended to cover situations in which uncertainties in program performance exist. Allowable costs are paid to the supplier, together with additional incentive fee payments based on designated accomplishments. At the time of negotiation, individual factors such as schedule milestones and specific performance measures may be identified as items in which incentives are to be specified in order to motivate suppliers to excel in these areas. Contract negotiation will result in a defined target cost, target fee, a minimum and maximum fee, and a fee-adjustment formula. On completion of the contract, the supplier's performance will serve as the basis for fee adjustment. An application of this type of contract can include the negotiation of incentives against each of the technical performance measures (TPMs) specified for the system, as they apply to the item being procured.

6. *Cost-sharing contract.* This is primarily designed for research and development work conducted with educational institutions and nonprofit organizations. Such work is jointly sponsored, and reimbursement to the supplier is in accordance with a predetermined sharing agreement. No fee is awarded; however, in lieu thereof, the supplier anticipates that the work accomplished will derive other benefits (e.g., a patentable item, acquisition of technical know-how, a good publication).

7. *Time and material contract.* This allows for the payment for actual materials and services expended in the performance of designated tasks. This type of contract is employed when the extent and duration of work cannot be determined ahead of

Major Program Activity	Conceptual Design	Preliminary System Design	Detail System Design and Development	Production and/or Construction
1. Need analysis, feasibility studies, operational requirements, maintenance concept				
2. System specification				
3. Test and evaluation master plan				
4. System engineering management plan				
5. Conceptual design review				
6. Functional analysis and requirements allocation				
7. System analysis, synthesis, and trade-off studies				
8. System design reviews				
9. System design integration				
10. System test and evaluation				
11. Equipment/software design reviews				
12. Critical design review				

Contractual payment schedule				
1. Progress payment				
2. Incentive/penalty determination				

FIGURE 6.36 Schedule of proposed contractual payments.

331

time and when costs cannot be estimated to any degree of accuracy. Appropriate applications include specific subcontracted research and development tasks, maintenance repair and overhaul services, and so on.

8. *Letter agreement.* Often used as a preliminary contractual document initiated with the intent of authorizing the supplier to start work on a project immediately. These agreements serve as an interim means for providing a rapid response to an identified need that otherwise might be delayed pending the negotiation of a definitive contract. Letter agreements usually do not include total pricing information; however, an upper-limit dollar amount is usually specified to preclude excessive spending. Under this type of agreement, all costs incurred by the supplier for work accomplished are fully reimbursed by the contractor.

Associated with each major type of contract is the question concerning schedule of payments. When will the supplier be reimbursed for the successful completion of contracted tasks? What is the magnitude of expected payments? For incentive contracting, what type of incentive/penalty plan should be applied? These and comparable questions are significant, particularly for the larger contracts, because the contractor is generally tied to a specific budgeting cycle and the supplier must offset operating costs without going too far into debt. Thus, a payment schedule of some type should be developed.

Figure 6.36 presents an example of one type of plan, in which progress payments are tied to the successful completion of formal design reviews; that is, the system design review, the last equipment/software design review, and the critical design review. These particular design reviews will include coverage of supplier activity, and tying progress payments to these events should motivate the supplier to produce effective results, ensuring success.

If incentive contracting is used, an incentive/penalty plan should be developed as a supplement to the schedule for progress payments. Such a plan should specify the application of incentive and penalty payments to significant project milestones and/or demonstrated system performance and effectives characteristics. See Figure 5.2 (Chapter 5); the TPMs applied at the system level should be allocated to the subsystem or to the level applicable to the item being provided by the supplier. Performance measures that are realistic for the item being procured may be appropriate factors for consideration in the development of an incentive/penalty plan for the supplier.

In developing an incentive/penalty payment plan, it is necessary to identify the parameters to which incentives and penalties are to be applied. In many instances there is more than one parameter, resulting in a multiple structure. The appropriate sum of money for each incentive is difficult to determine and will depend on the type of component (or service) and the importance of the item to which the incentive is to be applied. It is unlikely that all selected parameters will be equally important; therefore, it will be necessary to assign an "importance value" or "weighting" to each parameter and to estimate the magnitude of the incentive/penalty values accordingly. An example of the multiple approach, involving two component characteristics, is illustrated in Figure 6.37. A target value is established based on specification requirements, which may also be considered as a "contracted value." If, after test

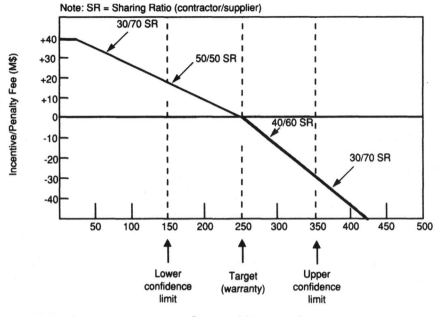

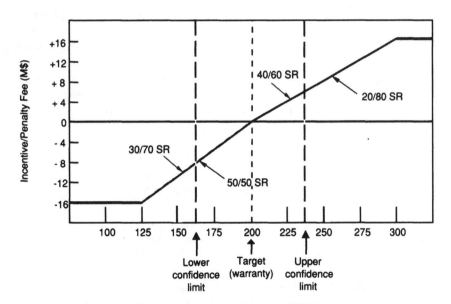

FIGURE 6.37 Multiple incentive/penalty plans.

and evaluation, the actual measured value is an improvement over the target value, an incentive fee is awarded to the supplier at the designated time, as indicated in the schedule in Figure 6.36. More specifically, if the measured MTBM exceeded the upper confidence limit of approximately 238 hours in Figure 6.37, the assigned fee would be split, with 20% going to the contractor and 80% going to the supplier. Conversely, if the measured MTBM fell below the target objective, a penalty of 50% of the indicated ordinate value would be paid by the supplier. Similar applications involving other key parameters may be covered through incentive/penalty contracting.

Although there are a variety of possible contract types, care must be taken in adapting the appropriate contract structure to the particular procurement action. For example, if design and development activity is required on the part of the supplier, then it may be appropriate to negotiate either a cost-plus-fixed-fee (CPFF) or a cost-plus-incentive-fee (CPIF) type of contract. In such instances, when considering the flexibility, the contractor may have to implement tighter monitoring and control activities to ensure the timely completion of tasks by the supplier. At the same time, the contractor needs to be careful not to impose (or cause) any design changes that may have an impact on the supplier. If the contractor even suggests a possible improvement to the supplier's product, or a change in direction relative to activity, then the supplier is likely to claim a change in the scope of work and charge the contract accordingly. In addition, a supplier knowledgeable of contracting may initially submit a proposal representing a "minimal effort" in order to keep the price low and win the competition. At the same time, the supplier is planning to initiate changes and/or additions at a later time to cover items that perhaps should have been included in the initial proposal. These changes are likely to be processed through a series of individual engineering change proposals (ECPs), and the ultimate costs will increase accordingly. In such situations, it is important that the system engineer not only be familiar with contracting methods in general, but also thoroughly familiar with the item(s) being proposed, its technical makeup and how it fits into the system hierarchy, and the various interface and support requirements that are applicable.

At the other end of the spectrum, there will undoubtedly be many different system components that are well defined, for which no additional design effort is required. In this instance, the implementation of a firm-fixed-price (FFP) contract may be preferred. For the performance of services, such as in the accomplishment of maintenance and repair actions, the basic time and materials type of contract may be the most appropriate.

The ultimate achievement of definitive contract terms and conditions is accomplished through formalized negotiations between the contractor and the supplier. Negotiations per se can assume a simplified approach involving several representatives from each side, meeting on a given day to discuss requirements in general. By contrast, for relatively large subsystems and/or major system components, the contract negotiation process can become quite complex. In a more formalized negotiation, the contractor, in response to the supplier's proposal, will interrogate the supplier relative to the validity of his or her proposed technical approach, management approach, and/or price. Questions along a technical line will attempt to ascertain whether the supplier has demonstrated that his or her technical approach is the best (based on the results of design trade-off studies), and that he or she has the technical expertise and experience to follow through in developing and producing the proposed item. Concerning cost, the object is to verify that the supplier's price is fair and reasonable and that it was developed through a logical cost analysis. From the supplier's standpoint, the negotiation initially takes the form of defending the proposal as submitted to the contractor. The supplier may be required to provide any amount of supporting material to help convince the contractor that he or she is thorough, honest, and offering the best deal possible.

Negotiation, in general, is an art and usually requires some strategy on both sides. Initially, a plan is developed that identifies the location where the negotiations are to be held and includes an agenda for each meeting that is scheduled. The contractor and the supplier each identify the personnel who will participate in the negotiations process. Both technical and administrative personnel will be included, and a representative from the contractor's system engineering organization should be present for technical discussions covering system-oriented requirements. During the formal negotiations at the bargaining table, both sides will assume a minimum-risk position, considering the contractual terms and conditions mentioned earlier. Interruptions will occur, short strategy meetings will be held to discuss pertinent events, attempts will be made to gain a measure of sympathy from the opposition, and, it is hoped, an agreement will be made after some compromises on both sides. This process may evolve through a number of iterations, perhaps consuming more time than initially anticipated. However, the final objective is to realize a signed formal contract between the contractor and the supplier.

6.3.5 Supplier Monitoring and Control

With the identification, approval, and establishment of formal contractual relationships with suppliers, the contractor's main activity now includes program coordination, evaluation, and control. This ongoing activity can be rather significant for the following reasons:

1. The magnitude of supplier activity and the number of individual product/component suppliers for a given system may be extensive. For some systems, as much as 50 to 75% of the planned development and production activity will be accomplished by suppliers.

2. In addition to the large number of suppliers involved in system acquisition, the geographic distribution of these suppliers may be worldwide. Many systems utilize components that are developed and manufactured in Pacific Rim countries, Europe, Africa, Canada, Mexico, South America, and so on. The requirements in system acquisition may dictate a truly international communications and distribution network.

3. In the acquisition of relatively large-scale systems, where there are many different component suppliers, the variety of tasks being accomplished at any given time can be rather extensive. Some suppliers may be undertaking a full-scale design and development effort, others may be performing manufacturing and production functions, and many suppliers may be providing standard off-the-shelf components in

response to routine purchase orders. There are some programs that are staggered and discontinuous, and there are other programs that are continuous over a long period of time. Figure 6.38 presents a sample plan of supplier project activities.

In this type of environment (i.e., many different suppliers, located worldwide, performing a wide variety of functions), the contractor is faced with a formidable and challenging task. As discussed earlier, specific supplier requirements must be carefully developed and clearly stated from the beginning, and an appropriate contracting structure must be established to ensure that the requirements will be met. The type of contract, of course, should be tailored to the supplier level of effort.

In Figure 6.38, suppliers A, C, D, F, and G are each involved in a project that includes some design and development activity. As part of this effort, trade-off studies are conducted, reliability and maintainability prediction reports are prepared, design reviews are scheduled, test and evaluation functions are accomplished, and so on. The process described in Chapter 2 and many of the activities discussed throughout this text are applicable, although the effort must be scaled down to be compatible with the particular needs of the supplier's program.

In regard to supplier program evaluation and control, the contractor must incorporate supplier activities as part of the overall design review process described in Chapter 5. For large design and development efforts, individual selected design reviews may be conducted at the supplier's facility, with the results of these reviews being included in the higher-level reviews conducted at the contractor's plant (refer to Figure 5.1). For smaller programs, the review process may not be as formal, with the results of the supplier's effort being integrated into the evaluation of a larger element of the system. When addressing projects involving the manufacture and production of components (e.g., each of the projects in Figure 6.38), the contractor's primary concern is that of incoming inspection and quality control. It is essential that the characteristics designed into the component, or as "advertised" in an off-the-shelf item, be maintained throughout.

In essence, supplier evaluation and control are merely extensions of the program review and control activities initiated by the customer and imposed on the contractor. The contractor, in turn, must impose certain requirements on the supplier. Large suppliers must impose the necessary controls on smaller suppliers in the event that a "layering of suppliers" exists. The objectives are to (1) ensure that system-level requirements are being properly allocated from the top down and (2) that compliance with these requirements is being realized from the bottom up. The SEMP must describe the necessary procedures, technical reviews, and so on, as related to supplier activities (refer to Section 6.2).

6.4   INTEGRATION OF DESIGN SPECIALTY PLANS

As indicated in the basic definition of system engineering in Chapter 1, a major objective is to ensure the proper integration of all applicable engineering disciplines into the total design effort (refer to Figure 2.31, Chapter 2). Although there are some

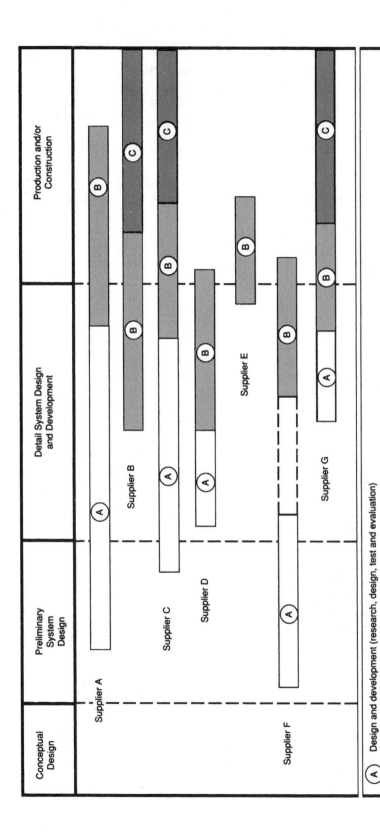

FIGURE 6.38 A sample of supplier project activities.

Conceptual Design | Preliminary System Design | Detail System Design and Development | Production and/or Construction

Supplier A
Supplier B
Supplier C
Supplier D
Supplier E
Supplier F
Supplier G

(A) Design and development (research, design, test and evaluation)
(B) Production (manufacturing and test of prime components)
(C) Production (manufacturing and test of support elements—spare/repair parts)

337

variations with each program, those disciplines identified in Figure 6.5 are considered critical.

Along with the hierarchy of specifications illustrated in Figure 6.15, there is also a hierarchy of program plans, with the Program Management Plan (PMP) at the top, the System Engineering Management Plan (SEMP) next, and a number of subordinate plans supplementing the SEMP. Figure 6.39 illustrates this hierarchical relationship.[18]

In Chapter 3 there is a description of the requirements for each of these supporting disciplines. An evaluation of the requirements illustrates the importance of reliability, maintainability, human factors, supportability/serviceability, producibility, quality, and so on, in design. Further, these requirements are closely interrelated in addition to being key to the system engineering process. Thus, these factors must be properly addressed in the design process, in a timely manner, and commencing from the beginning during the conceptual design phase.

From a review of the individual sections in Chapter 3, one can see that there are many tasks that are similar across the board. The first task in each of the design disciplines represented is the *preparation of a program plan,* a document that specifies the tasks to be accomplished in order to fulfill program requirements. Within each area of activity, there are analysis tasks, prediction tasks, design review and evaluation tasks, and test and demonstration tasks. In the accomplishment of design trade-offs, as part of the ongoing analysis effort, the net results must reflect a *balanced* approach. This, in turn, forces the proper mix of reliability characteristics in design, maintainability characteristics in design, and so on. In other words, these factors (as they impact the design process) must be carefully integrated throughout.

In the past, common practice has resulted in the preparation of these program plans on a separate and independent basis, and often at different times in the system development process. This has caused some differences in stated objectives, inconsistencies in schedules, redundancies in program task requirements, and conflicts in output. In view of the importance of these disciplines in meeting system engineering objectives, it is recommended that the respective plans be prepared and integrated into the SEMP. This is illustrated in Figure 6.39.

In the integration shown in Figure 6.39, the intent is not to hamper or in any way curtail the efforts of the individual disciplines in fulfilling program requirements. The purpose is to ensure the proper relationships among the many tasks that must be accomplished, as well as eliminate possible redundancies. For instance, the results of reliability prediction must feed into the accomplishment of maintainability prediction and the supportability analysis; the preparation of the failure mode, effect, and

---

[18]It should be emphasized that within the broad spectrum of system engineering, there may be a wide variety of different design plans covering not only the prime functional design disciplines (e.g.. civil engineering, electrical engineering, chemical engineering, mechanical engineering), but some of the basic engineering supporting disciplines as well. Relative to the latter, many of the supporting disciplines (e.g., reliability, maintainability, logistics) have been treated as separate entities, each requiring a stand-alone plan, and being implemented not as an integral part of the total engineering effort, but on an independent basis. The emphasis in Figure 6.39 is to ensure that these plans are integrated into the overall process.

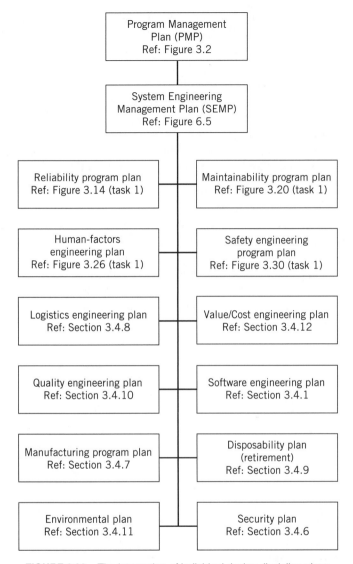

FIGURE 6.39   The integration of individual design discipline plans.

criticality analysis (FMECA) constitutes an input to other reliability tasks, the maintainability analysis, the supportability analysis, and the safety analysis; the fault-tree analysis (FTA) constitutes an input to the safety hazard analysis; the accomplishment of the human-factors operator task analysis (OTA) must be compatible with and directly support the maintenance task analysis (MTA); and the reliability analysis (model), maintainability analysis, operator task analysis, and logistic support analysis must evolve from the system-level functional analysis (refer to Section 2.7).

It is essential that the system engineer completely understand the many interrelationships among these disciplines and that such activities be properly integrated through the SEMP.

6.5 INTERFACES WITH OTHER PROGRAM ACTIVITIES

Although it is important to provide the proper integration of the individual design discipline plans, as conveyed in Figure 6.39, it is also necessary to ensure that the proper communications links exist between the SEMP and other related program plans. Of particular interest are those noted in Figure 6.40 and identified in the following paragraphs.

1. *Individual design plans.* For some programs, individual plans may be prepared by the traditional design disciplines such as civil engineering, electrical engineering, industrial engineering, mechanical engineering, and other such disciplines. It is essential that these plans be supportive of the material presented in the SEMP and be referenced accordingly. An objective of system engineering is to allow the integration of *all* engineering disciplines, requiring that these plans "communicate" with each other.

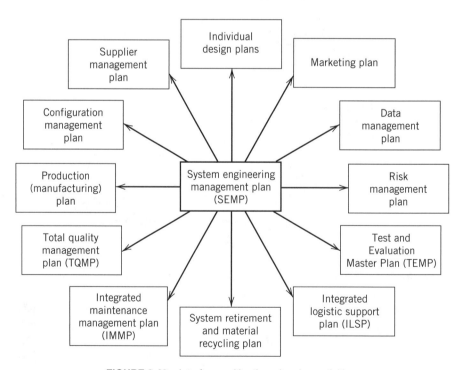

FIGURE 6.40   Interfaces with other planning activities.

2. *Marketing plan.* Individual company/agency marketing plans are often directed toward the "short term" and do not address the necessary long-term, *life-cycle* considerations that are essential in system/product acquisitions. Further, some marketing plans may be directed toward the establishment of certain *partnership* relations with organizations that may, or may not, be sympathetic to the concepts methods of system engineering. Thus, all marketing plans must be prepared to convey the *system engineering approach.*

3. *Data management plan.* The proper integration of all design and supporting data is necessary to ensure that the various elements are compatible (i.e., track where applicable) and available at the right place and in a timely manner, that data redundancies are minimized (if not eliminated), and that data costs are minimized to the extent possible. Although the data environment is rapidly changing with the advent of new technologies (i.e., the conversion of data to a digital format and using a shared database approach), there is still a need for some degree of consistency in presentation, data format, control of data/documentation changes, and so on (refer to Section 2.10 and Figure 2.33 in Chapter 2).

4. *Risk management plan.* Inherent within any system development effort is the aspect of *risk;* that is, risk due to technical decisions, risk due to management decisions, and so on. The objective is, of course, to minimize risk throughout, and a major goal in system engineering is to implement a *risk management plan* that will allow for the early identification of potential areas of risk, the assessment of risk, and risk abatement. As shown in Figure 6.5 (item 11.0), the SEMP should address the area of risk management. This subject is discussed further in Section 6.7.

5. *Test and Evaluation Master Plan (TEMP).* As indicated in Section 2.11 and Figure 2.34, there is a need for *integrated* test planning from the beginning. As system-level requirements are initially defined and TPMs are established, one has to determine how the system will ultimately be evaluated to ensure that the initially specified requirements will be met. This constitutes the *validation* loop within the system engineering process; hence, the preparation of the TEMP has been identified as one of the critical tasks in implementing a system engineering program (refer to Figure 6.6, task 4).

6. *Integrated logistic support plan (ILSP).* See Section 3.4.8; the ILSP covers all of the activities associated with the design of the prime elements of the system for *supportability,* design of the support infrastructure, the procurement and acquisition of the elements of support (maintenance personnel, spares and repair parts, test equipment, facilities, transportation and handling provisions, computer resources, and technical data), and for the sustaining maintenance and support of that system throughout its planned life cycle. Within the ILSP, there is a *logistics engineering* section that should either be incorporated within the SEMP or constitute a major reference required for the successful completion of system engineering requirements (refer to Figure 6.39). Further, as one progresses through the development process, there is an ongoing need for the evaluation of the overall support infrastructure, and for the data collection and feedback process as the system is being utilized by the consumer in the field.

7. *System retirement and material recycling plan.* Although the system retirement and material recycling and disposal part of the life cycle has not been properly addressed in many programs, the environmental concerns described in Section 3.4.11 are assuming an increasing degree of importance. The engineering aspects pertaining to the *design for the environment* should be addressed within the context of the SEMP, and the life-cycle activities associated with system development, construction and/or production, operations and support, and retirement should be monitored in terms of their impact on the environment. Further, as the system evolves through the consumer utilization and support phase, there must be an ongoing assessment relative to the possible impact of external environmental factors on the system; that is, the impact of ecological, technological, political, economic, and related factors on system operations. Is the system still performing as initially intended, or has there been some degradation due to external factors?

8. *Integrated maintenance management plan (IMMP).* Whereas the ILSP deals with the major issues pertaining to the system support infrastructure in the defense sector, the IMMP may serve in a comparable role in the commercial sector. Whether one is dealing with a transportation system, a communications system, a healthcare or hospital complex, or manufacturing plant, there is still a need to *plan for maintenance.* This includes the design for reliability and maintainability, the design of the maintenance and support infrastructure, the procurement of the elements of support, and the sustaining maintenance and support of the system throughout its planned life cycle.

9. *Total quality management plan (TQMP).* Within the context of total quality management (TQM), there are activities associated with *quality engineering* dealing with the design of products, the design of the manufacturing process, and the design of the maintenance and support infrastructure. These areas should either be incorporated within the SEMP or constitute a major reference required for the successful completion of system engineering requirements (refer to Figure 6.39). Further, as one progresses through the development, manufacturing, and support processes, there is an ongoing requirement to maintain the *quality* that has been built into the design.

10. *Production/manufacturing plan.* What may initially appear to be a well-designed and configured set of prime mission-oriented elements of a system may turn out not to be so after they are subjected to the production process. The production process is highly *dynamic,* with variances introduced throughout, and may have a significant impact on the products being manufactured. This would certainly be the case if *producibility* considerations were not initially incorporated in the design of the applicable product(s). The concurrency approach, whereby the product and the manufacturing life cycles must be properly integrated, is critical in the implementation of system engineering concepts and methods.

11. *Configuration management plan.* The importance of configuration management (CM), or *baseline* management, is critical to the fulfillment of system engineering objectives and has been emphasized throughout this text. Maintaining the design baseline and controlling design changes are essential for system evaluation and cost control.

12. *Supplier management plan.* With the increasing trend toward outsourcing and the selection of suppliers from various sources around the world, it may be feasible to develop a plan (in support of the marketing plan) incorporating the criteria for initial supplier selection and the follow-on procedures leading to some form of contracting, subsequent monitoring and control activities, and so on. For some programs, more than 50% of the components of a given system may be subcontracted, and it is imperative that the specifications being imposed on a contract are complete and well prepared, include performance-based requirements, and are supportive of the system specification (type A). The objective is to ensure that all of the components of the system that are procured from an outside source will properly *fit* or can be *integrated* into the system as an entity without unexpected problems occurring. Thus, it is essential that system engineering criteria be included in the supplier selection process, in the preparation of specifications for the purposes of subcontracting, and for the follow-on supplier monitoring and control activity.

Although the successful implementation of a system engineering program requires close coordination with all of the design-related activities, a special emphasis is required to ensure close working relationships with those organizations responsible for the activities covered in these plans.

6.6 MANAGEMENT METHODS/TOOLS

A major system engineering challenge is to be able to evolve through the process illustrated in Figure 1.12 (Chapter 1) following an organized, logical, and methodical approach and utilizing whatever techniques/tools are available at the time. Inherent within this process is the requirement for accomplishing synthesis, analysis, and evaluation efforts, commencing as early in the life cycle as practicable. The goal is to gain early visibility, reduce the time that it takes for system acquisition, be able to make good design and management decisions as system development evolves, and to minimize the risks associated with the decision-making process. However, too much activity too early may turn out to be meaningless and costly. The challenge is to be able to quickly assess *what is required, when, and to what extent.* At the same time, it is essential that one be familiar with the techniques/tools that are available and can be applied in helping to meet this goal. The system engineer must take a leadership role here, and knowledge of the latest technology that can be applied in facilitating this process is required.

Although it is impossible to mention all of the analytical techniques/tools that may be utilized to assist the system engineer in successfully fulfilling the aforementioned objective, it is recommended that such professionals become familiar with at least these six:

1. Possible applications of electronic commerce (EC), information technology (IT), electronic data interchange (EDI), and the Internet in system development and in the implementation of the system engineering process (refer to Section 4.3).

2. Application of CAD, CAM, and CAS methods in design (refer to Sections 4.5, 4.6, and 4.7).

3. The utilization of simulation methods in design (refer to Section 4.4.1).

4. The use of rapid prototyping methods in software design and development (refer to Section 4.4.2).

5. The use of scaled models and mock-ups in design evaluation (refer to Section 4.4.3).

6. Application of statistical and operations research methods in system analysis (refer to Section 4.2).

The challenge is to know what techniques/tools to use in solving certain design-related problems and in accomplishing various types of analyses. These design-enhancement capabilities should be described in the SEMP, in terms of their application. Care must be taken to ensure that their application is compatible with the capabilities of suppliers when they are required. For example, if the prime contractor utilizes a specific CAD software and depends on "live" inputs from one or more suppliers, then one needs to ensure that the supplier(s) utilizes software that is compatible. The compatibility of technologies across the spectrum of the design team members is essential.

6.7   RISK MANAGEMENT PLAN[19]

Risk is the potential that something will go wrong as a result of one or a series of events. It is measured as the combined effect of the probability of occurrence and the assessed consequence given that occurrence. The potential for risk becomes increasingly higher as complexities and new technologies are introduced in the design of systems. Risk, as used in the context described herein, refers to the potential of not meeting a specified technical and/or program requirement—for example, not meeting a requirement specified by a TPM, a schedule, or a cost projection.

Risk management is an organized method for identifying and measuring risk, and for selecting and developing options for handling risk. Risk management is not a separate program thrust by itself, but should be an inherent part of any sound management activity. Risk management includes three basic activities:

1. *Risk assessment*. This involves the ongoing review of technical design and/or program management decisions and the identification of potential areas of risk.

2. *Risk analysis*. This includes conducting an analysis to determine the probability of events and the consequences associated with their occurrence. The purpose of risk analysis is to identify the cause(s), the effects, and the magnitude of the risk perceived and to identify alternative approaches for risk avoidance. There are many tools

---

[19]"Risk" and "risk management" constitute a very important part of a system engineering program. Three good references for a more in-depth discussion of this area are (1) E. M. Hall, *Managing Risk: Methods for Software Systems Development* (Reading, MA; Addison-Wesley, 1998): (2) Y. Haimes, *Risk Modeling, Assessment, and Management*, 2nd ed. (Hoboken, NJ: John Wiley & Sons, Inc., 2004) and (3) EIA/IS-632, *Processes for Engineering a System* (Washington DC: Electronic Industries Association, EIA).

available that can be used as an aid in conducting risk analyses: for example, scheduling network analysis, life-cycle cost analysis, FMECA, the Ishikawa cause-and-effect or *fishbone* diagram, hazard analysis, and trade-off studies in varying forms.

3. *Risk abatement.* This involves the techniques and methods developed to reduce (if not eliminate) or control risk. A plan must be implemented for the handling of risk.

One of the first steps in risk management is the identification of the potential areas of risk. Although there is some degree of risk associated with any program area of activity where decisions are being made, one needs to identify those in which the potential consequences of failure can be significant. Program areas of risk may include funding, schedule, contract relationships, political, and technical. Technical risks relate primarily to the potential of not meeting a design requirement, not being able to produce an item in multiple quantities, and/or not being able to support a product in the field. Design engineering risks can be tied directly to the technical performance measures (TPMs) identified in Section 2.6 and in Figure 5.2. These TPMs, which reflect critical factors in design, can be prioritized to reflect relative degrees of importance.

Given the identification of performance characteristics to which the system is to be designed (i.e., those parameters that require monitoring on a regular basis), the next step is to evaluate these by indicating possible causes for failure. In the event of failure to meet a specific design requirement, one must ask, What are the possible causes and what are the probabilities of occurrence? Although the output measure being monitored may be a high-priority TPM, the cause of a possible failure may be the result of a misapplication of a new technology in design, a schedule delay on the part of a major supplier, a cost overrun, or a combination of these.

The causes are evaluated independently to determine the degree to which they can impact the TPM(s) being monitored. Sensitivity analyses are conducted, using various analytical models as appropriate, to determine the magnitude of the potential risk. This, in turn, will lead to the classification of factors in terms of "high," "medium," or "low" risk. These classifications of risk are then addressed within the program management review and reporting structure. High-risk items are monitored to a greater extent, with a higher priority relative to initiating a risk abatement plan, than low-risk items.

To facilitate the risk management implementation process, it is often feasible to develop a model of some type. One approach is to address risk in terms of two major variables: the probability of failure (P_f) and the effect or consequence of that failure (C_f). Consequences may be measured on the basis of technical performance, cost, or schedule. Mathematically, this model can be expressed as:[20]

$$\text{Risk factor}\,(RF) = P_f + C_f - (P_f)(C_f) \qquad (6.5)$$

---

[20]This model was adapted from the procedure included in 1986 edition of the *Systems Engineering Management Guide*, published by the Defense Systems Management College (DSMC), Fort Belvoir, VA. Although there are other models in use today, presentation of the material included herein will provide an idea as to an approach in the quantification of risk. It should be emphasized, however, that one needs to develop a model tailored to the system and the program in question.

where P_f is the probability of failure and C_f is the consequence of failure. The quantitative relationships of these parameters are described in Figure 6.41.

To illustrate the model application, with Figure 6.41 as the prime source of information, consider four system design characteristics:

1. System design uses off-the-shelf hardware with minor modifications to the software.
2. The design is relatively simple, involving the use of standard hardware.
3. The design requires software of somewhat greater complexity.
4. The design requires a new database to be developed by a supplier (subcontractor).

The characteristics of the system suggest that there is potential risk associated with the software development task. Using the criteria in Figure 6.41 (and applying the weighting factors as indicated), the probability of failure (P_f) is calculated as follows:

$$
\begin{aligned}
P_{Mhw} &= 0.1, \text{ or } (a)(P_{Mhw}) = (0.2)(0.1) = 0.02 \\
P_{Msw} &= 0.3, \text{ or } (b)(P_{Msw}) = (0.1)(0.3) = 0.03 \\
P_{Chw} &= 0.1, \text{ or } (c)(P_{Chw}) = (0.4)(0.1) = 0.04 \\
P_{Csw} &= 0.3, \text{ or } (d)(P_{Csw}) = (0.1)(0.3) = 0.03 \\
P_D &= 0.9, \text{ or } (e)(P_D) \quad = (0.2)(0.9) = \underline{0.18} \\
& \qquad\qquad\qquad\qquad\qquad\qquad\qquad 0.30
\end{aligned}
$$

Given the preceding criteria, P_f of this item is 0.30.

If the consequence of the item's failure due to technical factors causes problems of a correctible nature, but the correction results in an 8% cost increase and a two-month schedule slippage, the C_f is calculated as follows:

$$
\begin{aligned}
C_t &= 0.3, \text{ or } (f)(C_t) = (0.4)(0.3) = 0.12 \\
C_c &= 0.5, \text{ or } (g)(C_c) = (0.5)(0.5) = 0.25 \\
C_s &= 0.5, \text{ or } (h)(C_s) = (0.1)(0.5) = \underline{0.05} \\
& \qquad\qquad\qquad\qquad\qquad\qquad 0.42
\end{aligned}
$$

Based on the preceding (using the weighting factors indicated), the C_f factor is 0.42 and, from equation (6.5), the calculated risk factor (RF) is 0.594. This can be classified within the category of medium risk, as noted in Figure 6.42. In this instance, the risk is primarily associated with the system software and the reliance on a supplier.

A similar approach can be applied in performing a risk analysis on all other applicable parameters. The net result is the development of a list of critical items, presented in order of priority, that require special management attention. Risk reports are prepared at different times (i.e., frequency of distribution) depending on the

(1) Risk Factor = $P_f + C_f - P_f \bullet C_f$

(2) $P_f = (a)(P_{Mhw}) + (b)(P_{Msw}) + (c)(P_{Chw}) + (d)(P_{Csw}) + (e)(P_D)$

where

P_{Mhw} = Probability of failure due to degree of hardware maturity

P_{Msw} = Probability of failure due to degree of software maturity

P_{Chw} = Probability of failure due to degree of hardware complexity

P_{Csw} = Probability of failure due to degree of software complexity

P_D = Probability of failure due to dependency on other items

and where: a, b, c, d, and e are weighting factors whose sum equals 1.

(3) $C_f = (f)(C_t) + (g)(C_c) + (h)(C_s)$

where

C_t = Consequence of failure due to technical factors

C_c = Consequence of failure due to changes in cost

C_s = Consequence of failure due to changes in schedule

and where f, g, and h are weighting factors whose sum equals 1.

Magnitude	Maturity Factor (P_M)		Complexity Factor (P_C)		Dependency Factor (P_D)
	Hardware PMhw	Software PMsw	Hardware PChw	Software PCsw	
0.1	Existing	Existing	Simple design	Simple design	Independent of existing system, facility, or associate contractor
0.3	Minor redesign	Minor redesign	Minor increases in complexity	Minor increases in complexity	Schedule dependent on existing system, facility, or associate contractor
0.5	Major change feasible	Major change feasible	Moderate increase	Moderate increase	Performance dependent on existing system performance, facility, or associate contractor
0.7	Technology available, complex design	New software similar to existing	Significant increase	Significant Increase/major increase in # of modules	Schedule dependent on new system schedule, facility, or associate contractor
0.9	State of art some research complete	State of art never done before	Extremely complex	Extremely complex	Performance dependent on new system schedule, facility, or associate contractor

Magnitude	Technical Factor (C_f)	Cost Factor (C_c)	Schedule Factor (C_s)
0.1 (low)	Minimal or no consequences, unimportant	Budget estimates not exceeded, some transfer of money	Negligible impact on program, slight development schedule change compensated by available schedule slack
0.3 (minor)	Small reduction in technical performance	Cost estimates exceed budget by 1 to 5 percent	Minor slip in schedule (less than 1 month), some adjustment in milestones required
0.5 (moderate)	Some reduction in technical performance	Cost estimates increased by 5 to 20 percent	Small slip in schedule
0.7 (significant)	Significant degradation in technical performance	Cost estimates increased by 20 to 50 percent	Development schedule slip in excess of 3 months
0.9 (high)	Technical goals cannot be achieved	Cost estimates increased in excess of 50 percent	Large schedule slip that affects segment milestones or has possible effect on system milestones

FIGURE 6.41 A mathematical model for risk assessment. *Source:* Defense Systems Management College, *Systems Engineering Management Guide* (Fort Belvoir, VA: DSMC, 1986).

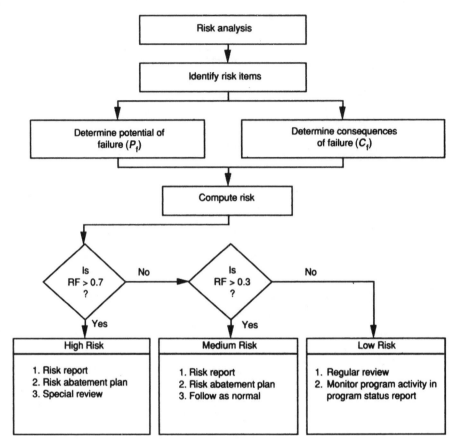

FIGURE 6.42 Risk analysis and reporting procedure.

nature of the risk. High-risk items require frequent reporting and special management attention, whereas low-risk items can be handled through the normal program review, evaluation, and reporting process.

For items classified under "high" and "medium" risk, a risk abatement plan should be implemented. This constitutes a formal approach for eliminating (if possible), reducing, and/or controlling risk. The accomplishment of such may involve one or a combination of the following:

- Provide increased management review of the problem area(s) and initiate the necessary corrective action through an internal allocation or shift in resources.
- Hire outside consultants or specialists to help resolve existing design problems.
- Implement an extensive testing program with the objective of better isolating the problem and eliminating possible causes.
- Initiate special research and development activities, conducted in parallel, in order to provide a fall-back position.

The purpose of a risk abatement plan is to *highlight* those areas where special management attention is required. The identification of technical risks is of particular interest in regard to system engineering, because the fulfillment of design objectives is highly dependent on the proper and expeditious handling of these risks. In this respect, risk management should be an inherent aspect of system engineering management.

6.8 GLOBAL APPLICATIONS/RELATIONSHIPS

As a final step in the planning process, one needs to refer to the "system operational and maintenance flow" in Figure 1.20 (Chapter 1) and ensure that all activities within both the *forward* and *reverse* flows are adequately covered. Such coverage must address customer activities, prime contractor (producer) activities, subcontractor activities, and supplier activities. Many of these activities may be assigned to various organizations throughout the world (refer to Figure 6.30). In addressing the overall spectrum, care must be taken to ensure proper compatibility and integration of these activities throughout the program in question.

More specifically, there are cultural differences, variations in political structures, variations in technical capabilities (technology applications), different methods for doing business, different geographical and environmental factors to consider, different logistics and maintenance support infrastructures, and so on, which must be addressed (and planned for) as suppliers are selected and partnerships are established. In addition, and related to these factors, are the communication processes that must effective and in place from the beginning. Given the cultural and language differences throughout the world, it would be relatively easy to misinterpret or not completely understand a given requirement, or to not understand certain assumptions and symbols used in the design process. In any event, one may wish to address the following questions:

1. Have the appropriate processes been established and put in place to ensure good communications throughout the project and across the spectrum of customer, contractor, subcontractor, and supplier activities? Has a focal point been identified in the organizational structure for the resolution of possible conflicts in this area?

2. Are the technologies and technology applications used by the various project organizations compatible across the board? For example, does every organization involved in the design process utilize the same CAD/CAM/CAS configurations?

3. Are the business processes compatible in regard to all of the organizations participating as members of the project team?

4. Are the various logistics and maintenance support infrastructures compatible with the system requirements? For example, does each participating organization have the required transportation capability to move people and materials as required?

5. Are the political structures for each participating organization compatible and supportive of program/project objectives?

Although there are many questions of this nature that can be addressed, the objective here is to ensure success by covering some of the issues that are important in a *global* environment. These issues must be inherent in the development of the SEMP.

6.9 SUMMARY

With the focus of this chapter primarily oriented to the subject of *planning,* the major emphasis has been on the development of the System Engineering Management Plan (SEMP). This planning document serves as the vehicle through which system engineering functions/tasks are initially defined and later implemented. As this plan is the key to describing the system engineering requirements for a typical program, a summary in the form of a checklist, as it applies to the SEMP's content, may be helpful to the systems engineer:

1. Does the SEMP include
   (a) A statement of work (SOW)?
   (b) A description of system engineering tasks?
   (c) A description of work packages and a work breakdown structure (WBS)?
   (d) A description of the program/project organization, the system engineering organization, the critical organizational interfaces (i.e., customer interfaces, producer/contractor interfaces, supplier interfaces) and applicable policies and procedures?
   (e) A specification/documentation tree?
   (f) A detailed program schedule?
   (g) Program/task cost projections?
   (h) A procedure for cost/schedule/technical performance measurement, review, evaluation, and control?
   (i) A description of program reporting requirements?
   (j) A risk management plan?
2. Does the SEMP adequately describe the system engineering process, including coverage of
   (a) A needs analysis?
   (b) Feasibility analysis?
   (c) System operational requirements?
   (d) Maintenance and support concept?
   (e) A procedure for identifying and prioritizing technical performance measures (TPMs)?
   (f) Functional analysis and allocation?
   (g) System synthesis, analysis, and design optimization?
   (h) Design integration and support?

(i) Design reviews?

(j) System test and evaluation (validation)?

(k) Production and/or construction?

(l) System utilization and sustaining support?

(m) System upgrades and modifications?

(n) System retirement and material recycling/disposal?

3. Does the SEMP cover the requirements for the integration of applicable engineering specialties into the total design process, including

(a) Software engineering?

(b) Reliability engineering?

(c) Maintainability engineering?

(d) Human-factors engineering?

(e) Safety engineering?

(f) Security engineering?

(g) Manufacturing and production engineering?

(h) Logistics and supportability engineering?

(i) Disposability engineering?

(j) Quality engineering?

(k) Environmental engineering?

(l) Value/cost engineering?

4. Does the SEMP describe the necessary communication links with other program planning documents such as

(a) Program Management Plan (PMP)?

(b) Individual functional design plans (as applicable)?

(c) Marketing and supplier management plan?

(d) Manufacturing/production plan?

(e) Integrated Logistic Support Plan (ILSP) and/or integrated maintenance management plan?

(f) Test and Evaluation Master Plan (TEMP)?

(g) Configuration management plan?

(h) Data management plan?

(i) Total quality management plan (TQMP)?

(j) System retirement and material recycling/disposal plan?

5. Does the SEMP support the requirements of the system specification (type A)?

6. Does the SEMP address the globalization and international environmental requirements (as applicable)?

7. Does the SEMP include adequate coverage of the current technology requirements?

8. Does the SEMP adequately support the objectives of system engineering?

QUESTIONS AND PROBLEMS

1. System engineering planning commences early at program inception with the definition of overall program requirements. Why is it essential that this planning activity start as soon as possible? What is likely to happen if system engineering planning is initiated later?

2. How do the system specification (type A) and SEMP relate to each other?

3. Who is responsible for preparing the SEMP—consumer, producer, contractor, subcontractor, or supplier? Describe some of the conditions and interfaces as applicable.

4. Select a system of your choice, describe the acquisition process, and develop a detailed outline of a SEMP for the program in question.

5. Select a program of your choice, and describe the system engineering tasks for that program (justify the tasks selected). Identify some of the key interfaces that exist.

6. For the tasks identified in Question 5, develop (a) a detailed schedule in the form of a program network and (b) a cost estimate for the proposed scheduled activity.

7. The following data are available (in Figure 6.43):
   (a) Construct a PERT/CPM chart from the data.
   (b) Determine the values for standard deviation, TE, TL, TS, TC, and P.
   (c) What is the critical path? What does this value mean?

8. When employing PERT/COST, the cost–time option applies. What is meant by the cost–time option? How can it affect the critical path?

9. What is the purpose of a WBS? What is the difference between a WBS, an SWBS, and a CWBS? How do work packages relate to the WBS? Construct a WBS for a program of your choice.

10. Describe in your own words the steps that should be followed in determining the "supplier requirements" associated with the acquisition of a new system.

Event	Previous Event	t_a	t_b	t_c
8	7	20	30	40
	6	15	20	35
	5	8	12	15
7	4	30	35	50
	3	3	7	12
6	3	40	45	65
	2	25	35	50
5	2	55	70	95
4	1	10	20	35
3	1	5	15	25
2	1	10	15	30

FIGURE 6.43 Problem 7 data.

11. Identify and describe some of the factors that should be considered in *make-or-buy* or *outsourcing* decisions.

12. Why is the development of a make-or-buy plan important? What is included?

13. Should the system engineering organization be involved in make-or-buy decisions? If so, in what capacity? If not, why not?

14. Development of an RFP in preparation for supplier proposals, evaluation, and selection is extremely critical from a system engineering standpoint. Identify and explain the reasons for such, and briefly describe key features that should be included.

15. How can political, social/societal, cultural, and economic factors influence the supplier selection process? Provide a few examples.

16. Describe in your own words some of the trends that are occurring in the world today as they relate to customer, contractor, and supplier activities.

17. What is meant by *postproduction support*? Provide some examples.

18. How does technical performance measurement fit into the system engineering planning process? What is the significance of the TPMs?

19. There are various types of contract structures that can be imposed through the contract negotiation process to include FFP, FP, CPFF, CPIF, cost sharing, and time and material. Describe each and include some discussion as to applications.

20. When establishing multiple incentives under incentive contracting, what steps would you follow? How will you determine the specific factors or characteristics on which to establish incentives?

21. Under incentive contracting, what is meant by an incentive/penalty sharing ratio? How is it applied? How does the SR relate to supplier/contractor risks?

22. Should the system engineer participate in the contract negotiation process? If so, in what capacity?

23. Describe some of the methods/tools that can be used to facilitate implementation of the system engineering process.

24. Why is it important to develop a risk management plan? What is included?

7

ORGANIZATION FOR SYSTEM ENGINEERING

The initial planning for system engineering commences during the early stages of conceptual design and evolves through the development of the System Engineering Management Plan (SEMP) described in Chapter 6. To implement this plan successfully requires an organizational structure that will promote, support, and generally enhance the application of system engineering principles and concepts. The proper organizational *environment* must be created that will allow for the accomplishment of system engineering requirements in an effective and efficient manner—that is, the implementation of a top-down, life-cycle-oriented, integrated approach in system design and development. In addition, the organization must be *dynamic* in response to the many changes that are taking place worldwide.

Figure 6.1 shows two sides of the spectrum; that is, the *technology* issues that can be applied to enhance and facilitate the implementation of the system engineering process and the *management* issues that are necessary to meet the objectives in this area. Inherent in this overall spectrum is the *organizational* element. *Organization* is the combining of resources in such a manner as to fulfill a certain need. Organizations constitute groups of individuals of varying levels of expertise, combined in a social structure of some form to accomplish one or more functions. Organizational structures vary with the functions to be performed, and the results will depend on the established goals and objectives, the resources available, the communications and working relationships between the individual participants, the motivation of personnel, and many other factors. The ultimate objective is to achieve the most effective and efficient utilization of human, material, and monetary resources through the establishment of communications and decision-making processes designed to accomplish specific objectives.

This chapter begins with a discussion of different types of organizational structures (their advantages and disadvantages from a generic perspective) and then emphasizes the system engineering organization, its functions, organizational interfaces, and the staffing needed to meet the objectives described throughout this text. Among the structures addressed are the *functional, product line, project, matrix*, and the *combined functional-project* approaches. Customer (consumer), producer (contractor), and supplier relationships are covered, along with their respective functions/tasks. Finally, the chapter discusses human resource requirements: the selection of personnel, the skill levels required, organizational leadership characteristics, personal motivational factors, and so on. The material presented herein is directly supportive of the planning process described in Chapter 6.[1]

7.1 DEVELOPING THE ORGANIZATIONAL STRUCTURE

When dealing with organizations, one must address a number of issues, including *structure, processes, culture, environment,* and various combinations of these. As an initial step, it is logical to consider *structure* first. Processes, culture, and the organizational environment are discussed later.

In the development of any type of an organizational structure, one must start by determining the goals and objectives for the overall company/agency/institution involved, along with the functions and tasks that must be accomplished. Depending on the complexity and size of a program, the structure may assume a pure *functional* model, a *project* or *product line* orientation, a *matrix* approach, or combinations thereof. Further, the structure may change in context as the system development evolves from the conceptual design phase through detail design and development, production, and so on. The ultimate goal, of course, is to achieve the most effective utilization of human, material, and monetary resources in accomplishing the functions that are required at the time.

In regard to system engineering, a prime objective during the early stages of conceptual design is to ensure the proper development of system-level requirements; that is, the needs analysis, feasibility analysis, operational requirements, maintenance concept, identification of technical performance measures (TPMs), and the preparation of the system specification (type A). These activities are highly *customer/user* focused and directed toward the *system* as an entity, and their accomplishment does not require a large organization per se. However, the selection of a few key personnel with the appropriate skills, backgrounds, and experience levels is essential.

---

[1] The level (depth) of discussion of organizational concepts in this chapter is very cursory and is intended to provide the reader with an overview of some of the key points in respect to system engineering. Three good references for additional material are (1) J. L. Gibson, J. H. Donnelly, J. M. Ivancevich, and R. Konopaske, *Organizations: Behavior, Structure, Processes,* 12th ed. (New York: McGraw-Hill/Irwin, 2005); (2) H. Kerzner, *Project Management: A Systems Approach to Planning, Scheduling, and Controlling,* 9th ed. (Hoboken, NJ: John Wiley & Sons, Inc., 2005); and (3) A. P. Sage, *Systems Management for Information Technology and Software Engineering* (New York: John Wiley & Sons, Inc., 1995). Also check Appendix F for additional references.

As the program evolves into the preliminary and detail design and development phases, the number of assigned personnel may increase as the design requirements at the subsystem level (and below) may dictate the necessity for including the expertise of various design disciplines—for example, reliability, maintainability, human factors, safety, and logistics. In this context, the organizational structure may change from a pure *project* configuration to a mixed *functional-project* or *matrix* approach. As the system and its components enter the production phase, the organizational structure may shift once again.[2]

In addressing the organizational issue overall, the emphasis herein is intended to stress the many and varied tasks, described in Section 6.2.2 (Figure 6.6), that must be accomplished, regardless of which organizational element (department or group of personnel) performs the work. Experience indicates that there are organizational departments/groups located within industrial firms and/or government agencies that have been designated as *system engineering* and assigned the appropriate responsibilities, but are not performing the tasks required. Conversely, there are organizational elements with different identities that are, in actuality, performing the desired functions. Further, for small projects, where a single individual must assume many different roles, the system engineering responsibilities may be accomplished by an electrical engineer, a mechanical engineer, or someone with equivalent background and experience. For instance, the chief engineer or project manager may serve as the "system engineer," or there may be a designated group performing the required tasks.

Whereas there may be variations in approach, Sections 7.2 through 7.5 provide a more in-depth discussion on the various types of organizational structures, the advantages and disadvantages of each, personnel staffing issues, and so on.

7.2 CUSTOMER, PRODUCER, AND SUPPLIER RELATIONSHIPS

To properly address the subject of "organization for system engineering," one needs to view this in the context of the total flow of activities, evolving from the customer and down to the producer (prime contractor) and suppliers. Although this top-down flow may vary in detail, depending on the size of the project and the stage of design and development, this discussion is primarily directed to a large project activity, applicable in the acquisition of many large-scale systems. By addressing large projects, it is hoped that a better understanding of the role of system engineering in a somewhat complex environment will be provided. The reader must then adapt and structure an approach for his or her own program requirements.

For a relatively large project, the system engineering function may appear at several levels, as shown in Figure 7.1. The requirements for system engineering and the responsibility for implementing the tasks described in Chapter 6 lie with the customer. The customer may establish a system engineering organization to accomplish

---

[2]It should be noted that, from time to time, a shift in organizational structure may occur as a result of "outside" or "external" influences: for example, resulting from changes in technology applications, changes in supplier requirements, changes in political and economic conditions, and the like.

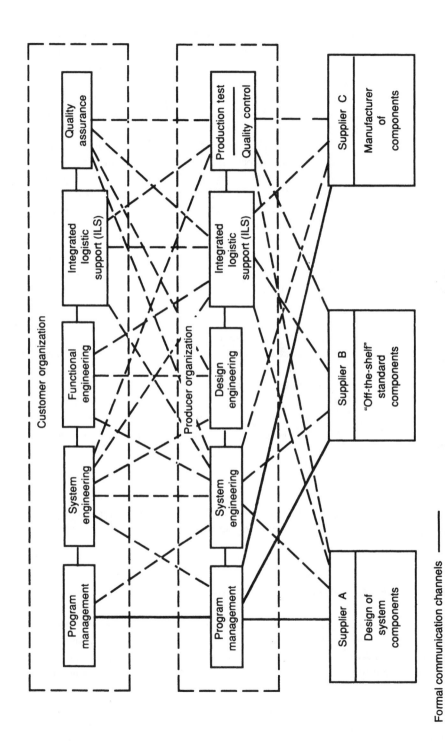

FIGURE 7.1 Consumer/producer/supplier interfaces.

Formal communication channels ———
Informal communication channels – – – –

357

the required tasks, or these tasks may be relegated (in part or in total) to the producer through some form of contractual arrangement. In any event, the responsibility, along with the authority, for accomplishing system engineering functions must be clearly defined from the beginning.

In some instances, the customer may assume full responsibility for the overall design and development, production, and installation of the system and its elements for operational use. The needs analysis, feasibility studies, definition of operational requirements and the maintenance concept, identification and prioritization of TPMs, preparation of the system specification (type A), and preparation of the SEMP are accomplished by the customer. Top-level functions are defined and specific program requirements are allocated to individual producers, subcontractors, and suppliers.

In other cases, whereas the customer provides the overall guidance in terms of is-suing a general statement of work (SOW) or a contractual document of an equivalent nature, the producer (or *prime contractor*) is held responsible for the entire system design and development effort and for completing the tasks described in Chapter 6. In other words, although both the customer and the producer have established sys-tem engineering organizations, the basic responsibility for fulfilling the objectives described throughout this text lies with the producer's organization, with supporting tasks being accomplished by individual suppliers as required. To accomplish this, the customer must delegate the appropriate level of *responsibility* for completing the functions specified, and the necessary *authority* as well. Further, the customer must make available all of the input data necessary for the producer to successfully com-plete the conceptual design tasks noted earlier.[3]

In Figure 7.1, it should be noted that there is an extensive amount of communica-tion required, not only within each of the customer and producer organizations, but also between the various customer, producer, and supplier organizations. Although the solid lines pertain primarily to the more formal program management direction of contractual nature, there are many informal channels of communication that must ex-ist to ensure that the proper dialogue is established between the numerous and varied entities involved in the system development effort. The successful implementation of a *teaming* or *partnership* approach, along with the fostering of concurrent engineer-ing principles, is heavily dependent on good communications (both downward and upward) from the beginning.

7.3 CUSTOMER ORGANIZATION AND FUNCTIONS

The customer/consumer organization may vary, ranging from one or a small group of individuals to an industrial firm, a commercial business, an academic institution, a government laboratory, the Department of Defense, or a military service. The cus-tomer may be the ultimate "user" of the system or may be the procuring agency for

---

[3]It is not uncommon for the customer to perform a requirements analysis, prepare a report describing the requirements for a new system, place it in a file somewhere, and then fail to pass on the necessary information later to the responsible producer. Thus, the producer has to generate a new set of requirements that may, or may not, be consistent with those initially developed by the customer.

a user. An example of the latter is found within the defense sector, where the Air Force, Army, and Navy each have acquisition agencies that are responsible for the contracting and procurement of systems, and the "user" is the operating command in the field/fleet responsible for the utilization and sustaining maintenance and support of the system throughout its planned life cycle.

In Figure 7.1, the acquisition agency may be represented by the top block, with a chain of industrial firms, small businesses, and component suppliers providing the materials and services necessary for the development of the system and its elements. In such instances, it is incumbent on the procuring agency to ensure that the early contracting and acquisition process will result in satisfying the needs of the ultimate "user," not just to respond to the short-term desires of the procuring agency. In this case, the procuring agency must be responsive to the "user" organization (as the customer), the producer or industrial firm (in Figure 7.1) must be responsive to the acquisition agency (as the customer), and the suppliers must be responsive to the producer (as the customer). The question is, Who is the *ultimate customer,* who is *your customer,* and do the requirements associated with the latter support the objectives specified for the first? It is essential that this overall "chain" of organizational entities be addressed in the planning and development of systems.

There are a variety of approaches and associated organizational relationships involved in the design and development of new systems. The objective is to identify the overall "program manager" and to pinpoint the responsibility for *system engineering management.* In the past, there have been numerous instances in which the procuring agency (e.g., the "customer" in Figure 7.1) has initiated a contract with an industrial firm (e.g., the "producer") for the design and development, and/or reengineering, of a large system, but has not delegated the complete responsibility (or corresponding *authority*) for system engineering management. The industrial firm has been held responsible for the design, development, production, and delivery of a system in response to certain specified requirements. However, the customer has not always provided the producer with the necessary data and/or controls to allow the development effort to proceed in accordance with good system engineering practices. At the same time, the customer has not performed the necessary functions of system engineering management. The net result has been the development of systems without the consideration of many of the characteristics discussed throughout Chapter 3; that is, systems that are unreliable, not maintainable, not supportable, not cost-effective, and not responsive to the needs of the ultimate users.

The fulfillment of system engineering objectives is highly dependent on a *commitment* from the top down. These objectives must be recognized from the beginning by the customer, and an organizational entity needs to be established to ensure that these objectives are met. The program manager must first understand and believe in the concepts and principles of system engineering, and then must create the appropriate environment and take the lead by initiating either of the following courses of action:

1. Accomplish the system engineering functions within the customer's organizational structure (see Figure 7.1). This may include completing the basic activities reflected in Figure 1.12 and described in Figure 6.6; that is, the needs

analysis and feasibility studies, development of operational requirements and the maintenance concept, the identification and prioritization of TPMs, functional analysis and allocation, synthesis, design optimization, and so on. In other words, the customer (or procuring agency) will prepare the system specification (type A), will perform all of the tasks required at the *system level,* and will delegate requirements for the subsystem level and below.

2. Accomplish the system engineering functions within the industrial firm or the producer's organizational structure (shown in Figure 7.1). These may include the completion of the system engineering tasks reflected in Figure 1.12 and described in Figure 6.6; that is, development of operational requirements and the maintenance concept, functional analysis and allocation, synthesis, design optimization, and so on. Although the customer will define the program requirements in the form of a statement of work (SOW), all of the system engineering tasks and associated management functions will be delegated to and be accomplished by the producer.

Although these two options represent the extremes, there may be any combination of models in which the responsibilities for accomplishing system engineering management functions have been split. In such cases, it is essential that the responsibility for system engineering be established from the beginning. The customer must clarify system objectives and program functions, and the requirements for system engineering must be well defined. It is critical that the process described in Chapter 2 be implemented properly, independent of organizational splits, the sharing of responsibilities, or any other conditions.

In the event that the system engineering responsibility is delegated to the producer (i.e., the preceding second option), the customer must completely support this decision by providing the necessary top-down guidance and managerial backing. Responsibilities must be properly delineated, system-level data generated through earlier customer activities and studies made available to the producer (e.g., the results of feasibility analyses, the documentation of operational requirements), and the producer must be given the necessary leeway relative to making decisions at the system level. The challenge for the customer is to prepare a good, comprehensive, well-written, and clear SOW to be implemented by the producer. The emphasis should be on the issues of producer *performance,* specifying *what* needs to be accomplished and *when,* versus telling the producer *how* to perform the job. In addition, the various lines of communication between the customer and producer shown in Figure 7.1 must support a unified and consistent approach throughout.

7.4 PRODUCER ORGANIZATION AND FUNCTIONS (THE CONTRACTOR)

For the purposes of discussion, it is assumed that the producer (or contractor) in Figure 7.1 will undertake the bulk of the system engineering activities associated with the design and development of a large-scale system. The customer will specify the necessary system-level and program requirements through the preparation of a *request for proposal* (RFP) or an *invitation for bid* (IFB), and various industrial firms will respond by submitting a formal proposal. The response may represent the results of a *teaming* arrangement involving a designated number of industrial firms and component suppliers. As there may be a number of responding proposals, a formal competition is initiated, individual proposals are reviewed and evaluated, contractual negotiations are consummated, and a selection is made. The successful contractor (i.e., producer) will then proceed with the proposed level of effort.

In addressing program requirements, it is essential that the successful contractor have access to all information and data leading to the requirements specified in the technical portions of the RFP/IFB. In some instances, the RFP will include a system specification covering the technical aspects of system development, along with a statement of work (SOW) directed toward project tasks and the management aspects of a program. The preparation of the specification will result from the completion of those activities described in Sections 2.1 through 2.7. In other words, by the time that the contractor gets involved in this case, the customer will have completed the first three tasks shown in Figure 6.6. The main objective here is to ensure continuity in the transition from the activities accomplished by the customer to those to be performed by the contractor.

This transition process is one of the most critical points in a program. First, the process described in Chapter 2 must be maintained, and a thorough understanding of this process by both customer and contractor personnel is essential. Second, the specification and SOW prepared by the customer must be complete and easily understandable; they must "talk to each other," and they must jointly *promote* the system engineering process. Often, in attempting to meet a schedule, specifications and SOW are hurriedly put together without the benefit of a complete review and the proper level of integration. The results are usually diasterous, and the follow-on activities reflect inconsistencies and the lack of the proper integration of those activities described in Chapter 3. Finally, given a good specification and SOW, the key system engineering activities must not be *negotiated out* in the development of a contractual agreement between the customer and the contractor (i.e., the development of a contract work breakdown structure; refer to Section 6.2.4). Sometimes there is a tendency to eliminate system engineering tasks to save money, which reflects a lack of understanding of the process and its objectives. This must not be allowed to happen.

Given that system-level requirements have been properly defined and that a prime contractor has been selected to accomplish the design and development effort, the next step is to address the subject of system engineering in the context of the contractor's organizational structure. Organizational structures vary from the pure *functional*, to the *project*, the combined *project-functional*, the *matrix*, and so on. These organizational patterns are discussed in the sections to follow, as they relate to the objectives of system engineering.

7.4.1 Functional Organization Structure

The primary building block for most organizational patterns is the *functional structure* reflected in Figure 7.2. This approach, sometimes referred to as the "classical" or "traditional" approach, involves the grouping of specialties or disciplines into

362

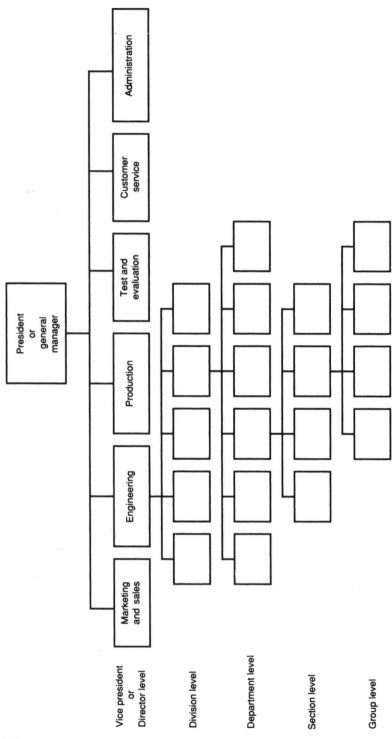

FIGURE 7.2 Producer organization (traditional functionally oriented structure).

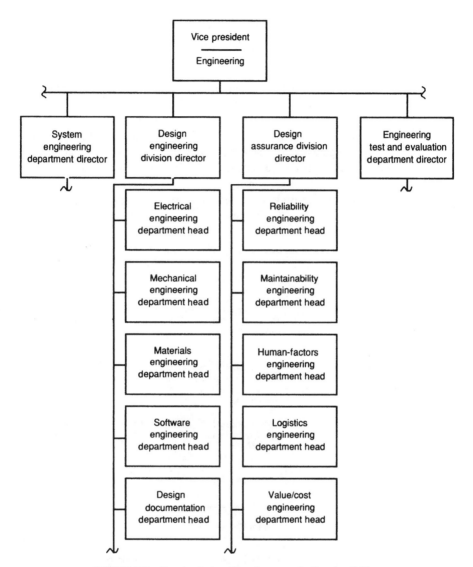

FIGURE 7.3 Breakout of engineering organizational activities.

separately identifiable entities. The intent is to perform similar activities within one organizational component. For example, all engineering work would be the responsibility of one executive, all production or manufacturing work would be the responsibility of another executive, and so on. Figure 7.3 shows a further breakout of engineering activities for illustrative purposes.

As shown in the figures, the depth of the individual elements of the organization will vary with the type of project and level of emphasis required. For projects involving the conceptual and/or preliminary design of new systems, there will be a great deal of emphasis on marketing and engineering. Within engineering, the system engineering organization should be highly influential in the design decision-making process, as compared with some of the individual design disciplines. Later, as the development process phases into detail design, the individual design disciplines will assume a greater degree of importance, and the interest in production and manufacturing increases.

As with any organizational structure, there are advantages and disadvantages. Figure 7.4 identifies some of the pros and cons associated with the pure functional approach illustrated in Figure 7.2. As shown, the president (or general manager) controls all the functional entities necessary to design and develop, produce, deliver, and support a system. Each department maintains a strong concentration of technical expertise, and thus a project can benefit from the most advanced technology in the field. In addition, levels of authority and responsibility are clearly defined,

---

Advantages

1. Enables the development of a better technical capability for the organization. Specialists can be grouped to share knowledge. Experiences from one project can be transferred to other projects through personnel exchange. Cross-training is relatively easy.

2. The organization can respond more quickly to a specific requirement through the careful assignment (or reassignment) of personnel. There are a larger number of personnel in the organization with the required skills in a given area. The manager has a greater degree of flexibility in the use of personnel and a broader manpower base with which to work. Greater technical control can be maintained.

3. Budgeting and cost control are easier because of the centralization of areas of expertise. Common tasks for different projects are integrated, and it is easier not only to estimate costs but also to monitor and control costs.

4. The channels of communication are well established. The reporting structure is vertical, and there is no question as to who is the "boss."

Disadvantages

1. It is difficult to maintain an identity with a specific project. No single individual is responsible for the total project or the integration of its activities. It is hard to pinpoint specific project responsibilities.

2. Concepts and techniques tend to be functionally oriented with little regard for project requirements. The "tailoring" of technical requirements to a particular project is discouraged.

3. There is little customer orientation or focal point. Response to specific customer needs is slow. Decisions are made on the basis of the strongest functional area of activity.

4. Because of the group orientation relative to specific areas of expertise, there is less personal motivation to excel and innovation concerning the generation of new ideas is lacking.

FIGURE 7.4 A functional organization–advantages and disadvantages.

communication channels are well structured, and the necessary controls over budgets and costs can be easily established. In general, this organizational structure is well suited for a single project operation, large or small.

However, the pure functional organization may not be as appropriate for large multiproduct firms or agencies. Where there are a large number of different projects, each competing for special attention and the appropriate resources, there are some disadvantages. The main problem is that there is no strong central authority or individual responsible for the total project. As a result, the integration of activities that cross functional lines becomes difficult. Conflicts occur as each functional activity struggles for power and resources, and decisions are often made on the basis of what is best for a functional group rather than what is best for the project. Further, the decision-making processes are sometimes slow and tedious because all communications must be channeled through upper-level management. Basically, projects may fall behind and suffer in the classical functional organization structure.

7.4.2 Product-Line/Project Organization Structure

As industrial firms grow and there are more products being developed, it is often convenient to classify these products into common groups and to develop a product-line organization structure, as shown in Figure 7.5. A company may become involved in the development of communication systems, transportation systems, and electronic test and support equipment. Where there is functional commonality, it may be appropriate to organize the company into three divisions, one for each product line. In such instances, each division will be self-sufficient relative to system design and support. Further, these divisions may be geographically separated, and each may serve as a functional entity with operations similar to those described in Section 7.4.1.

In divisions in which large systems are being developed, the product-line responsibilities may be subdivided into projects, as illustrated in Figure 7.6. In such cases, the project will be the lowest independent entity.

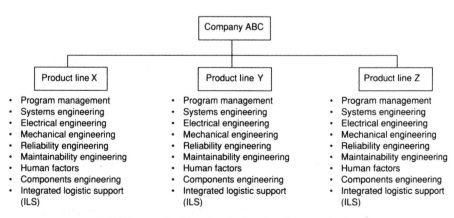

FIGURE 7.5   Traditional project/product-line organization.

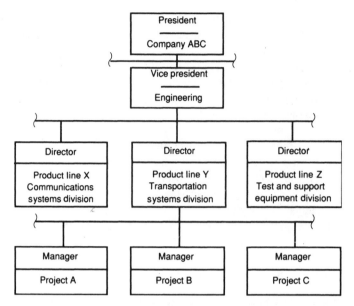

FIGURE 7.6 Product-line organization with project subunits.

A project organization is one that is solely responsible for the planning, design and development, production, and support of a single system or a large product. It is time-limited, directly oriented to the life cycle of the system, and the commitment of personnel and material is purely for the purposes of accomplishing tasks peculiar to that system. Each project will contain its own management structure, its own engineering function, its own production capability, its own support function, and so on. The project manager has the authority and the responsibility for all aspects of the project, whether it is a success or a failure.

In the case of both product-line and project structures, the activities are organized as presented in Figure 7.5. On the one hand, the lines of authority and responsibility for a given project are clearly defined, and there is no question as to priorities. On the other hand, there is potential for the duplication of activities within a firm, which can be quite costly. The emphasis is on individual projects in this structure, as compared with the overall functional approach illustrated in Figure 7.2. Some of the advantages and disadvantages of product-line/project structures are presented in Figure 7.7.

7.4.3 Matrix Organizational Structure

The matrix organizational structure is an attempt to combine the advantages of the pure functional organization and the pure project organization. In the functional organization, technology is emphasized and project-oriented tasks, schedules, and time constraints are often sacrificed. In the pure project structure, technology tends to suffer, because there is no single group responsible for its planning and

---

Advantages

1. The lines of authority and responsibility for a given project are clearly defined. Project participants work directly for the project manager, communication channels within the project are strong, and there is no question as to priorities. A good project orientation is provided.

2. There is a strong customer orientation, a company focal point is readily identified, and the communication processes between the customer and the contractor are relatively easy to maintain. A rapid response to customer needs is realized.

3. Personnel assigned to the project generally exhibit a high degree of loyalty to the project, there is strong motivation, and personal morale is usually better with product identification and affiliation.

4. The required personnel expertise can be assigned and retained exclusively on the project without the time sharing that is often required in the functional approach.

5. There is greater visibility relative to all project activities. Cost, schedule, and performance progress can be easily monitored, and potential problem areas (with the appropriate follow-on corrective action) can be identified earlier.

Disadvantages

1. The application of new technologies tends to suffer without strong functional groups and the opportunities for technical interchange between projects. As projects go on and on, those technologies that are applicable at project inception continue to be applied on a repetitive basis. There is no perpetuation of technology, and the introduction of new methods and procedures is discouraged.

2. In contractor organizations where there are many different projects, there is usually a duplication of effort, personnel, and the use of facilities and equipment. The overall operation is inefficient and the results can be quite costly. There are times when a completely decentralized approach is not as efficient as centralization.

3. From a managerial perspective, it is difficult to effectively utilize personnel in the transfer from one project to another. Good, qualified workers assigned to projects are retained by project managers for as long as possible (whether they are being effectively utilized or not), and the reassignment of such personnel usually requires approval from a higher level of authority, which can be quite time-consuming. The shifting of personnel in response to short-term needs is essentially impossible.

4. The continuity of an individual's career, growth potential, and opportunities for promotion are often not as good when he or she is assigned to a project for an extended period of time. Project personnel are limited in terms of opportunities to be innovative relative to the application of new technologies. The repetitiveness of tasks sometimes results in stagnation.

FIGURE 7.7   A project/product-line organization–advantages and disadvantages.

development of such! Matrix management is an attempt to acquire the greatest amount of technology, consistent with project schedules, time and cost constraints, and related customer requirements. Figure 7.8 presents a typical matrix organization structure.

Each project manager reports to a vice president, has the overall responsibility, and is accountable for project success. At the same time, the functional departments

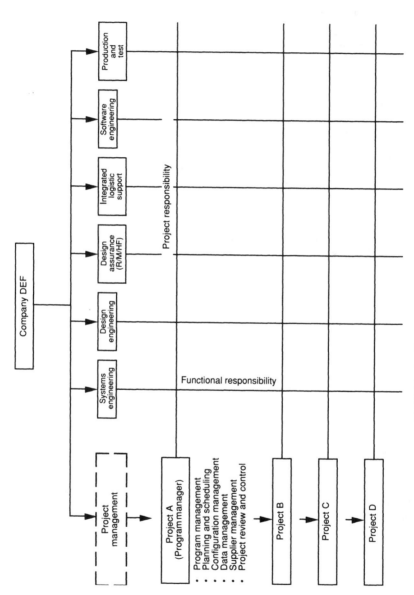

FIGURE 7.8 Pure matrix organization structure.

are responsible for maintaining technical excellence and for ensuring that all available technical information is exchanged between projects. The functional managers, who also report to a vice president, are responsible for ensuring that their personnel are knowledgeable of the latest accomplishments in their respective fields.

The matrix organization, in its simplest form, can be considered as being a two-dimensional entity, with the projects representing potential profit centers and the functional departments identified as cost centers. For small industrial firms, the two-dimensional structure may be the preferred organizational approach because of the flexibility allowed. The sharing of personnel and the ability to shift back and forth are often inherent characteristics. For large corporations with many product divisions, the matrix becomes a multidimensional structure.

As the number of projects and functional departments increases, the matrix structure can become quite complex. To ensure success in implementing matrix management, a highly cooperative and mutually supportive environment must be created within the company. Managers and workers alike must be committed to the objectives of matrix management. Four key points follow:

1. Good communication channels (vertical and horizontal) must be established to allow for a free and continuing flow of information between projects and the functional departments. Good communications must also be established from project to project.

2. Both project managers and functional department managers should participate in the initial establishment of companywide and program-oriented objectives. Further, each must have an input and become involved in the planning process. The purpose is to help ensure the necessary commitment on both sides. In addition, both project and functional managers must be willing to negotiate for resources.

3. A quick and effective method for conflict resolution must be established, to be used in the event of disagreement. A procedure must be developed with the participation and commitment of both project and functional managers.

4. For personnel representing the technical functions and assigned to a project, the project manager and the functional department manager should agree on the duration of assignment, the tasks to be accomplished, and the basis on which the individual(s) will be evaluated. The individual worker must know what is to be expected of him or her, the criteria for evaluation, and which manager will be conducting the performance review (or how the performance review will be conducted). Otherwise, a "two-boss" situation (each with his or her own objectives) may develop, and the employee will be caught in the middle.

The matrix structure provides the best of several worlds: that is, a composite of the pure project approach and the traditional functional approach. The main advantage pertains to the capability of providing the proper mix of technology and project-related activities. At the same time, a major disadvantage relates to the conflicts that arise on a continuing basis as a result of a power struggle among project and functional managers, changes in priorities, and so on. Further advantages and disadvantages are noted in Figure 7.9.

Advantages

1. The project manager can provide the necessary strong controls for the project while having ready access to the resources from many different function-oriented departments.

2. The functional organizations exist primarily as support for the projects. A strong technical capability can be developed and made available in response to project requirements in an expeditious manner.

3. Technical expertise can be exchanged between projects with a minimum of conflict. Knowledge is available for all projects on an equal basis.

4. Authority and responsibility for project task accomplishment are shared between the project manager and the functional manager. There is mutual commitment in fulfilling project requirements.

5. Key personnel can be shared and assigned to work on a variety of problems. From the company top-management perspective, a more effective utilization of technical personnel can be realized and program costs can be minimized as a result.

Disadvantages

1. Each project organization operates independently. In an attempt to maintain an identity, separate operating procedures are developed, separate personnel requirements are identified, and so on. Extreme care must be taken to guard against possible duplication of efforts.

2. From a company viewpoint, the matrix structure may be more costly in terms of administrative requirements. Both the project and the functional areas of activity require similar administrative controls.

3. The balance of power between the project and the functional organizations must be clearly defined initially and closely monitored thereafter. Depending on the strengths (and weaknesses) of the individual managers, the power and influence can shift to the detriment of the overall company organization.

4. From the perspective of the individual worker, there is often a split in the chain of command for reporting purposes. The individual is sometimes "pulled" between the project boss and the functional boss.

FIGURE 7.9 A matrix organization–advantages and disadvantages.

7.4.4 Integrated Product and Process Development (IPPD)

With the objectives of concurrent engineering in mind, the Department of Defense (DOD) initiated the concept of *integrated product and process development* (IPPD) in the mid-1990s. IPPD can be defined as "a management technique that simultaneously integrates all essential acquisition activities through the use of multidiscipline teams to optimize the design, manufacturing, and support processes."[4] The concept promotes the communications and integration of the key functional areas as they

---

[4]DOD 5000.2-R, *Mandatory Procedures for Major Defense Acquisition Programs (MDAPS) and Major Automated Information System (MAIS) Acquisition Programs* (Washington DC: Office of the Secretary of Defense).

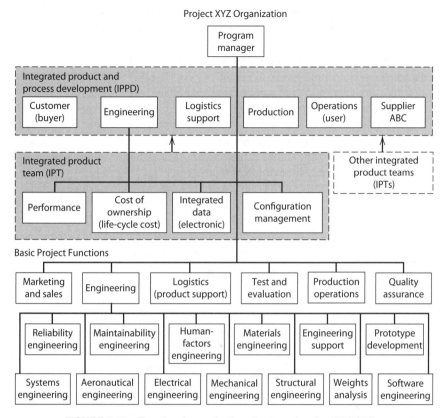

FIGURE 7.10   Functional organization structure showing IPPD/IPTs.

apply to the various phases of program activity from conceptual design through detail design and development. Although the specific nature of the activities involved and the degree of emphasis exerted will change somewhat as system design and development evolves, the concept conveyed in Figure 7.10 is maintained throughout to ensure the appropriate integration. In this regard, the concept of IPPD is directly in line with system engineering objectives; that is, to cause the integration of the various features of design and the organizations involved in the design process.

7.4.5   Integrated Product/Process Teams (IPTs)

Inherent within the IPPD concept is the establishment of *integrated product teams* (IPTs), with the objective of addressing certain designated and well-defined issues.[5] An IPT, constituting a selected team of individuals from the appropriate disciplines, may be established to investigate a specific segment of design, a solution for some

---

[5]"IPT" is also used as a designator for *Integrated Process Team.* Another term used in a similar context is *Process Action Team* (PAT).

outstanding problem, design activities that have a great impact on a high-priority TPM, and so on. The objective is to create a *team* of qualified individuals that can effectively work together to solve some problem in response to a given requirement. Further, there may be a number of different teams established to address issues at different levels in the overall system hierarchical structure; that is, issues at the system level, subsystem level, and/or component level. As shown in Figure 7.10, an IPT may be established to concentrate on those activities that significantly impact selected *performance factors, cost of ownership,* and *configuration management.* There may be another IPT assigned to "track" the *integrated data* environment issue. The objective is to provide the necessary emphasis in critical areas and to reap the benefits of a *team* approach in arriving the best solution possible.

IPTs are often established by the program manager or by some designated high-level authority in the organization. The representative team members must be well qualified in their respective areas of expertise, empowered to make on-the-spot decisions when necessary, proactive relative to team participation, success-oriented, and resolved to addressing the problem assigned. The program manager must clearly define the objectives for the team, the expectations in terms of results, and the team members must maintain a continuous "up-the-line" communications channel. The longevity of an IPT will depend on the nature of the problem and the effectiveness of the team in progressing toward meeting its objective. Care must be taken to avoid the establishment of too many teams, as the communication processes and interfaces become too complex when there are many teams in place. In addition, there often are conflicts when it comes to issues of importance, and a critical issue may be traded off as a result. Further, as the team ceases to be effective in accomplishing its objectives, it should be disbanded accordingly. An established team that has outlived its usefulness can be counterproductive.

7.4.6 System Engineering Organization

Sections 7.4.1 through 7.4.5 provide an overview of the major characteristics of the functional, project, and matrix organization structures. Some of the advantages and disadvantages of each are identified. It is important to thoroughly understand and have these characteristics in mind in developing an organizational approach involving system engineering. More specifically, in considering system engineering objectives, the following points should be noted:

- The function of system engineering must be oriented to the objective of bringing a system into being in an effective and efficient manner. In this regard, there is a natural close association with the project type of organizational structure. System engineering is heavily involved in the initial establishment of requirements and in the follow-on integration of design engineering and supporting activities throughout system development, production, and operational use. System engineering influences design to a significant degree, and this is best accomplished through a project organizational structure.

- The nature of the system engineering function, its objectives in terms of design integration, its many interfaces with other program activities, and so on, require the existence of good communication channels (both vertically and horizontally). The personnel within the system engineering organization must maintain effective communications with all other project organizational elements, with many different functional departments, with a variety of suppliers, and with the customer. These requirements are facilitated through the project organization approach.

- The successful fulfillment of system engineering objectives requires the specification of technical requirements for the system, the conductance of trade-off studies, the selection of appropriate technologies, and so on. Personnel within the system engineering organization must be current (i.e., up to date) relative to the latest technology applications and/or must have access to technical expertise in the appropriate disciplines. A strong technical thrust is required, and good communications must be established with the functional departments (as applicable). Thus, the preferred organization structure should include selected functional elements, in addition to the project orientation.

Although the implementation of system engineering requirements can actually be fulfilled through any one of a number of organizational structures, the preferred approach should respond to these three major considerations. It appears that the best organizational structure includes a combination of project requirements and functional requirements. Although a major *project* orientation is required in response to customer needs, a *functional* orientation is necessary to ensure consideration of the latest technology applications. The combined project-functional organization approach may vary somewhat, depending on the size of the industrial firm. For a large firm, the organization structure illustrated in Figure 7.11 may be appropriate. Project activities are relatively large in scope (and in personnel loading), with supporting functional activities covering selected areas of expertise where centralization is justified. For smaller firms, the functional departments are relatively large, and they provide support to individual projects according to demand. This support is assigned on a task-by-task basis. Figure 7.12 illustrates an organizational structure in which the emphasis is on the functional end of the spectrum. In essence, the degree of "project" emphasis and "functional" emphasis often shifts back and forth, depending on both the size of the firm and the nature of the activity; that is, whether conceptual design, preliminary system design, or detailed design and development activities are in progress.

As shown in the figures, there may be a variety of approaches within the same firm. One or two large projects may exist along with numerous smaller projects. The large projects will tend to support an organizational structure similar to that presented in Figure 7.11, whereas the smaller projects will likely follow the format in Figure 7.12. Where the larger projects can afford to support significant numbers of personnel on a full-time basis, the smaller projects may be able to support a select number of individuals on only a part-time basis. The specific requirements are dictated through the generation of program tasks by the project organization; that is, a request for assistance is initiated by the project manager, with the task(s) being completed within the functional department.

374

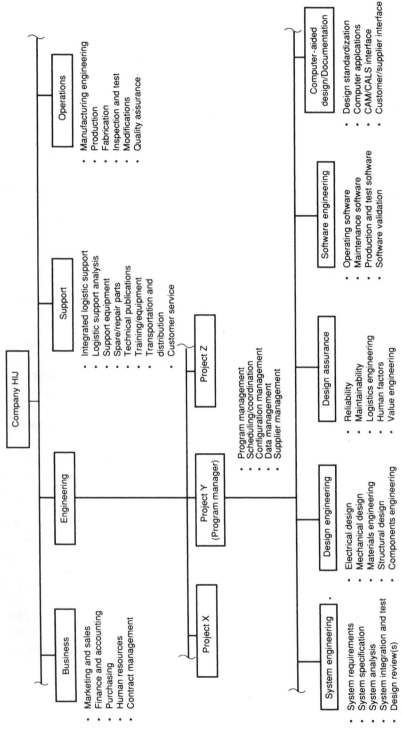

FIGURE 7.11 Producer organization (combined project-functional structure).

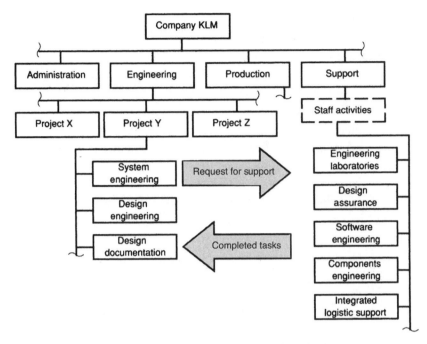

FIGURE 7.12   Producer organization (work flow).

Project size will vary not only with the type and nature of the system being developed, but also with the specific stage of development. A large-scale system in the early stages of conceptual design may be represented by a small project organization, as shown in Figure 7.12. As system development progresses into the phases of preliminary system design and detail design and development, the organization structure may shift somewhat, replicating the configuration in Figure 7.11. In other words, the characteristics and structures of organizations are often dynamic by nature. The organization structure must be adapted to the needs of the project at the time, and these needs may shift as system development evolves.

In regard to system engineering, the tasks identified in Figure 6.6 (Chapter 6) can be allocated by phase, as follows:[6]

1. *Conceptual design phase:*
   (a) Perform needs analysis and conduct feasibility studies.
   (b) Define operational requirements, the system maintenance concept, technical performance measures (TPMs), and accomplish system-level functional analysis.

---

[6]It is not intended to imply that the system engineering organization does everything. The emphasis here is on providing a *technical* thrust and assuming a *technical* leadership role in the design and development of the system. The project/program manager must, of course, provide the necessary leadership from the overall organizational standpoint.

    (c) Accomplish system integration.

    (d) Prepare the system specification (type A).

    (e) Prepare the Test and Evaluation Master Plan (TEMP).

    (f) Prepare the System Engineering Management Plan (SEMP).

    (g) Plan, coordinate, and conduct the conceptual design review.

2. *Preliminary system design phase:*

    (a) Accomplish functional analysis and the allocation of requirements.

    (b) Accomplish system analysis, synthesis, and trade-off studies.

    (c) Accomplish system integration; that is, the integration of design disciplines, supplier activities, and data.

    (d) Plan, coordinate, and conduct system design reviews.

3. *Detail design and development phase:*

    (a) Accomplish system analysis, synthesis, and trade-off studies.

    (b) Accomplish system integration; that is, the integration of design disciplines, supplier activities, and data.

    (c) Monitor and review system test and evaluation activities.

    (d) Plan, coordinate, implement, and control design changes.

    (e) Plan, coordinate, and conduct equipment/software design reviews and the critical design review.

    (f) Initiate and maintain production and/or construction liaison and customer service activities.

4. *Production and/or construction phase:*

    (a) Monitor and review system test and evaluation activities.

    (b) Plan, coordinate, implement, and control design changes.

    (c) Maintain production and/or construction liaison and conduct customer service activities (i.e., field service).

5. *System utilization and life-cycle support:*

    (a) Monitor and review system test and evaluation activities.

    (b) Plan, coordinate, implement, and control design changes.

    (c) Conduct customer service activities (i.e., field service).

    (d) Collect, analyze, and process field data. Prepare reports on system operations in the user environment.

6. *System retirement and material disposal:*

    (a) Prepare plan for system retirement.

    (b) Monitor material disposal and recycling activities.

    (c) Prepare reports on environmental impact(s).

As noted in Section 6.2.2, these tasks reflect what is envisioned as being critical to the successful implementation of the system engineering process. This does not mean to imply a tremendous level of effort. The tasks must be tailored to the particular applications, and the accomplishment of many of these requires a significant input from other organizations. A major objective is to provide a mechanism for *integration,* and the interfaces and relationships between system engineering and other organizations are extensive. For example, the conductance of a life-cycle cost analysis as part of the overall system analysis task involves a data exchange with all project-related engineering organizations, finance and accounting, logistic support, production, the customer, and so on. The goal of the system engineering organization is to provide the necessary technical management and guidance to ensure that these activities are completed in a timely manner.

To further illustrate the many interfaces that exist, the combined project-functional organization structure in Figure 7.11 has been extended to include a sampling of the required communication channels that must be established between the system engineering function and other organizational elements. These communication links are shown in Figure 7.13, and an abbreviated description covering the nature of the necessary communications is presented in Figure 7.14. The system engineering function, as an integrating agency, must not only provide the technical leadership throughout the system development activity, but must initially establish and subsequently maintain open and free-flowing communications across the board.

Ultimate success in meeting these system engineering objectives is, of course, highly dependent on managerial support from the top down. The president (or general manager), the vice president of engineering, the project Y manager, and other high-level managers must each *understand* and *believe in* the concepts and objectives of system engineering. If the system engineering manager is to be successful, these higher-level managers must be directly supportive all the time. There will be many occasions when individual design engineers and/or middle managers will go off on their own, making decisions that will conflict with system engineering objectives. When this occurs, the system engineering manager must have the necessary support to ensure that actions are taken to "get things back on track."

7.5 SUPPLIER ORGANIZATION AND FUNCTIONS

As defined in Section 6.3, the term *supplier* refers to a broad category of organizations that provide various materials and/or services to the producer (i.e., prime contractor). These items may range from the design and development of a major subsystem to the delivery of a small off-the-shelf part from an existing inventory or the performance of some service. The process for identifying outsourcing requirements, identifying potential sources of supply and the selection of a qualified supplier, and the follow-on contracting activity is described in Chapter 6. The discussion is extended here with a consideration of the supplier organization.

For relatively large projects (involving design and development, manufacturing and production, etc.), the system engineering function may be established at different levels. As shown in Figure 7.1, the customer may establish a system engineering group and a similar function may be included in the producer's organization (i.e., the contractor). Where the outsourcing requirements are significant and suppliers

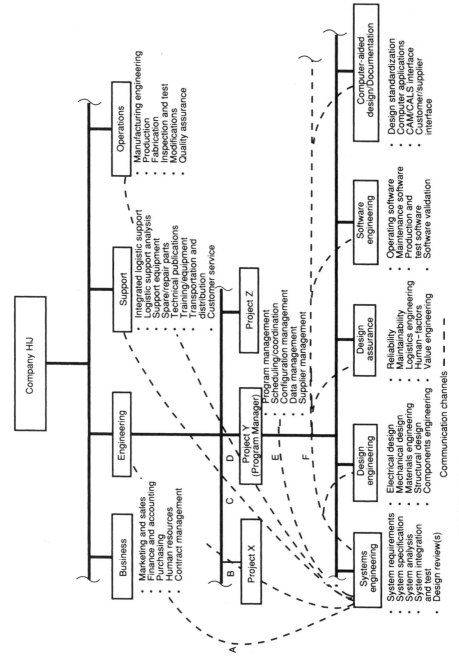

FIGURE 7.13 Major system engineering communication links (producer organization).

Company HIJ

Business
- Marketing and sales
- Finance and accounting
- Purchasing
- Human resources
- Contract management

Engineering

Support
- Integrated logistic support
- Logistic support analysis
- Support equipment
- Spare/repair parts
- Technical publications
- Training/equipment
- Transportation and distribution
- Customer service

Operations
- Manufacturing engineering
- Production
- Fabrication
- Inspection and test
- Modifications
- Quality assurance

Project X

Project Y (Program Manager)
- Program management
- Scheduling/coordination
- Configuration management
- Data management
- Supplier management

Project Z

Systems engineering
- System requirements
- System specification
- System analysis
- System integration and test
- Design review(s)

Design engineering
- Electrical design
- Mechanical design
- Materials engineering
- Structural design
- Components engineering

Design assurance
- Reliability
- Maintainability
- Logistics engineering
- Human-factors
- Value engineering

Software engineering
- Operating software
- Maintenance software
- Production and test software
- Software validation

Computer-aided design/Documentation
- Design standardization
- Computer applications
- CAM/CALS interface
- Customer/supplier interface

Communication channels — — —

Communication Channel (Figure 7.13)	Supporting Organization (Interface Requirements)
A	1. *Marketing and sales*—to acquire and sustain the necessary communications with the customer. Supplemental information pertaining to customer requirements, system operational and maintenance support requirements, changes in requirements, outside competition, etc., is needed. This is above and beyond the formal "contractual" channel of communications. 2. *Accounting*—to acquire both budgetary and cost data in support of economic analysis efforts (e.g., life-cycle cost analysis). 3. *Purchasing*—to assist in the identification, evaluation, and selection of component suppliers with regard to technical, quality, and life-cycle cost implications. 4. *Human resources*—to solicit assistance in the initial recruiting and hiring of qualified project personnel for system engineering, and in the subsequent training and maintenance of personnel skills; to conduct training programs for all project personnel across the board relative to system engineering concepts, objectives, and the implementation of program requirements. 5. *Contract management*—to keep abreast of contract requirements (of a technical nature) between the customer and the contractor; to ensure that the appropriate relationships are established and maintained with suppliers as they pertain to meeting the *technical* needs for system design and development.
B	To establish and maintain ongoing liaison and close communications with other projects with the objective of transferring knowledge that can be applied for the benefit of project Y; to solicit assistance from other companywide functionally oriented engineering laboratories and departments relative to the application of new technologies in support of system design and development.
C	To provide an input relative to project requirements for system support, and to solicit assistance in terms of the functional aspects associated with the design, development, test and evaluation, production, and sustaining maintenance of a support capability through the planned system life cycle.
D	To provide an input relative to project requirements for production (i.e., manufacturing, fabrication, assembly, inspection and test, and quality assurance), and to solicit assistance relative to the design for producibility and the implementation of quality engineering requirements in support of system design and development.
E	To establish and maintain close relationships and the necessary on going communications with such project activities as scheduling (the monitoring of critical program activities through a network scheduling approach); configuration management (the definition of various configuration baselines and the monitoring and control of changes/modifications); data management (the monitoring, review, and evaluation of various data packages to ensure compatibility and the elimination of unnecessary redundancies); and supplier management (to monitor progress and ensure the appropriate integration of supplier activities).
F	To provide an input relative to *system-level* design requirements, and to monitor, review, evaluate, and ensure the appropriate integration of system design activities. This includes providing a *technical* lead in the definition of system requirements, the accomplishment of functional analysis, the conductance of system-level trade-off studies, and the other project tasks presented in Figure 6.6.

FIGURE 7.14   Description of major project interface requirements.

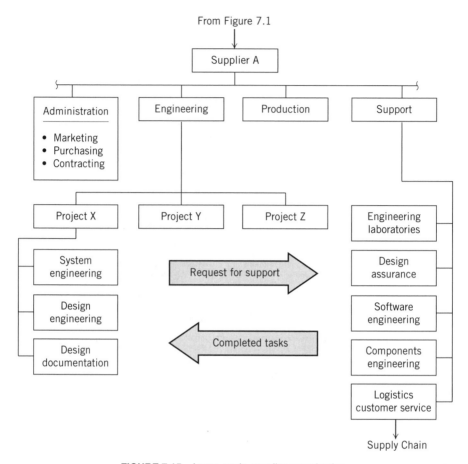

FIGURE 7.15   Large-scale supplier organization.

are selected to accomplish the design, development, and manufacture of subsystems and/or major components, the system engineering capability must be extended and included as an identifiable function within the supplier's organization as well. For the purposes of further discussion, assume that the organizational structure for a major supplier will take the form illustrated in Figure 7.15 (an extension of Figure 7.1). It is important to keep in mind that system engineering requirements must be traceable from the customer on down to the major supplier.

Referring to Figure 7.15, the system engineering activity need not be large and may include only one or two key individuals. In any event, there must be a *focal point* within the supplier's organization to ensure that the applicable system engineering tasks are performed in a timely and efficient manner. Such tasks may include the following:

1. Conduct feasibility studies and define specific design criteria for the system component(s) being developed. This information is based on system

operational requirements, the maintenance concept, functional analysis, and the allocation of requirements as they apply to the supplier-produced item. Supplier requirements in this instance are included in a development specification (type B)—see Figures 3.2 and 6.15.

2. Prepare a supplier engineering plan (or equivalent). This plan should constitute an expansion of the SEMP and must reflect those requirements that support system engineering objectives.

3. Accomplish synthesis, analysis, and trade-off studies in support of component design decisions, and as they impact higher-level system requirements.

4. Accomplish design integration activities; that is, ongoing coordination and integration of design disciplines, activities, and data.

5. Prepare and implement a test and evaluation plan covering the system component(s) being developed. Integrate testing activities into system-level test requirements where feasible. Monitor, review, and evaluate component testing activities conducted at the producer's facility.

6. Participate in equipment/software design reviews and critical design reviews as applicable; that is, formal design reviews that cover the system component(s) being developed and its interfaces with other elements of the system.

7. Review and evaluate proposed design changes as they apply to the component(s) and impact the overall system.

8. Initiate and maintain liaison with production/manufacturing activities that support the system component(s).

9. Initiate and maintain liaison with the producer (i.e., prime contractor) through out all phases of the program when supplier activities are in progress.

Although large projects may require the supplier to complete all of these tasks, the level of activity will obviously be scaled down for smaller projects. It must be emphasized (once again) that the system engineer is not be expected to personally complete all of these activities, but will be required to assume a leadership role to ensure that they are accomplished within the supplier's overall organizational structure.

For suppliers involved in production or manufacturing only, the system engineering focus is directed primarily to total quality management (TQM). It is important to ensure that the characteristics initially designed into system components are maintained through the subsequent production of multiple quantities of these items. As the production process can be highly dynamic and material substitutions often occur, it is necessary to guarantee that each of the components produced does indeed reflect the quality characteristics initially specified through design (refer to Section 3.4.10). Thus, close communications must be established with the supplier's quality control or quality assurance organization.

In dealing with suppliers of standard off-the-shelf components, it is important to prepare a good specification for the initial procurement of these items. Input-output parameters, size and weight, shape, density, and other key performance parameters must be covered in detail, along with allowable tolerances. Uncontrolled variances in component characteristics can have a significant impact on total system effectiveness and quality. Complete electrical, mechanical, physical, and functional interchangeability must be maintained where applicable. Although there are many different components that are currently in the inventory and fall within the category of "commercial off-the-shelf" (COTS), extreme care must be exercised to ensure that like components (i.e., those with the same part number) are actually manufactured to the *same standards*. Moreover, the allowable variances around key parameters must be minimized.

It must also be kept in mind that there are many different categories of suppliers and that there is often a layering effect, as illustrated in Figure 6.31. In addition, many of these suppliers may be located across the United States, Canada, Mexico, Europe, Africa, Asia, Australia, and South America (see Figure 6.30). It is anticipated that supplier organizational structures will vary from one instance to the next. Thus, an objective is to identify the particular organizational group (or individual) who has been assigned responsibility for the completion of the required system engineering tasks and to establish the necessary communications and working relationships that will produce effective results.

7.6 HUMAN RESOURCE REQUIREMENTS

In considering organizational elements as they pertain to system engineering, it is necessary to address the human resource requirements, the staffing of a system engineering organization, leadership characteristics and motivational factors, and personnel development. Although there will be variations in each situation, there are certain common objectives that should be met from the employer/employee standpoint.

7.6.1 Creating the Proper Organizational Environment

As discussed in Gibson, Donnelly, Ivancevich, and Konopaske, the study of organizations should address *structure, processes,* and *culture.*[7] Accordingly, "structure is the formal pattern of how an organization's people and jobs are grouped." Structure is usually illustrated in the form of an organizational chart, such as those described in Sections 7.4 and 7.5. Further, "processes are activities that give life to the organization chart. Communications, decision making, and organization development are examples of processes in organizations." Such processes have been highlighted and covered throughout the earlier chapters of this text. The third element is *culture,* which deals with an organization's personality, atmosphere, environment, and the like: "The culture of an organization defines appropriate behavior and bonds, it motivates individuals, and governs the way a company processes information, internal relations, and values. It functions at all levels from the subconscious to the visible. . . . Cultures of organizations can be positive or negative. An organization's culture is positive if it helps to improve productivity. A negative culture can hinder behavior, disrupt group effectiveness, and hamper the impact of a well-designed organization."

---

[7]J. L. Gibson, J. H. Donnelly, J. M. Ivancevich, and R. Konopaske, *Organizations: Behavior, Structure, Processes,* 12th ed. (New York: McGraw-Hill Irwin, 2005).

At this point, it is important to address the issue of "culture" through the design and development of an organization and establishing its human resource requirements, leadership characteristics, staffing, motivation factors and the development of its personnel, and so on. With regard to system engineering, the successful accomplishment of the objectives and specific tasks described throughout this text is highly dependent on the overall capabilities and environment that have been created within an organization. On the one hand, there are numerous instances in which organizations have been established and designated as the "system engineering department" or the "system engineering group," in which the requirements are not being met. On the other hand, there are other instances in which such organizations have been very successful. The key is to establish a structure and its processes and to create an environment that is positive. This requires promoting communications throughout a given project structure, acquiring the necessary respect in terms of technical capability and assuming a leadership role for the project, and being able to influence the system design and development process on a continuing basis. This may be particularly challenging because the system engineering organization must accomplish much without having control of all of the resources required to complete the job.

The nature of system engineering activities requires consideration of the following characteristics in developing an organizational structure:

- The personnel selected for the system engineering group must, in general, be highly professional senior-level individuals with varied backgrounds and a wide breadth of knowledge—for example, an understanding of research, design, manufacturing, and system support applications. The emphasis is on overall system-level design and technology applications, with knowledge of user operations and sustaining life-cycle maintenance and support in mind.

- The system engineering group must have "vision" and be creative in the selection of technologies for design, manufacturing, and support applications. Group personnel must constantly search for new opportunities and must be innovative, and applied research is often required in order to solve specific technical problems.

- A teamwork approach must be initiated within the system engineering group. The personnel assigned must be committed to the objectives of the organization; a certain degree of interdependence is required, and there must be mutual trust and respect.

- A high degree of communication must prevail, both within the system engineering group and with the many other related functions associated with a given project (refer to Figure 7.13). Communication is a two-way process and may be accomplished via written, verbal, and/or nonverbal means. Good communications must first exist within the system engineering organization. With that established, it is then necessary to develop two-way communications externally, using both vertical and horizontal channels as required.

Given the objectives described throughout this text, and with these considerations in mind, the appropriate "environment" must be created to allow for the accomplishment of system engineering tasks in an effective manner. *Environment* in this instance refers to both (1) the working environment external to the system engineering function but within the contractor's organizational structure and (2) the working environment within the system engineering group itself.

The creation of a favorable environment within the contractor's organization, or within any other organizational structure being addressed, must start from the *top*. The president, or general manager, must initially believe in and subsequently support the concepts and objectives of system engineering. On numerous occasions, power struggles will occur, conflicting organization goals and objectives will develop, and there will be a lack of communication between the key organizational entities. This, in turn, usually leads to redundancies, the hiring of unqualified personnel, and the expenditure of unnecessary resources, resulting in waste. A mechanism must be established for quick conflict resolution, and all project personnel must know that the system engineering philosophy *will* prevail, in spite of individual interests. Top management must create this understanding from the beginning.

In addition, the appropriate level of responsibility and authority must be delegated from the top, through the vice president of engineering and the program manager, to the system engineering department manager (refer to Figure 7.11). Responsibilities must be defined, and the commensurate level of authority to control and direct the means by which the activity is to be accomplished must be delegated accordingly. Quite often, a manager is willing to delegate responsibility for a particular activity to a subordinate, but will retain the authority for controlling the resources necessary for task completion. In this situation, the individual assigned the responsibility is powerless when it comes to the full utilization of available resources and, when things go wrong, is unable to respond relative to initiating the appropriate corrective action. He or she becomes discouraged, loses motivation, and the level of productivity ultimately falls off. In essence, the system engineering manager must be given the *responsibility,* the *authority,* and the *resources* to do the job assigned.

It is also important that the proper relationships be established between the vice president of engineering, the program manager, and the system engineering manager, as identified in Figure 7.11. Such relationships are, of course, heavily dependent on the managerial styles of the individuals in these positions. Although there are many variations, the two managerial styles most often discussed are the *autocratic* approach and the *democratic* approach.

The autocratic approach is basically dictatorial in nature and is restrictive in that decisions are generally unilateral; that is, decisions are made from the top down without input from those who are required to carry out these decisions. Managers control, direct, coerce, and even threaten employees to force them to work toward specific organizational goals. However, the democratic concept is participative and nonthreatening, and organizational interests are group-centered. The general theme is that individuals in the group have some voice in matters that directly affect them.[8]

---

[8]These concepts are related to the managerial views described as "Theory X" and "Theory Y" by Douglas McGregor in his classic book, *The Human Side of Enterprise* (New York: McGraw-Hill, 1960). This is recommended reading for a more in-depth understanding of the human relations movement developed in the 1940s.

Although both styles of management are prevalent for certain situations, the democratic style has been accepted as being more effective from a motivational standpoint. In general, people work harder, are more cooperative, and are more willing to accept changes if they feel that they have some influence on the results. Democratic leadership implies an organizational environment in which employees have a chance to grow and develop their skills, where formal supervision is considerate and the application of dictates is not arbitrary, and where individual opinions are solicited and respected. With the democratic approach, as compared with the autocratic approach, management is committed to the recognition of employees as high-level professionals and not merely as factors in the production scheme.[9]

In regard to system engineering, an environment must be created that will allow for individual initiative, creativity, flexibility, personnel growth, and so on. A democratic and participative approach seems appropriate in meeting the objectives stated. Although the manager must maintain authority and provide the necessary direction and control to effectively accomplish the organization's goals and objectives, he or she can introduce some practices that directly support the democratic style. The selected style, it is hoped, will help to create a favorable environment for the accomplishment of system engineering tasks. This is influenced not only by the managerial style employed within the system engineering group itself, but the approach employed by higher-level management that has an impact on the group. Creating such an environment is critical to the objectives stated throughout this text. There are examples in which two (or more) similar organizations have the same basic objectives, the same structure, the same position titles, and so on—but one is productive and the other is not. High-level productivity is a function of the working environment.

7.6.2 Leadership Characteristics

The system engineering organization is composed of a group of individuals with varying abilities, different roles and expectations, diverse personal goals, and distinct behavioral patterns. Although individuals within the organization are highly dependent on one another, they often push in different directions because of some of these factors. The challenge for the system engineering manager is to integrate these various characteristics into a cohesive force, leading to the accomplishment of organizational objectives. The manager must not only ensure that the job is completed in a satisfactory manner, but it is hoped that he or she will inspire and motivate his or her subordinates to *excel* in the fulfillment of organizational objectives. It is apparent that the manager must create a climate that presents challenges (not threats) for the individuals involved.

In facilitating this objective, the manager should initiate certain practices that are responsive to both the organization and the individual needs of people within the

---

[9]A review of the literature on human motivation is recommended at this point. Three good references are (1) A. H. Maslow, "A Theory of Human Motivation," *Psychological Review* (July 1943); (2) F. Herzberg, B. Mausner, and B. Snyderman, *The Motivation to Work* (Hoboken, NJ: John Wiley & Sons, Inc., 1959); (3) M. S. Myers, "Who Are Your Motivated Workers," *Harvard Business Review* (1970).

organization. As a start, the democratic style of leadership, discussed in Section 7.6.1, tends to promote the necessary environment for the organization. Within this framework, the manager should encourage the participation of individuals in goal setting and in decision making, and should promote communications. By doing so, he or she will not only tend to improve motivation from within, but will also acquire a better understanding of the individuals in the organization. This understanding can be enhanced by taking the following steps:

1. Recognize the personal characteristics of each individual in the organization in order to better match the individual with the job requirements. A person may excel in one situation, but do only a mediocre job in another situation, even though the ability to do both and the overall organization climate remain relatively constant. In essence, the manager must assign employees to the type of work they do most effectively. A high-quality output is essential if the system engineering organization is to gain respect and retain a leadership role on a project.

2. Inspire each individual to excell in his or her job by creating an atmosphere of personal interest. An employee tends to perform better if he or she knows that the supervisor is personally interested. Personal interest is developed through involvement at the employee level.

3. Be sensitive to employee problems related to their work such so that each can be addressed on personal terms. The solution to a problem should, if at all possible, consider the effects on the individual employee.

4. Evaluate employees on a personal basis and initiate rewards promptly when warranted. Promotions and merit raises should not be given to the organizational hierarchy alone, but should be directed to the best performers.

Good communications and good rapport with employees must be established from the beginning. Obtaining the desired organization environment not only depends on the intent of the manager in initially establishing such practices, but is highly dependent on the manager's personal leadership ability in directing the activities of the organization over time. The best planning in the world will have little benefit unless the actions that follow in the implementation phase are directly supportive. As a goal, the manager should strive to exhibit the characteristics listed in Figure 7.16.

7.6.3 The Needs of the Individual

With the discussion thus far primarily covering the organization environment and the desired characteristics of leadership, it is now important that some attention be given to the needs of the individual employee. If the manager is to inspire, motivate, and deal successfully with subordinates, then a good understanding of these needs is necessary.

As a starting point, the reader should first review A. H. Maslow's theory dealing with the hierarchy of needs. The theory was developed to identify relatively

Leadership Characteristics
1. Acceptance: Earns respect and has the confidence of others.
2. Accomplishment: Effectively uses time in meeting goals and objectives.
3. Acuteness: Mentally alert and readily comprehends instructions, explanations, and unusual circumstances.
4. Administration: Organizes his or her own work and that of his or her subordinates; delegates responsibility and authority; measures, evaluates, and controls position activities.
5. Analysis and judgment: Performs critical evaluation of potential and current problem areas; breaks problem into components; weighs alternatives; relates, and arrives at sound conclusions.
6. Attitude: Enthusiastic; optimistic; and loyal to firm/agency, superior, position, and associates.
7. Communication: Promotes communication within and between organization elements.
8. Creativeness: Has inquiring mind; develops original ideas; and initiates new approaches to problems.
9. Decisiveness: Makes prompt decisions when necessary.
10. Dependability: Meets schedules and deadlines in consistent manner and adheres to firm/ agency policies and procedures.
11. Developing others: Develops competent successors and replacements.
12. Flexibility: Adaptable; quickly adjusts to changing conditions; and copes with the unexpected.
13. Human relations: Is sensitive to and understands personnel interactions; has "feel" for individuals and recognizes their problems; considerate of others; ability to motivate and get people to work together.
14. Initiative: Self-starting; prompt to take hold; seeks and acts on new opportunities; exhibits high degree of energy in work; not easily discouraged; and possesses basic urge to get things done.
15. Knowledge: Possesses knowledge (breadth and depth) of functional skills needed to fulfill position requirements; uses information and concepts from other related fields of knowledge; and generally understands the "big picture."
16. Objectivity: Has an open mind and makes decisions without the influence of personal or emotional interests.
17. Planning: Looking ahead; developing new programs; preparing plans; and scheduling requirements.
18. Quality: Accuracy and thoroughness of work; maintains high standards consistently.
19. Self-confidence: Self-assurance; inner security; self-reliance; takes new developments in stride.
20. Self-control: Calm and poised under pressure.
21. Self-motivation: Has well-planned goals; willingly assumes greater responsibilities; realistically ambitious; and generally eager for self-improvement.
22. Sociability: Makes friends easily; works well with others; and has sincere interest in people.
23. Verbal ability: Articulate; communicative; and is generally understood by persons at all organizational levels.
24. Vision: Possesses foresight; sees new trends and opportunities; anticipates future events; and is not bound by tradition or custom.

FIGURE 7.16   Checklist of leadership characteristics.

separate and distinct drives that motivate individuals in general. The following five basic needs are identified:[10]

1. *The physiological needs, such as thirst, hunger, sex, sleep, and activity.* These constitute the needs of the body, and unless these needs are basically satisfied, they remain the prime influencing factors in the behavior of an individual.

[10] A. H. Maslow, "A Theory of Human Motivation," *Psychological Review* (July 1943).

2. *The safety and security needs, which include protection against danger, threat, and deprivation.* Having satisfied the bodily needs, the safety and security needs become a dominant goal.

3. *The need for love and esteem by others, or social needs.* This includes belonging to groups, giving and receiving friendship, and the like.

4. *The need for self-esteem or self-respect, and for the respect of others (i.e., ego need).* An individual wishes to consider him- or herself strong, able, competent, and basically worthy by his or her own standards.

5. *The need for self-fulfillment or the achieving of one's full potential through self-development, creativity, and self-expression.* This relates to the human desire to grow, develop to the point of full potential, and ultimately attain the highest level possible.

The needs are related to each other, and are ordered in a manner that will stimulate consciousness and activity. As a need becomes satisfied, activity emphasis shifts to the next need category. In other words, a satisfied need is no longer a motivator, and the next need becomes a driving factor.

Whereas Maslow addressed the overall hierarchy of needs from a general standpoint, Herzberg's research resulted in his *motivation-hygiene* theory, which identifies factors commonly known as "satisfiers" and "dissatisfiers." The theory was developed from research pertaining to the job attitudes of 200 engineers and accountants, and was based on two questions: (1) "Can you describe, in detail, when you felt exceptionally *good* about your job?" and (2) "Can you describe, in detail, when you felt exceptionally *bad* about your job?" The results were classified in two categories as follows:[11]

1. *Satisfiers:*

   (a) Achievement—personal satisfaction in job completion and problem solving.

   (b) Recognition—acknowledgment of an accomplishment (e.g., a job well done)

   (c) Work itself—actual content of the job and its positive/negative effect on the employee

   (d) Responsibility—both responsibility and authority in relation to the job

   (e) Advancement—promotion on the job

   (f) Growth—learning new skills offering greater possibility for advancement

2. *Dissatisfiers:*

   (a) Company policy and administration—feelings about the adequacy or inadequacy of company organization and management, policies and procedures, and so on.

---

[11]F. Herzberg, "Work and Motivation," *Studies in Personnel Policy Number 316: Behavioral Science, Concepts and Management Application* (New York: National Industrial Conference Board, 1969).

(b) Supervision—competency or technical ability of supervision

(c) Working conditions—physical environment associated with the job

(d) Interpersonal relations—relations with supervisors, subordinates, and peers

(e) Salary—pay and fringe benefits

(f) Status—miscellaneous items such as size of office, having a secretary, and a private parking place

(g) Job security—tenure, company stability or instability

(h) Personal life—personal factors that affect the job (e.g., family problems, social problems)

Most of these factors have some bipolar effects. For instance, "advancement" certainly is a *satisfier* when it happens and may be somewhat of a *dissatisfier* when it does not occur. However, this category weighs much more heavily as a satisfier. The item "salary" is definitely a dissatisfier when pay scales and fringe benefits are poor, and is a mild satisfier when the compensation is good. In this case, the predominant classification for salary is that of a dissatisfier. In any event, all of the factors listed represent needs of the individual and should be considered by management.

Myers conducted research involving 282 subjects (including engineers, scientists, and technicians) interviewed at Texas Instruments commencing in 1961. The categories used were "motivators" and "dissatisfiers," and the results as they pertain to engineers are noted in the following list. The items listed and identified with an "M" are clearly motivators, and those with a "D" are definitely dissatisfiers. For instance, the item "pay" is clearly a dissatisfier if considered as being inadequate, and "advancement" is definitely a motivator when it occurs. Once again, there are bipolar effects; however, the categorization does indicate where the greatest impact occurs.[12]

1. Work itself (M)
2. Responsibility (D)
3. Company policy and administration (D)
4. Pay (D)
5. Advancement (M)
6. Recognition (D)
7. Achievement (M)
8. Competence of supervision (D)
9. Friendliness of supervision (D)

In summary, the needs of the individual employee will vary somewhat, depending on his or her situation. If one need is satisfied, then another need becomes predominant, and so on. In addition, these needs are often related to the business position

[12]M. S. Myers, "Who Are Your Motivated Workers," *Harvard Business Review* (1970). This study employs factors comparable to those used by Herzberg.

of the company in which the person is employed (or the success of the organization overall). If the firm is in a growth posture, the individual's perceived needs may be somewhat different than if the firm is experiencing a business decline and the prospect of laying off employees is apparent. Finally, the manager's job is twofold. He or she must (1) be aware of the individual needs in the organization and create the necessary conditions for employee motivation and (2) satisfy those needs on a continuing basis to the extent possible. Human motivation is a key to organizational success, and an understanding of the concepts in this section should help in meeting this objective.[13]

7.6.4   Staffing the Organization

The requirements for staffing an organization initially stem from the results of the system engineering planning activity described in Chapter 6. Tasks are identified from both short- and long-range projections (refer to Figure 6.25), combined into work packages and the work breakdown structure (WBS), and the work packages are grouped and related to specific position requirements. The positions are, in turn, arranged within the organizational structure considered to be most appropriate for the need (refer to Figures 7.2 through 7.15).

With regard to determining specific position requirements for a system engineering organization, one should first have a good understanding of the basic functions of the organization. These are discussed throughout the earlier chapters of this text and, more specifically, in Chapter 6. A review of the assigned tasks, the nature and challenges of the organizational structure, and so on, indicate that, in general, an entry-level "system engineer" should have the following:

- A basic formal education at the undergraduate and graduate levels in some recognized field of engineering; that is, a master's degree in engineering or equivalent.[14]
- A high level of general technical competence in the engineering fields being pursued by the organization, project, and so on.
- Relevant design experience in the appropriate areas of activity. For example, if the company is involved in the development of electrical/electronic systems, then it is desirable for the candidate to have had some prior design experience in electrical/electronic systems. A different type of experience would be required for aeronautical systems, civil systems, hydraulic systems, and so on.

---

[13]A comparison of four "content theories of motivation" is presented in J. L. Gibson, J. H. Donnelly, J. M. Ivancevich, and R. Konopaske, *Organizations: Behavior, Structure, Processes,* 12th ed. (New York: McGraw-Hill Irwin, 2005). This includes a discussion of Maslow's Need Hierarchy, Alderfer's ERG Theory (existence, relatedness, and growth), Herzberg's Two-Factor Theory, and McClelland's Learned Needs Theory (achievement, power, and affiliation). Only two of these are covered in this chapter.

[14]Recognized accredited programs in engineering are defined by the Accreditation Board for Engineering and Technology (ABET), United Engineering Center, 345 East 47th Street, New York, NY 10017. Refer to the latest *Annual Report.*

- A basic understanding of the design requirements in areas such as reliability engineering, maintainability engineering, human factors, safety engineering, logistics engineering, software engineering, quality engineering, and value/cost engineering.
- An understanding of the system engineering process and the methodologies/ tools that can be effectively employed in bringing a system into being; for example, the definition of system requirements and functional analysis and allocation.
- An understanding of the relationships between functions, including marketing, contract management, purchasing, integrated logistic support, configuration management, production (manufacturing), quality control, customer and supplier operations, and so on.

As the specific definition of a "system engineer" often varies from one organization to the next, individual perceptions as to the requisites will differ. Based on experience, it is believed that a good solid technical engineering education is a necessary foundation, some design experience is essential, knowledge of the customer (user) environment a thorough understanding of the system life cycle and its elements is required, and knowledge of the many design interfaces is appropriate. If an individual is to successfully implement the functions identified in Chapter 6 (Figure 6.6), then some prior experience in these areas is highly recommended.

Given the basic requisites, the system engineering department manager will prepare an individual position description for each open slot in the organization. A sample position description format is illustrated in Figure 7.17. The position title, responsible supervisor, areas of responsibility and job objectives, background requirements, and the date of need should be clearly identified. The system engineering position requirements are completed and forwarded to the human resources department (or equivalent) in order to proceed with the necessary steps for recruiting and employment.

In staffing an organization, possible sources include (1) qualified personnel from within the company and ready for promotion and (2) personnel from outside and available through the open market. It is the responsibility of the system engineering department manager to work closely with the human resources department in establishing the initial requirements for personnel, in developing position descriptions and advertising material, in recruiting and the conducting of interviews, in the selection of qualified candidates, and in the final hiring of individuals for employment within the system engineering organization. In the process of conducting interviews and selecting system engineering personnel, the characteristics identified in Figure 7.16 should be kept in mind.[15]

---

[15]The human resources department in most companies is responsible for establishing job classifications and salary structures, for the recruiting and hiring of personnel, for initiating employee benefit coverage, for providing employee opportunities for education and training, and so on. It is incumbent upon the system engineering department manager to ensure that his or her organizational requirements are initially understood and subsequently met through recruiting, employment, and training activities.

```
                                              Date of need:

Position title:                               Supervisor:

Senior system engineer-communications        System Engineering
                                              Department Head


Broad Function:

Responsible for the performance of system engineering functions in the design and development of
communication products.

Functional Objectives:

1.  Perform system feasibility studies and evaluate alternative technology applications.
2.  Develop operational requirements and maintenance concepts for new communication systems/
    equipment.
3.  Interpret and translate system-level requirements into functional design requirements.
4.  Prepare system and subsystem specifications and plans.
5.  Accomplish system integration activities (including supplier functions).
6.  Determine requirements and conduct formal design reviews for all system elements.
7.  Prepare system test and evaluation requirements, monitor test functions, and evaluate test
    results to determine system performance and effectiveness. Make recommendations for
    corrective action and/or improvement as appropriate.
8.  Provide assistance to marketing in product sales activities, and fulfill customer service
    requirements as necessary.

Requirements:

Degree in electrical engineering (master's degree or higher and some training in managment skills
and practices), plus at least 10 years of experience in communication systems design.
```

FIGURE 7.17 Sample position description.

7.6.5 Personnel Development and Training

Nearly every engineer wants to know how he or she is doing on a day-to-day basis and what are the opportunities for growth. Response to the first part of the question is derived from a combination of the "formal performance review," which is often conducted on a regularly scheduled basis (either semiannually or annually), and the ongoing day-to-day "informal communications process" with the supervisor. The engineer is given responsibility and seeks recognition and approval from the supervisor. As discussed in Section 7.6.2, there must be close communication, and the supervisor must provide some reinforcement that the employee is doing a good job. The employee also needs to know as soon as possible when his or her work is unsatisfactory and improvement is desired. Waiting until the formal performance review is conducted to indicate that the employee's work is not satisfactory is a poor practice. It is also demoralizing because, by virtue of not having heard any comments to the contrary, the employee assumes that all is well. In a system engineering organization, it is particularly important for the appropriate close level of communication to be established from the beginning.

The second part of the question, pertaining to opportunities for growth, depends on (1) the climate provided within the organization and the actions of the manager that allow for individual development, and (2) the initiative on the part of the engineer to take advantage of the opportunities provided. Within a system engineering department, it is *essential* that individual personal growth take place if that department is to function effectively. The climate (or environment) must allow for individual development, and the individual system engineer must seek opportunities accordingly. The system engineering department manager should work with each employee in preparing a tailored *development plan* for that employee. The plan, adapted to each person's specific needs, should allow for (and promote) personal development by providing a combination of the following four factors:

1. Formal internal training designed to familiarize the engineer with the policies and procedures applicable to the overall company as a whole, as well as the detailed operating procedures of his or her own organization. This type of training should enable the individual to function more successfully within the framework of the total organization through familiarization with the many interfaces that he or she will encounter on the job.

2. On-the-job training through selective project assignments. Although extensive shifting of personnel from job to job (or project to project) can be detrimental, it is sometimes appropriate to reassign an individual to work where he or she is likely to be more highly motivated. Every employee needs to acquire new skills, and occasional transfers may be beneficial as long as the overall productivity of the organization does not suffer.

3. Formal technical education and training designed to upgrade the engineer relative to the application of new methods and techniques in his or her own field of expertise. This pertains to the necessity for the engineer to maintain currency (and avoid technical obsolescence) through a combination of (a) continuing education short courses, seminars, and workshops, (b) formal off-campus graduate engineering programs provided at the local level (leading to an advanced degree), and (c) long-term training involving opportunities for research and advanced education on a university or college campus. The opportunities for acquiring continuing education while on the job are greater now than ever before, with the availability of the Internet, satellite TV, two-way compressed video (VTEL/PICTEL), videotapes, and computer-based delivery capabilities. If an individual is motivated, he or she can acquire a great deal of support in this area.[16]

4. A technical exchange of expertise with others in the field through participation in technical society activity, industry association activity, symposia and congresses, and the like.

---

[16]With the current computer technology, one can take courses and acquire a graduate degree (at the master's level) in system engineering, and in other fields, through the Internet without having to travel or leave home. Two of a number of software systems currently in place, and through which course material can be offered, are WebCT and Blackboard (for example).

The system engineering manager must recognize the need for the ongoing development of personnel in his or her organization and should encourage each individual to seek a higher level of performance by offering not only challenging job assignments, but opportunities for growth through education and training. The long-term viability of such an organization is highly dependent on personnel development. This, in turn, should enhance individual motivation and result in the fulfillment of system engineering functions to the highest quality standards.

7.7 SUMMARY

It should be noted that the successful implementation of system engineering requirements is not dependent on any one specific organizational structure. Although the various organizational "structures," and their advantages and disadvantages, are identified in Figures 7.2–7.13, a successful system engineering program can be realized through any one of these structures. However the accomplishment of such is dependent on: (1) providing the proper "environment" from the top-down which will allow system engineering principles and concepts to be implemented effectively and efficiently; (2) having the proper leadership that understands and believes in system engineering and the benefits that can be realized as a result of its implementation; (3) the establishment of a good communications capability throughout the entire organization, with the customer, and among the suppliers; and (4) incorporating an effective feedback and control capability that will permit periodic evaluation and allow for continuous process improvement.

From experience, there have been numerous occurrences where a designated "system engineering" organization has been established and assigned the responsibility for implementation of system engineering requirements, but where the results turned out to be a failure. In other words, there have been system engineering organizations established, but which have been unsuccessful for one reason or another (e.g., have not had the proper understanding and support from a higher level of management, have not established the appropriate level of communications across the board, have not "educated" others in the project as to the objectives and benefits of system engineering, and so on). On the other hand, there have been situations where a separately identified organizational entity has not been too apparent, but where the requirements have been very successfully implemented, primarily due to upper management support, good communications, a desire for a good quality product, and an appreciation throughout the project/company as to the benefits that can be derived through the implementation of good system engineering practices.

Finally, it should be emphasized that the successful implementation of system engineering requirements is not dependent on any single organizational entity, but is a project-wide (or program-wide) responsibility. Referring to Figure 7.13, while there is an identified "lead" system engineering group, the successful accomplishment of system engineering requirements is highly dependent on the communications and cooperation of all of the other organizational entities as shown. In this instance, the system engineering group will serve in a leadership role, but much of the "doing" takes place in other organizational groups. Again, it's a team approach that is essential.

QUESTIONS AND PROBLEMS

1. Describe what is meant by *organization*. What are its characteristics and objectives? What factors must be addressed to ensure the successful accomplishment of its objectives?

2. There are various types of organizational structures. Identify at least four different types, briefly describe each, and discuss some of the advantages and disadvantages of each.

3. Refer to Figure 7.1. Where is system engineering accomplished? Who is responsible for the accomplishment of system engineering functions? Identify some of the major concerns associated with the organizational relationships shown in the figure.

4. Based on your own experience, what type of an organizational structure is preferred from a system engineering perspective? Why? Construct an organization chart showing the system engineering group/department/section within the context of a company's (i.e., producer/contractor's) organizational structure. Identify the major organizational interfaces between system engineering and other major activities within the company.

5. Refer to Question 4. For the organizational structure developed, describe the system engineering tasks to be accomplished and identify "input" requirements and expected "output" results for each task.

6. From an organizational perspective, identify and describe some of the conditions that must exist in order to accomplish system engineering objectives in an effective and efficient manner.

7. In regard to organizational "environment," what factors must be considered to ensure the successful implementation of system engineering requirements?

8. Describe some of the major challenges associated with the management of supplier organizations and related activities.

9. Relative to the styles of management, what is meant by "Theory X" and "Theory Y?" Which is preferred for a system engineering organization?

10. What is meant by *organizational culture*? Why is it important? Why is it particularly important when operating in an international environment?

11. Assume that you, as the vice president of engineering, are planning to hire a new system engineering department manager. What leadership characteristics would you identify as being critical (identify in order of importance)?

12. Refer to Figure 7.16. Select and list in order of importance the top 10 characteristics, based on your own experience.

13. What are IPPD and IPT? Describe the purpose and objectives of each.

14. Refer to the results of Herzberg's research in Section 7.6.3. Based on your own experience, list the "satisfiers" and the "dissatisfiers" in order of importance. List some additional factors as you see fit. A bar chart showing the bipolar relationships is a good way to present your thoughts.

15. From your own perspective, describe the characteristics of a system engineer (background, experience, personal characteristics, motivational factors, etc.).

16. As a manager of a system engineering department, what steps would you take to ensure that your organization maintains a lead position relative to technical competency?

17. Refer to Figure 7.13. What steps would you take to ensure maximum cooperation with (and support from) design engineering? logistics support? software engineering? business? operations?

8

SYSTEM ENGINEERING PROGRAM EVALUATION

Inherent within and part of the overall system engineering management activity is a four-step process: (1) the initial definition of system requirements, (2) the ongoing activity of fulfilling these requirements through a good and effective system design and development effort, (3) the measurement, evaluation, and assessment of the results, and (4) providing feedback and taking any necessary corrective action to achieve or exceed the initially specified objectives.

The subject of requirements has been emphasized to a great extent in Chapters 2 and 3. More specifically, the development of system operational requirements, the maintenance concept, and the identification and prioritization of technical performance measures (TPMs), described in Sections 2.4 through 2.6, constitute the steps involved in defining the requirements for the system. These requirements (which are also described in the various design-related sections in Chapter 3) are then allocated and apportioned downward to the various subsystems and below, and are included in the appropriate specifications (refer to Section 3.1). Through this allocation process, and resulting from decisions pertaining to outsourcing, the requirements for each of the different suppliers are then determined. In essence, this is where the process starts; that is, in the *definition of requirements*.

The next step is to identify the tasks that must be accomplished and the organizational approach that must be implemented to meet the overall objectives that have been identified through the requirements definition process. The tasks that must be completed, and the available technology applications to facilitate this effort, are discussed throughout Chapters 3, 4, 5, and 6. Initially, there is a planning process, which is covered rather extensively in Chapter 6. Program task schedules and cost projections are initiated, supplier requirements are identified, contractual requirements are negotiated, and program review and reporting requirements are established. In Chapter 7 the emphasis is on "organization" and the approach proposed for implementation of a program designed to fulfill the desired objectives as stated. Having identified the "whats" (i.e., What must be accomplished?), the next question relates to the "hows" (i.e., How can this be best accomplished?). Basically, the planning and organizational activities are discussed in depth in Chapter 6 and 7, primarily addressing the first two items in the four-step process mentioned earlier.

The next issue is "measurement, evaluation, feedback, and taking corrective action as required" (i.e., the third and fourth steps in the process). *Measurement* means determining, through both informal and formal reporting, the degree to which progress toward the meeting the objectives (requirements) is being made. The evaluation and reporting of technical performance measurement (TPM) status in Figure 5.6 and the cost-schedule reporting in Figure 6.28 are examples of formal reports. In addition, reports covering the results from the formal design reviews are another source for determining status. *Evaluation* is determining cause and possible steps to take when there are significant deviations from the planned performance. *Feedback and corrective action* include the development and implementation of a plan to correct any deficiencies that may exist. Such a plan must be coordinated with the development of the risk management plan described in Section 6.7.

The overall process involving the initial definition of system requirements, identification of the work (tasks) to be accomplished, development and implementation of the SEMP (or SEP), and the subsequent measurement and evaluation of program work (task) accomplishment(s) is illustrated in Figure 8.1. The requirements definition functions and the program planning and implementation activities are discussed

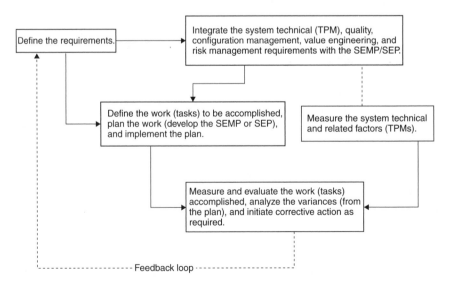

FIGURE 8.1 System requirements, review and evaluation, and feedback and corrective-action process.

in earlier chapters. This chapter addresses primarily *measurement and evaluation* as it pertains to the implementation of a system engineering program. The primary area of emphasis is the organization and management of a system engineering department/group in fulfilling its intended functions and meeting desired objectives. More specifically, the subjects of organizational requirements and benchmarking, evaluation of a system engineering organization overall, and individual program reporting, feedback, and control are covered.

8.1 EVALUATION REQUIREMENTS

Although the introduction to this chapter points to the requirements associated with the design and development of a single system, or the requirements determined for a specific project, it should be noted that an established system engineering organization may be involved in many different projects concurrently; for example, the design and development of a large-scale system, the design of many different subsystems, the manufacture and testing of a large system element, and/or the monitoring of many varieties of supplier activities such as illustrated in Figure 6.38. As the requirements and system engineering tasks are varied, the system engineering organization must be able to respond to all of the functions described in Chapter 6 (Figure 6.6), across the board and at the same time.

Thus, the emphasis should be on "organizational development" and building a capability so that the system engineering organization can be responsive to a wide mix of situations. As a start, the system engineering manager (with the support of key senior personnel both within and external to his or her organization) needs to define organizational objectives, goals, and responsibilities. To this end, it would be appropriate to establish a *benchmarking* capability and a model for the measurement and evaluation of the organization and its operations. The basic questions are, Where are we today? How do we compare with the competition (relative to both product and organization)? Where would we like to be in the future?

8.2 BENCHMARKING

The term *benchmark* may be defined in different ways, depending on one's individual background and experience. *Webster's Collegiate Dictionary* (10th edition) defines *benchmark* as "a point of reference from which measurements may be made; something that serves as a standard by which others may be measured." Although this definition primarily refers to a surveyor's mark or point of reference, the term has also been used in the context of setting and measuring standards related to product characteristics and organizational performance. In the early 1970s, the Xerox Corporation (and others) promoted the concept of benchmarking as a "business practice." According to Camp, benchmarking can be defined as "the continuous process of measuring products, services, and practices against the toughest competitors or those companies recognized as industry leaders."[1] Balm provided a more comprehensive definition, whereby benchmarking is "the ongoing activity of comparing one's own process, product, or service against the best known similar activity, so that by challenging attainable goals a realistic course of action can be implemented to efficiently become and remain the best of the best in a reasonable time."[2] This definition includes the element of *time*, which is critical if improvement is to be made in a competitive manner.

In regard to system engineering, there have been a number of benchmarking studies, and a few companies that practice the concepts and principles described throughout this text have implemented an active benchmarking effort internally.[3] The emphasis in most of these instances has been oriented directly to organizations and the processes that they use in accomplishing their day-to-day functions. Although this is appropriate, care must be taken to first define the company's objectives in terms of *product output* and then address the organizational characteristics that are considered to be essential in order to successfully meet the overall product goals. It is often tempting to launch into an evaluation of organizational effectiveness, employing some measures that may or may not be relevant to the ultimate objectives, and then initiating changes. Such changes may turn out to provide negative results because the proper goals were not defined at the beginning.

As shown in Figure 8.2, the general approach to benchmarking commences with the development of a plan for implementation (see block 2). This is based on a definition of the organization's objectives as they pertain to product goals. Product goals may be specified in terms of the technical performance measures (TPMs) for a given system, or some equivalent set of measures for one or more products. For example, Figure 2.11 identifies the TPMs for a system/product that resulted from a quality function deployment (QFD) analysis. Assuming that the quantitative requirements in the third column represent current status and that the immediate objectives include progressing to the requirements specified in the second column, then a plan needs to be developed covering the steps that must be accomplished in progressing from the current status to the level of performance ultimately desired. These steps relate to the organizational structure and the processes that are currently being implemented to support the product-oriented goals. For the purposes of this text, these include the system engineering functions and tasks described in Section 6.2.2.

[1]Robert C. Camp, *Benchmarking—The Search for Industry Best Practices That Lead to Superior Performance* (Milwaukee: ASQ Quality Press, 1993).

[2]Gerald J. Balm, *Benchmarking: A Practitioner's Guide for Becoming and Staying the Best of the Best* (Schaumburg, IL: QPMA Press, 1992).

[3]Kenneth Jones, *Benchmarking Systems Engineering in United States Industry,* Systems Engineering Design Laboratory, Virginia Poltechnic Institute and State University (Blacksburg, VA: 1994). Forty individuals from 21 different companies who had previously indicated that they were implementing the concepts and principles of systems engineering participated in this project study. Several additional references are (1) G. C. Thomas, and H. R. Smith, *Using Structured Benchmarking to Fast-Track CMM Process Improvement*, IEEE Computer Society Press, 2001; and (2) M. Zair, and P. Leonard, *Practical Benchmarking: The Complete Guide* (Chapman & Hall, 1994).

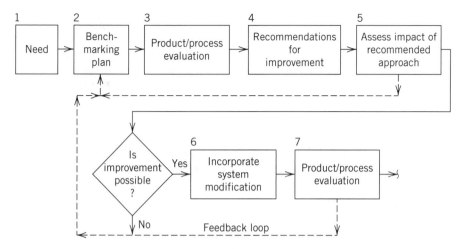

FIGURE 8.2 Benchmarking process.

In block 3 of Figure 8.2, one of the first steps is to define what is meant by *system engineering,* what is included, and what tasks must be accomplished in order to properly implement the concepts and principles described herein as they pertain to the product goals. This may lead to the development of a questionnaire, or series of checklists, used to facilitate the evaluation process. An assessment of the current processes is accomplished, possible problem areas are noted, recommendations for process/product improvement are developed (block 4), the potential impact of these proposed changes is assessed (block 5), and, if feasible, modifications are incorporated as appropriate. This may be a continuous process until the desired level of performance is attained.

Figure 8.3 illustrates a benchmarking plan showing the current status in terms of some level of performance, the status of the major competition, and the desired objective. It can be assumed that the competitor is also involved in a benchmarking effort and has established some higher-level goals. Thus, for the system/product in question, a plan must be developed that will enable one to follow path A–B in lieu of path C–D.[4]

8.3 EVALUATION OF THE SYSTEM ENGINEERING ORGANIZATION

Certain company/agency/institution goals having been established, the next step is to discover the extent to which the system engineering organization has progressed

---

[4]It should be reemphasized that the first step is to establish organizational "capability" goals, which stem from the projects (and their respective requirements) that the organization wishes to take on, and then to identify the steps required to develop the organization so that it can respond to such goals both effectively and efficiently.

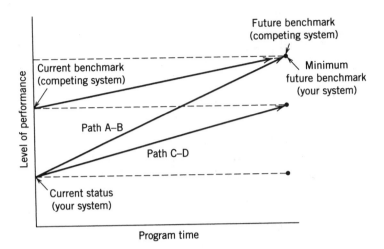

FIGURE 8.3   Benchmarking.

toward meeting these goals; that is, the measure of the organization's capability to meet the desired level of performance. Given the objectives of system engineering and the recommended tasks that must be performed, there are some questions that should be addressed: To what extent is the organization completing these tasks effectively and efficiently? Does the management understand the principles and concepts of system engineering? Is there a commitment from the top down toward the implementation of the system engineering process? If so, what policies are currently being implemented to support this? Have standards, measurable goals, and the appropriate processes been established for the successful accomplishment of system engineering objectives? Has the organization developed a plan for continuous improvement?

Although there are many questions of this nature that can be asked, the objective is to determine the organization's *level of maturity,* where it may "fit" in the hierarchical structure as compared with other organizations functioning in a similar area of activity, and where there are weaknesses that need to be addressed. In other words, although the benchmarking process aids in establishing specific goals, there is a need to develop a model to assist in the evaluation of an organization's current capability.

In response, there has been a concerted and continuing effort since the late 1980s to develop a model that will address the organizational assessment issue. Although there have been numerous models used to varying degrees through the years, a series of recently developed specific projects/models is noteworthy. Through the early efforts of the Software Engineering Institute (SEI) at Carnegie Mellon University, a process improvement model oriented to software development, Software Capability Maturity Model (SW-CMM), was first introduced in 1989. As a result of experience and continuous upgrading, Version 1.1 of SW-CMM was released in 1993. Based on this experience, and through the combined efforts of many in industry, government, and academia, the System Engineering Capability Maturity Model (SE-CMM) was developed and released for use in 1994.[5] At the same time, and with the coordination and support of the International Council on Systems Engineering (INCOSE), the Systems Engineering Capability Assessment Model (SECAM) was released in 1994.[6] These two models were then successfully merged into EIA/IS-731 in 1998 as a result of a collaborative effort involving EIA (Electronic Industries Alliance), EPIC (Enterprise Process Improvement Collaboration), and INCOSE.[7]

Subsequent to the initial release of EIA/IS-731, there have been a number of individual efforts to develop comparable models for different purposes. In addition to the models covering software development and systems engineering, an effort was initiated to develop a model for *integrated product and process development* (IPPD). Further, there have been efforts to address other critical areas where a measure of organizational maturity is desired. Given the trend relative to developing a series of different models for individual purposes, an effort was initiated in 1998 to study the feasibility of developing one comprehensive model that would represent an integrated approach and combine the capabilities of the SW-CMM, SE-CMM, SECAM, and the IPPD model. The result of this effort has produced a new product, Capability Maturity Model Integration (CMMI). The objective is to eliminate the "stovepipe" models and to adapt CMMI as the ultimate measurement tool for the various areas of concern.[8]

To get some idea of the detailed approach for implementation, given that the emphasis throughout this text is on systems engineering, it would be appropriate at this point to consider the Systems Engineering Capability Model (SECM), discussed in EIA/IS-731. One of the first steps in its development was, of course, to define the goals and objectives of a system engineering organization. Having accomplished this, essential systems engineering and management tasks that an organization must perform to ensure a successful effort were identified and included in three basic focus-area categories; that is, a *technical category focus area,* a *management category focus area,* and an *environment category focus area.* Establishment of the focus-area categories then led to the identification of specific *focus areas,* which led to *themes,*

---

[5]Software Engineering Institute (SEI), *A Systems Engineering Capability Maturity Model (SE-CMM),* Version 1.1, SECMM-95-01, (Pittsburgh, PA: Carnegie Melon University, 1995).

[6]A good reference that provides a historical basis for the SECAM and its applications is B. A. Andrews and E. R. Widmann, "A Synopsis of Metrics and Observations from Systems Engineering Process Assessments Conducted Using the INCOSE SECAM," in *Proceedings of the Sixth Annual International Symposium of the INCOSE,* Vol. 1 (Seattle, WA: INCOSE, 1996), p. 1071. Additional references are included in the *Proceedings* from earlier INCOSE symposia.

[7]GEIA (Government Electronics and Information Technology Association), EIA/IS 731: *Systems Engineering Capability Model (SECM)* (Washington, DC, 2001) (Web site: http://www.geia.org/sstc/G47/page6.htm, October 2001).

[8]A good reference covering the history and background leading to the development of the CMMI is *Systems Engineering: The Journal of the International Council on Systems Engineering* Vol. 5, no. 1 (2002) published by John Wiley & Sons, Inc. There are a series of articles in this *Journal* issue that deal with CMMI, the status of EIA/IS-731, and related topics. Further, there are a number of issues of *CrossTalk: The Journal of Defense Software Engineering,* published by the Software Technology Support Center, Hill AFB, Utah, that discuss the CMMI model and objectives. The development of the CMMI process continues through the joint efforts of government, industry, and the Carnegie Melon University.

Focus-Area Categories		
Technical	Management	Environmental
• Define stakeholder and system-level requirements • Define the technical problem • Define the solution • Assess and select • Integrate system • Verify system • Validate system	• Plan and organize • Monitor and control • Integrate design disciplines • Coordinate with suppliers • Manage risk • Manage data • Manage configurations • Ensure quality	• Define and improve the system engineering process • Manage competency • Manage technology • Manage system engineering support environment

FIGURE 8.4 SECM Focus areas and categories (EIA/IS-731).

which led to a description of specific *practices*. The results of this progression are summarized in a presentation of the major topics shown in Figure 8.4.

Given a description of the desired practices, the next step was to identify different *capability levels* (levels of maturity), or the degree(s) of capability an organization should strive to meet, evolving from current capability to a future level, indicating growth potential. Six capability levels were established and related to individual focus areas. A focus area includes a list of practices describing the activities that an organization must successfully perform. In Figure 8.5, the levels of capability (from 0 to 5) are identified as *initial, performed, managed, defined, measured,* and *optimized*. These levels are supported by a description of specific practices that are desired in order to meet the requirements for a given level. The objective, of course, is to progress to level 5.

In applying this model in the appraisal (or assessment) of an organization's capability, there are different phases: *preassessment, on-site assessment,* and *postassessment*. During the preassessment phase, it is necessary to solicit management support of the organization to be evaluated and to develop the process for evaluation. Included in this phase is the development of a rather extensive questionnaire (which contains many different questions for the EIA/IS-731 requirement, or a minimum of 40 questions pertaining to level one, 91 questions for level two, 156 questions for level three, 56 questions for level four, and 83 questions for level five).[9] The "on-site assessment" phase includes the following steps: administering the questionnaire, analyzing the results, developing some additional exploratory questions, conducting interviews with focus groups, analyzing exploratory data, summarizing the results and coordinating with management, and preparing the final evaluation report. This phase is usually conducted during a one-week period, by a team of three to five people

---

[9]S. Alessi, "A Simple Statistic for Use with Capability Maturity Models," *Systems Engineering: The Journal of the International Council on Systems Engineering*, vol. 5, no. 3 (2002): 242–252 (published by John Wiley & Sons, Inc. New York).

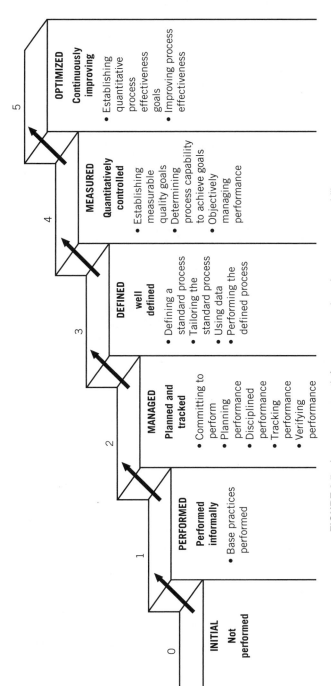

FIGURE 8.5 Improvement path for system engineering process capability.

0 INITIAL Not performed

1 PERFORMED Performed informally
- Base practices performed

2 MANAGED Planned and tracked
- Committing to perform
- Planning performance
- Disciplined performance
- Tracking performance
- Verifying performance

3 DEFINED well defined
- Defining a standard process
- Tailoring the standard process
- Using data
- Performing the defined process

4 MEASURED Quantitatively controlled
- Establishing measurable quality goals
- Determining process capability to achieve goals
- Objectively managing performance

5 OPTIMIZED Continuously improving
- Establishing quantitative process effectiveness goals
- Improving process effectiveness

working with a combination of department managers, project leaders, and workforce practitioners, and results in rapid feedback and minimizing any impact on internal projects and the day-to-day scheduled work. The postassessment phase involves management briefings and the preparation of a plan for future action as required.

The results of the assessment, utilizing the SECM, should include a summary chart/graphic showing the different focus areas and the degrees to which each has achieved a given "level of capability." In Figure 8.6, it can be seen that focus area 1 in the "Managed" category has achieved "capability" at level 3, and that focus area 2 (in the same category) is only at level 1. Given these results, the final assessment report (and plan for future action) should include some specific recommendations for improvement, particularly in regard to focus area 2, and the action(s) that need to be initiated in order to progress to the next higher level. The objective is, of course, to make progress in all of the focus areas with the proper balance being achieved across the board.

The preceding description provides only a rough idea as to the objectives and content of the Systems Engineering Capability Model (SECM). For more in-depth coverage, a detailed review of EIA/IS-731 is recommended. Relative to the future, although this model will, in all probability, continue to be applied in selected areas and oriented to the assessment of a systems engineering organization as an entity, acquiring a good understanding of the Capability Maturity Model Integration (CMMI) is also recommended, as this is a more comprehensive model and gaining in popularity.

In comparing the SECM with the CMMI, it is clear that the basic architectures are quite similar.[10] The SECM includes focus areas and categories; the CMMI takes the same basic approach, although the specific topics and nomenclature are different. In Figure 8.7, there are four process area categories: process management, project management, engineering, and support. Within each of these categories, there are a number of specific process areas, for which detailed questions have been prepared for purposes of assessment. Note that the activities in CMMI are much broader in scope than those in SECM. In regard to "levels of capability," the CMMI also has established six levels (i.e., "level 0" to "level 5"), including *incomplete, performed, managed, defined, defined, quantitatively managed,* and *optimizing.*

For purposes of assessment, the Standard CMMI Assessment Method for Process Improvement (SCAMPI) is accomplished through the application of questionnaires, local visits and interviews, and the like. Specific scoring rules for each capability level are used, and the highest resulting score reflects the "level of capability" attained for the process area being evaluated. In any event, the overall approach here is similar to that described earlier for SECM.

It should be noted that the development effort for CMMI continues and there are additional critical areas of activity that are being considered for inclusion (as this text goes to press). Thus, it is recommended that the reader pursue additional research in this area to ensure currency. This is particularly important, as the CMMI will likely

---

[10]I. Minnich, "EIA/IS-731 Compared to CMMI-SE/SW," *Systems Engineering: The Journal of the International Council on Systems Engineering,* vol. 5, no. 1 (2002) 62–72 (published by John Wiley & Sons, Inc.). Hoboken, NJ.

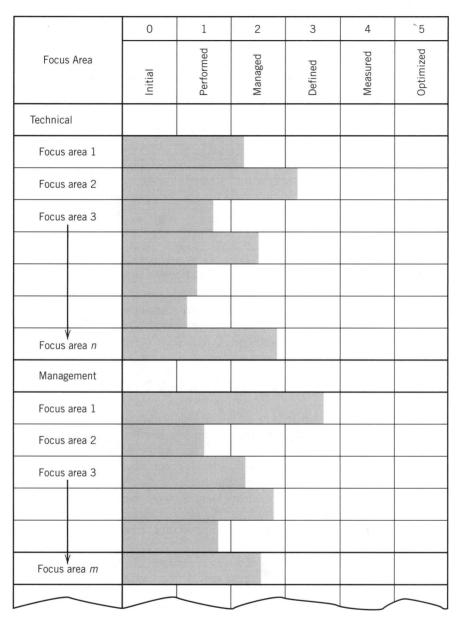

FIGURE 8.6   Focus area capability assessment.

Process Area Categories			
Process Management	Project Management	Engineering	Support
• Organizational process focus • Organizational process definition • Organizational innovation and deployment	• Project planning • Project monitoring and control • Supplier management • Integrated project management • Risk management • Quantitative project management	• Requirements management • Requirements developement • Technical solution • Product integration • Verification • Validation	• Configuration management • Process and product quality assurance • Measurement and analysis • Decision analysis and resolution • Causal analysis and resolution

FIGURE 8.7   CMMI process areas and categories. *Source:* I. Minnich, "EIA/IS-731 Compared to CMMI SE/SW," *Systems Engineering: The Journal of the International Council on Systems Engineering*, vol. 5, no. 1 (2002), Table II. (Published by John Wiley & Sons, Inc.)

be applied in the evaluation of all program organizations in the future. In any event, it is believed that the approach described throughout this section is excellent and certainly valid in the evaluation of a systems engineering organization.

8.4   PROGRAM REPORTING, FEEDBACK, AND CONTROL

The discussion in the earlier sections of this chapter applies primarily to the evaluation of a systems engineering activity operating within a large producer's organization (i.e., the prime contractor). As with any activity, the processes described in Sections 8.2 and 8.3 must be tailored to the specific organization being evaluated. As shown in Figure 7.1, the successful implementation of system engineering objectives depends not only on the producer's activities, but also on the related activities of the customer's organization and the activities of the various major suppliers participating in the program in question. Thus, there are both "upward" and "downward" impacts that must be considered.

In regard to an SECM, CMMI, or equivalent evaluation, the results highlight specific areas of weakness and where improvements in the applicable processes can be realized. With potential areas for improvement having been identified, two steps need to be addressed:

1. Determining ways for improvement of internal processes *within* the system engineering organization.[11] This encompasses evaluating alternative methods of doing business, determining the requirements for changing the existing

---

[11]In determining ways for improvement, reference should be made to the "benchmarks" that were established in Section 8.2. The objective in initiating change is to meet (if not exceed) a specified benchmark goal. In addition, one needs to assess the impact of change in terms of risk and the risk management plan (refer to Section 6.7). The goal is, of course, to reduce risk as a result of change.

procedures and processes, and assessing the impact of such changes on the other processes. A change in any one process should not have a negative impact on any other process.

2. Determining the possible impact(s) of changes in the processes being implemented by the systems engineering organization on any *external* and related organizational structures—the customer, other organizational groups within the producer's operation, major suppliers, and so forth. The proper environment must be established within the overall organizational infrastructure for the proposed changes described in item 1 to result in an improvement.

Proposed changes within the system engineering organization cannot be initiated in a vacuum. There must be a mutual commitment throughout the organization and, in particular, by the program manager and his or her staff. In any case, there must be a vehicle through which organizational improvement can be initiated.

Given the approval and incorporation of a "change" (or group of changes), the revised processes/procedures must be documented and reported and must serve as a baseline for the next organizational evaluation. Although there is no established frequency of evaluation, it is recommended that the approach and procedures discussed herein be included as a "continuing activity" within the overall spectrum of system engineering organization activities.

8.5 SUMMARY

In approaching a subject such as system engineering management, it is essential to first address the activity that is to be managed. The first five chapters of this text accomplish this by describing system engineering principles and concepts, the system engineering process, and supporting requirements as they apply in the design and development of major systems. The next step is to cover the necessary functions/tasks that must be implemented in order to realize the objectives described in the beginning. Thus, Chapters 6, 7, and 8 discuss the necessary steps for implementation; that is, the *planning* for system engineering, the *organization* for system engineering, and the subsequent *evaluation* (and *feedback*) in regard to how well we have planned from the beginning and how well we have performed in the organization and follow-on implementation of system engineering requirements. Accomplishing planning and organization activities alone, without having the benefit of subsequent evaluation and feedback, constitutes only part of the overall process and is certainly inhibiting when it comes to capturing the experiences from the past and realizing growth for the future.

This chapter emphasizes the importance of evaluation and feedback. Much of the material included herein addresses a popular set of models being developed and applied in the evaluation of system engineering organizations today (i.e., SECM and CMMI). As we evolve further into the future, there will (in all likelihood) be a new set of tools available for the purposes of evaluation. In any event, the important issue is to ensure that there is an *evaluation and feedback capability* built into any type of a system engineering program.[12]

QUESTIONS AND PROBLEMS

1. Why is system engineering evaluation and feedback important? Describe some of the benefits that can be realized through the implementation of such a capability. What is likely to occur should such a capability not be implemented?

2. What is meant by *benchmarking?* If you were assigned to develop and implement a benchmarking capability for your program (as program manager), what steps would you take in accomplishing this assignment?

3. In developing a benchmarking capability, what specific factors would you, as program manager, select in attempting to establish the appropriate goals for your program?

4. Review the literature pertaining to the SECM and CMMI tools and their application. What are the basic objectives of each (how do they differ)? What factors are measured? Briefly describe the steps to be followed in the implementation of each.

5. Assuming that you, as manager of the system engineering department, have just completed an assessment of your organization utilizing the SECM approach, what steps would you initiate next?

6. Assume that you, as manager of the system engineering department, need to gain some good visibility as to how well your organization is performing. What type of reports (or reporting requirements) would you require of your organization? How often would they be required?

7. Assume that you, as manager of the system engineering department, are dependent on the performance of a number of major suppliers. What steps would you take (and what should be included) in establishing the requirements for the evaluation of the suppliers?

8. As part of a supplier evaluation effort, you are planning to visit a major supplier's facility. What would you do in preparation, and what information would you solicit during the on-site visit?

9. In your opinion, how often should the evaluation of a system engineering organization be accomplished? Why?

---

[12] Again, it should be reemphasized that the approach to system engineering evaluation, feedback, and control, described in this chapter, is primarily oriented to the overall *system engineering organization* and its effectiveness in meeting the objectives described in Chapter 1 through 5. Specific program/project evaluation, feedback, and control are covered in Sections 6.2.7 through 6.2.10. Additionally, refer to the bibliography in Appendix F, Section 11, for several excellent references on *project management.*

10. What should be considered in recommending process changes resulting from an evaluation?

11. Refer to Figure 6.33 and Appendix E. Develop a supplier checklist for the purposes of evaluation (prepare the checklist in the format shown in Figure 6.33 and provide a breakout of factors for each item in your checklist, as illustrated in Appendix E).

APPENDIX A

FUNCTIONAL ANALYSIS (CASE-STUDY EXAMPLES)

A critical step in implementing the system engineering process is the accomplishment of the *functional analysis* and definition of the system in "functional" terms. Functions are initially identified as part of defining the need and the basic requirements for the system (Figure 1.12, block 0.1). System operational requirements and the maintenance concept are defined, and the functional analysis is expanded to establish a *functional baseline*, from which resource requirements for the system are identified (i.e., equipment, software, facilities, people, data, the various elements of maintenance and support, etc.). The functional analysis is initiated during the conceptual design phase, and is described in detail in Chapter 2, Section 2.7.

Referring to the discussion in Section 2.7, accomplishment of the functional analysis can best be facilitated through the development of functional flow block diagrams (FFBDs), as illustrated in Figure 2.12. Such FFBDs are developed primarily for the purposes of structuring and transitioning system requirements into "functional" terms. The process for accomplishing such is covered in Section 2.7.1, and a few examples of FFBDs are presented in Figures 2.13 through 2.18. With the objective of further illustrating the development of these function flow block diagrams, Figures A.1 through A.6 are included to show a few simple applications:

1. Figure A.1 provides an example of the basic format used in the development of functional flow diagrams in general. This is an extension of Figure 2.12.
2. Figure A.2 shows two levels of operational flow diagrams and two levels of maintenance flow diagrams for a basic lawn-mowing system.
3. Figure A.3 shows three levels of operational flow diagrams and one level of a maintenance flow diagram for an automotive system.
4. Figure A.4 shows two levels of operational flow diagrams and two levels of maintenance flow diagrams for a radar system.

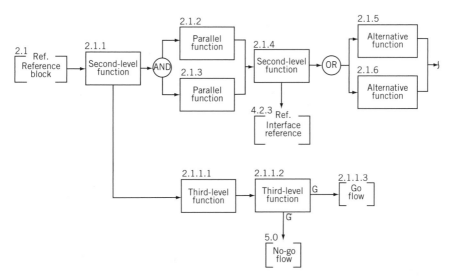

FIGURE A.1   General format for the development of functional flow block diagrams (FFBDs).

5. Figure A.5 shows two levels of operational flow diagrams for a space system.
6. Figure A.6 shows a maintenance functional flow diagram for the space system that evolves from the operational flow diagram in Figure A.5.

Although these sample block diagrams do not cover the selected systems entirely, it is hoped that the material is presented in enough detail to provide an appropriate level of guidance for the development of FFBDs for other applicable systems.

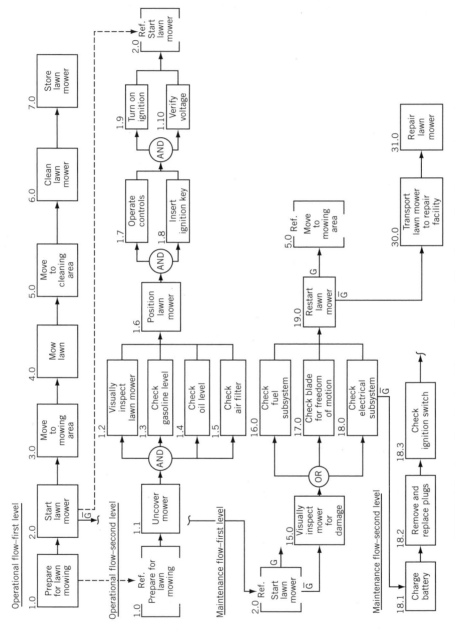

FIGURE A.2   Operational and maintenance functional flow diagram for a lawn-mowing system.

Operational functional flows—three levels

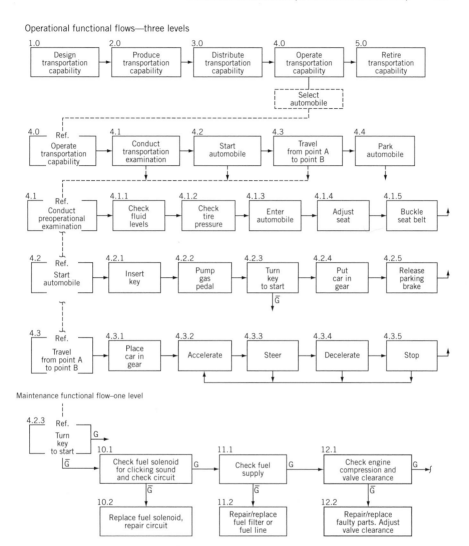

Maintenance functional flow—one level

FIGURE A.3 Operational and maintenance functional flow diagram for an automobile.

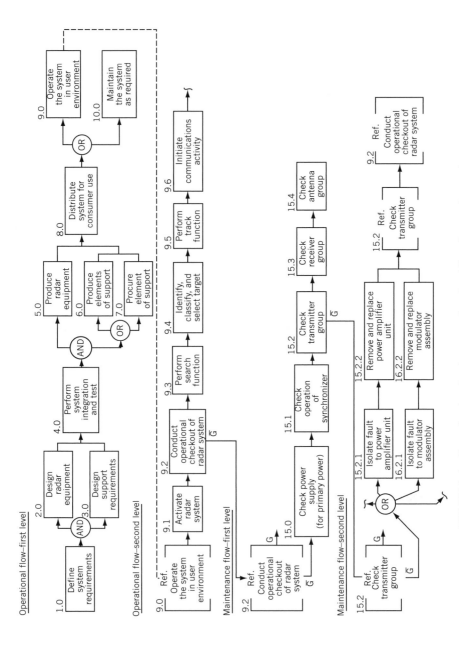

FIGURE A.4 Operational and maintenance functional flow diagram for a radar system.

416

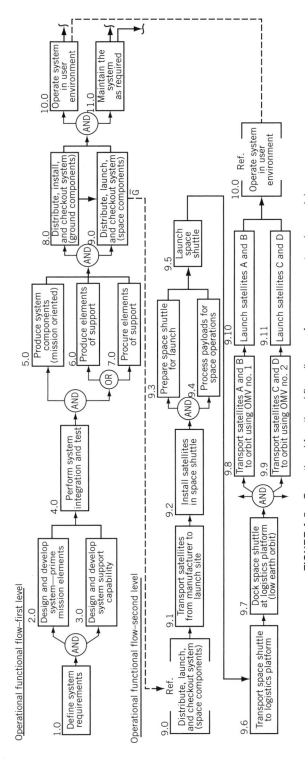

FIGURE A.5 Operational functional flow diagram for a space system (example).

417

Maintenance functional flow

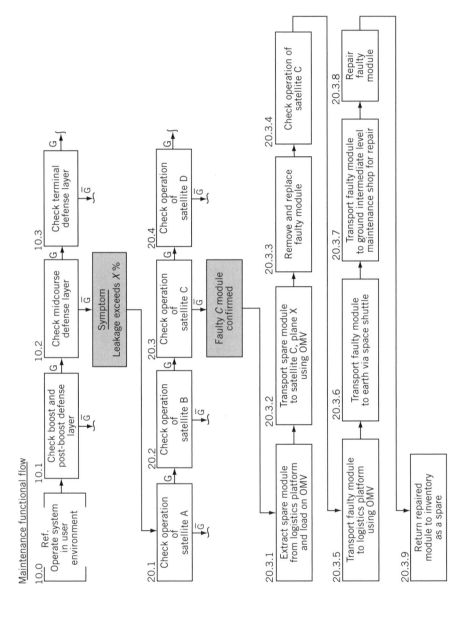

FIGURE A.6 Maintenance functional flow diagram for a space system (example).

APPENDIX B

LIFE-CYCLE COST-ANALYSIS PROCESS

Many of our day-to-day decisions, as they pertain to the design and development of new systems and the reengineering of existing systems, are based on *technical* performance-related factors alone. *Economic* considerations, if addressed at all, have dealt primarily with initial, procurement and acquisition costs only, and not the "downstream" costs associated with system operation and maintenance support. Yet these downstream costs, which often constitute a significant portion of the total life-cycle cost of a system, are highly influenced by the decisions made in the early phases of system development. In other words, the early decision-making process must consider the *total* spectrum of costs if economic benefits are to be gained in the long term. The consequences of the short-term approach often practiced in the past have been rather detrimental overall, as conveyed in Section 1.2 (Chapter 1). Total cost visibility, as illustrated in Figure B.1, is a *must* if the risks associated with the decision-making process are to be properly assessed.

Life-cycle costing includes the consideration of *all* future costs associated with research and development (i.e., design), construction, production, distribution, system operation, sustaining maintenance and support, system retirement, and material disposal and/or recycling. It involves the costs of all technical and management activities throughout the system life cycle; that is, customer activities, producer and/or contractor activities, supplier activities, and consumer or user activities. Although the influencing of these costs can best be realized during the early phases in the development of a *new* system, as conveyed in Figure B.2, benefits can also be gained through the identification and evaluation of high-cost contributors for *existing* systems already in use. In other words, the applications and benefits that can be gained through the accomplishment of life-cycle cost analyses are numerous, as shown in Figure 3.40 (Chapter 3).

In performing a life-cycle cost analysis, there is a series of steps to follow. These steps are briefly described in Figure 3.38 and are conveyed in the context of the overall process in Figure 3.41. The purpose of this appendix is to provide some additional explanation covering each of the steps identified in Figure 3.38.

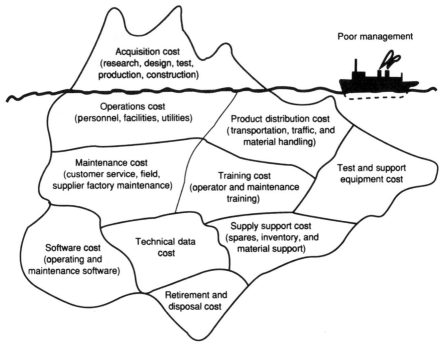

FIGURE B.1   Total cost visibility.

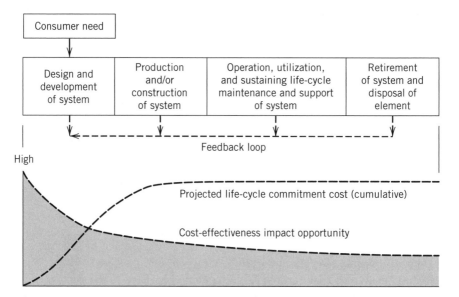

FIGURE B.2   Opportunity for impacting cost-effectiveness in the system life cycle.

B.1   DEFINE SYSTEM REQUIREMENTS

The first step in the performance of a life-cycle cost analysis is to define the problem, identify the proposed technical solution, describe the operational requirements and the maintenance concept for the system, identify the critical technical performance measures (TPMs), and describe the system configuration in functional terms; that is, the process described in Sections 2.1 through 2.7 (Chapter 2). Depending on where one is in the system life cycle, the definition may be rather cursory or more in-depth. In any event, the basic system requirements must be defined in order to provide the necessary structure for the analysis, and the assumptions that are made at this point may have a significant impact on the results.

In Figure B.3, it is assumed that a ground vehicle in development requires the incorporation of a communications capability. Multiple quantities of the vehicle will be deployed to three different geographical locations, (i.e., 20, 20, and 25 at each location, respectively), performing a variety of missions. Although there are variations from one location to the next, it is assumed that each vehicle will be utilized on the average of 4 hours per day, 360 days per year. The equipment must enable communication with other vehicles at a range of at least 200 miles, overhead aircraft at an altitude of up to 10,000 feet, and with a centralized area communications facility. The system must have a reliability mean time between failure (MTBF) of 450 hours, a corrective maintenance downtime ($\overline{\text{M}}$ct) of 30 minutes or less, a maintenance labor hours per system operating hour (MLH/OH) requirement of 0.2 or less, and a unit life-cycle cost not to exceed $20,000. The equipment will be functionally packaged in units (i.e., units A, B, and C) and, in the event of failure, the problem will be isolated to the unit level, faulty units will be removed and replaced with spares and sent back to the intermediate level of maintenance for corrective action, and so on.

In Figure B.3, the system operational requirements and the maintenance concept have been defined to the depth that will allow for the accomplishment of a life-cycle cost analysis during the late conceptual design or early preliminary design phase. The next step is to describe the system, and the mission(s) that is to be performed, in *functional* terms by accomplishing a top-level functional analysis. See Figure 2.13 (Chapter 2); the communication system can be described in a similar manner, followed with an evaluation of each functional block to determine the resource requirements that will provide the basis for functional costing (see Figure 2.18).

B.2   DESCRIBE THE SYSTEM LIFE CYCLE AND IDENTIFY THE MAJOR ACTIVITIES IN EACH PHASE

Given the definition of system requirements and the identification of functions, it is appropriate to provide a time line for these requirements in terms of the life cycle. In Figure B.3, the planned life cycle is 12 years. In other words, it is assumed that there is a need for the communication system and the functions that are to be performed for a 12-year period. Although this planning horizon may change (as requirements

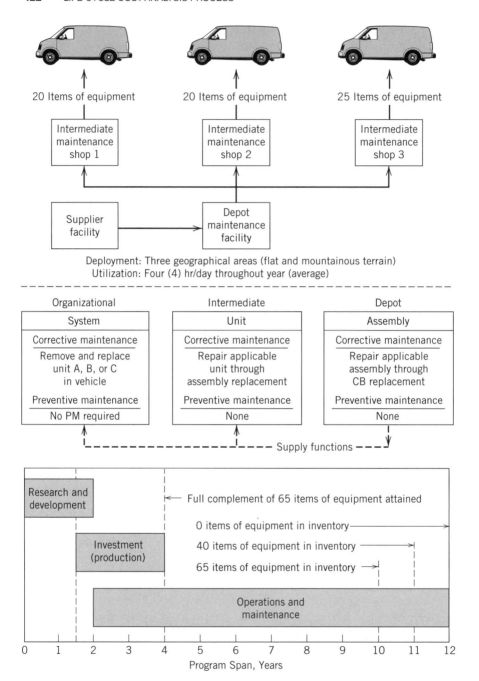

FIGURE B.3 Communication system requirements.

change), a baseline must be established. Thus, the 12-year period and the major activities identified in the figure will be assumed herein. The activity categories identified in the figure (i.e., research and development, investment/production, and operations and maintenance) form the basis for the development of a cost breakdown structure (CBS).

B.3 DEVELOP A COST BREAKDOWN STRUCTURE (CBS)

The functions described through the functional analysis can be broken down into subfunctions, categories of work, work packages, and, ultimately, the identification of physical elements. From a planning and management perspective, it is necessary to establish a top-down framework that will allow for the initial allocation and subsequent collecting, accumulating, organizing, and computing of costs. For a typical project, this may lead to the development of a work breakdown structure (WBS) prepared to show, in a hierarchical manner, all of the elements of work that are necessary to complete a given program. As shown in Section 6.2.4 (Figure 6.12), a summary work breakdown structure (SWBS) may be developed initially, followed by one or more individual contract work breakdown structures (CWBS) designed to address specific elements of work that are covered through some form of a contractual arrangement. It is the SWBS that provides a good basis for the development of a cost breakdown structure (CBS) used in life-cycle cost analyses, primarily because its intent is to cover *all* future activities and associated costs; that is, research and development, construction/production, distribution, operation and maintenance support, and retirement activities.

The CBS is intended to show all future functions/activities, broken down to the depth necessary to provide the appropriate level of visibility and tailored to the system configuration in question. Ultimately, the CBS will lead to the identification of a product and/or a process, with the objective of establishing a structure that can be initially used for the top-down allocation of costs during the conceptual design phase (refer to Section 2.8) and subsequently for the bottom-up collection of costs for the purposes of accomplishing a life-cycle cost analysis. Figure B.4 provides an illustration of a sample cost breakdown structure (CBS), and Figure B.5 provides an abbreviated example showing how each category of the CBS should be described in terms of what is included, how the costs are calculated, and the basis for accomplishing such. The CBS provides a vehicle for looking at costs from a *functional* perspective. As one proceeds with the life-cycle cost analysis, costs are estimated for each year in the planned life cycle and are summarized for each category in the CBS.[1]

---

[1] The cost breakdown structure (CBS) should be tailored to the system in question. In Figure 3.39, another example is presented. If the system is very "software-intensive," then Category Crs should be broken down to show more detail. If the system is very "operator-intensive" (e.g., a ground radar tracking station requiring a large number of operating personnel), then Category Cop should be expanded. On the other hand, if Category Cin is too detailed for the purposes of a given analysis, then one can summarize the costs accordingly. The objective is to provide *visibility* relative to key functional activities.

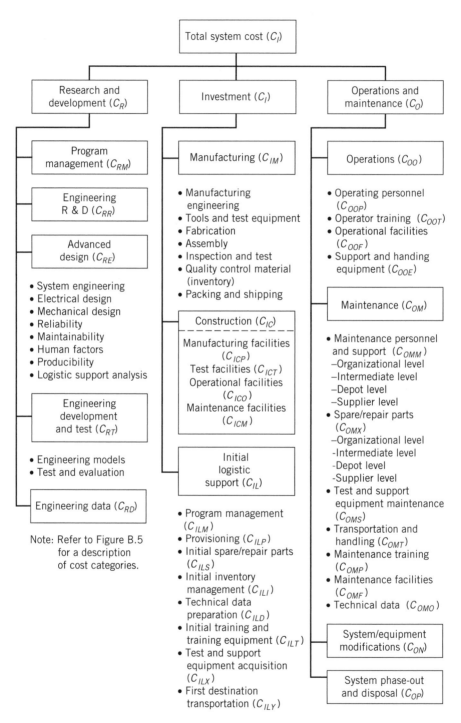

FIGURE B.4   Cost breakdown structure (CBS). *Source:* Benjamin S. Blanchard, *Logistics Engineering and Management*, 6th edition, © 2004, p. 468. Reprinted by permission of Pearson Education, Inc., Upper Saddle River, NJ.

Cost Category (Figure B.4)	Method of Determination (Quantitive Expression)	Cost Category Description and Justification
Total system cost (C)	$C = C_R + C_I + C_O$ $C_R = R$ and D cost $C_I =$ Investment cost $C_O =$ Operations and maintenance cost	Includes all future costs associated with the acquisition, utilization, and subsequent disposal of system/equipment.
Research and development (C_R)	$CR = C_{RM} + C_{RR} + C_{RE} + C_{RT} + C_{RD}$ $C_{RM} =$ Program management cost $C_{RR} =$ Advanced R&D cost $C_{RE} =$ Engineering design cost $C_{RT} =$ Equipment development/ test cost $C_{RD} =$ Engineering data cost	Includes all costs associated with conceptual/feasibility studies, basic research, advanced research and development, engineering design, fabrication and test of engineering prototype models (hardware) and associated documentation. Also covers all related program management functions. These cost are basically nonrecurring.
Investment (C_I)	$C_I = C_{IM} + C_{IC} + C_{IL}$ $C_{IM} =$ System/equipment manufacturing cost $C_{IC} =$ System construction cost $C_{IL} =$ Cost of initial logistic support	Includes all costs associated with the acquisition of systems/ equipment (once design and development have been completed). Specifically, this covers manufacturing (recurring and nonrecurring), manufacture-ing management, system construction, and initial logistic support.
Operations and maintenance (C_O)	$C_O = C_{OO} + C_{OM} + C_{ON} + C_{OP}$ $C_{OO} =$ Cost of system/equipment life-cycle operations $C_{OM} =$ Cost of system/equipment life-cycle maintenance $C_{ON} =$ Cost of system/equipment life-cycle modifications $C_{OP} =$ Cost of system/equipment phase-out and disposal	Includes all costs associated with the operation and maintenance support of the system throughout its life cycle subsequent to equipment delivery in the field. Specific categories cover the cost of system operation, maintenance, sustaining logic support, equipment modifications, and system/ equipment phase-out and disposal. Costs are generally determined for each year throughout the life cycle.

FIGURE B.5 Description of cost categories (partial).

Cost Category (Reference Figure B.4)	Method of Determination (Quantitive Expression)	Cost Category Description and Justification

Transportation and handling cost (C_{OMT})	$C_{OMT} = [(C_T)(Q_T) + (C_P)(Q_T)]$
	  C_T = Cost of transportation
	  C_P = Cost of packing
	  Q_T = Quantity of one-way shipments
	$C_T = [(W)(C_{TS})]$
	  W = Weight of item (lb)
	  C_{TS} = Shipping cost ($/lb)
	  C_{TS} will, of course, vary with the distance (in miles) of the one-way shipment.
	$C_P = [(W)(C_{TP})]$
	  C_{TP} = Packing cost ($/lb)
	  Packing cost and weight will vary depending on whether reuseable containers are employed.
		Initial (first destination) transportation and handling costs are covered in C_{ILY}. This category includes all sustaining transportation and handling (or packing and shipping) between organizational, intermediate, depot, and supplier facilities in support of maintenance operations. This includes the return of faulty material items to a higher echelon; the transportation of items to a higher echelon for preventive maintenance (overhaul, calibration); and the shipment of spare/repair parts, personnel, data, etc., from the supplier to forward echelons.

Maintenance training cost (C_{OMP})	$C_{OMP} = [(Q_{SM})(T_T)(C_{TOM})]$
	  Q_{SM} = Quantity of maintenance students
	  C_{TOM} = Cost of maintenance training ($/student-week)
	  T_T = Duration of training program (weeks)
		Initial maintenance training cost is included in C_{ILT}. This category covers the formal training of personnel assigned to maintain the prime equipment, test and support equipment, and training equipment. Such training is accomplished on a periodic basis throughout the system life-cycle to cover personnel replacements due to attrition. Total costs include instructor time; supervision; student pay and allowances while in school; training facilities (allocation of portion of facility required specifically for formal training); training aids and data; and student transportation as applicable.

Operational facilities cost (C_{OOF})	$C_{OOF} = [(C_{PPE} + C_U)(\% \text{ Allocation}) \times (N_{OS})]$
	  C_{PPE} = Cost of operational facility support ($/site)
	  C_U = Cost of utilities ($/site)
	  N_{OS} = Number of operational sites
	Alternative Approach
	$C_{OOF} = [(C_{PPF})(N_{OS})(S_O)]$
	  C_{PPF} = Cost of operational facility space ($/square foot/site). Utility cost allocation is included.
	  S_O = Facility space requirements (square feet)
		Initial acquisition cost for operational facilities is included in C_{ICO}. This category covers the annual recurring costs associated with the occupancy and maintenance (repair, paint, etc.) of operational facilities throughout the system life-cycle. Utility costs are also included. Facility and utility costs are proportionately allocated to each system.

FIGURE B.5   (*Continued*)

B.4 ESTIMATE THE COSTS FOR EACH PHASE OF THE LIFE CYCLE

The next step is to estimate the costs, by category in the CBS, for each year in the system life cycle. Such estimates must consider the effects of inflation, learning curves when repetitive processes or activities occur, and any other factors that are likely to cause changes in cost, either upward or downward. Cost estimates may be derived

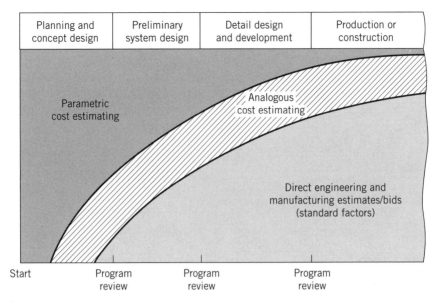

FIGURE B.6   Cost estimation by program phase.

from a combination of accounting records, cost projections, supplier proposals, and predictions in one form or another.

In Figure B.2, the early stages in the system life cycle is the preferred time to commence with the estimation of costs, because it is at this point when the greatest impact on total system life-cycle cost can be realized. However, the availability of good historical cost data at this time is almost nonexistent in most organizations, particularly the type of data that pertain to the downstream activities of operations and support for similar systems in the past. Thus, one must depend heavily on the use of various cost-estimating methods in order to accomplish the end objectives.

As shown in Figure B.6, as the system configuration becomes better defined in a developmental effort, the use of direct engineering and manufacturing standard factors based on past experience can be applied, as is the case for any "cost-to-complete" projection on a typical project today (e.g., cost per labor hour). On the other hand, in the earlier stages of the life cycle when the system configuration has not been well defined, the analyst must rely on the use of a combination of analogous and/or parametric methods developed from experience with similar systems in the past. The objective is to collect data on a "known entity," identify the major functions that have been accomplished and the costs associated with these functions, relate the costs in terms of some functional or physical parameter of the system, and then use this relationship in attempting to estimate the costs for a new system. A goal is to identify the applicable technical performance measures (TPMs) for the system in question and estimate the cost per a given level of performance (e.g., cost per unit of product output, cost per mile of range, cost per unit of weight, cost per volume of capacity used, cost per unit of acceleration, cost per functional output, etc.). Costs can be related to

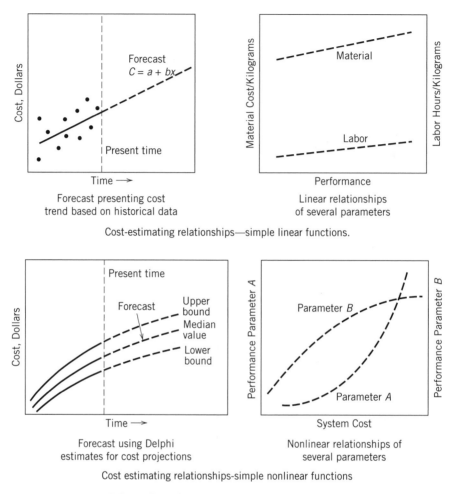

Forecast presenting cost
trend based on historical data

Linear relationships
of several parameters

Cost-estimating relationships—simple linear functions.

Forecast using Delphi
estimates for cost projections

Nonlinear relationships of
several parameters

Cost estimating relationships-simple nonlinear functions

FIGURE B.7   Cost-estimating relationships (CERs).

the appropriate blocks in the functional description of the system. Figures B.7 and
B.8 provide some simple illustrations of considerations in cost estimating. However,
care must be exercised to ensure that the historical information used in the development of cost-estimating relationships (CERs) is relevant to the system configuration
being evaluated today. CERs based on the mission and performance characteristics
of one system may not be appropriate for another system configuration, even if the
configuration is similar in a physical sense. Thus, costs must be related from a *functional* perspective.

To be effective in total cost management (and in the accomplishment of cost-
effectiveness analyses) requires full-cost visibility allowing for the traceability of
all costs back to the activities, processes, and/or products that generate these costs.
In the traditional accounting structures employed in most organizations, a large

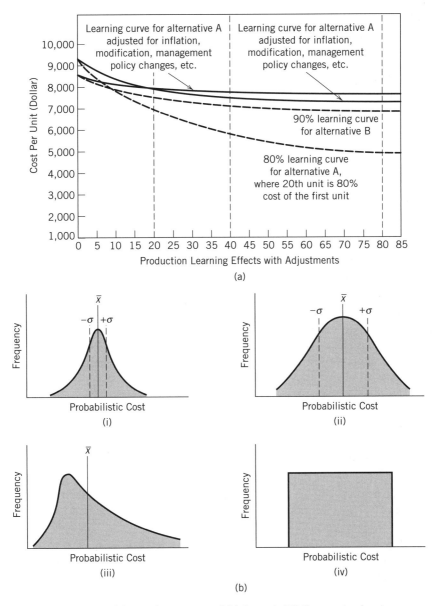

FIGURE B.8   (a) Learning curves and (b) the probabilistic aspects of costs.

percentage of the total cost cannot be traced back to the "causes." For example, overhead or indirect costs, which often constitute more than 50 percent of the total, include a lot of management costs, supporting organization costs, and other costs that are difficult to trace and assign to specific objects (refer to the overhead costs in Figure 6.27). With these costs being allocated across the board, it is impossible to identify the actual "causes" and to pinpoint the *true* high-cost contributors. As a result, the concept of *activity-based costing* (ABC) has been introduced.[2]

Activity-based costing is a methodology directed toward the detailing and assignment of costs to the items that cause them to occur. The objective is to enable the "traceability" of *all* applicable costs to the process or product that generates these costs. The ABC approach allows for the initial allocation and later assessment of costs by function. It was developed to deal with the shortcomings of the traditional management accounting structure whereby large overhead factors are assigned to all elements of the enterprise across the board without concern for whether they directly apply or not. More specifically, the principles of ABC include the following six factors:

1. Cost are directly traceable to the applicable cost-generating process, product, and/or a related object. Cause-and-effect relationships are established between a cost factor and a specific process or activity.

2. There is no distinction between direct and indirect (or overhead) costs. Generally, 80 to 90% of all costs are traceable, and nontraceable costs are not allocated across the board, but are allocated directly to the organizational unit(s) involved in the project.

3. Costs can be easily allocated on a *functional* basis; that is, according to the functions identified in Figures 2.14 and 2.17 (Chapter 2). It is relatively easy to develop cost-estimating relationships in terms of the cost of activities per some activity measure (i.e., the cost per unit output).

4. The emphasis in ABC is on *resource consumption* (versus spending). Processes and products consume activities, and activities consume resources. With resource consumption being the focus, the ABC approach facilitates the evaluation of day-to-day decisions in terms of their impact on resource consumption downstream.

5. The ABC approach fosters the establishment of cause-and-effect relationships and, as such, enables the identification of the "high-cost contributors." Areas of risk can be identified with some specific activity and the decisions that are being made associated with this activity.

6. The ABC approach tends to eliminate some of the cost doubling (or double counting) that occurs in attempting to differentiate as to what should be included as a "direct" cost or as an "indirect" cost. Without the necessary visibility, there is the potential for including the same costs in both categories.

It is essential to implement the ABC approach, or something of an equivalent nature, in order to do a good job of total cost management. Costs are tied to objects and viewed over the long term, and such a perspective facilitates the life-cycle

[2]D. T. Hicks, *Activity-Based Costing: Making It Work for Small and Mid-Sized Companies*, 2nd ed. (New York: John Wiley & Sons, 1999); and G. Conkins, *Activity-Based Cost Management: An Executive's Guide* (Hoboken, NJ: John Wiley & Sons, 2001).

cost-analysis process. An objective for the future is to persuade the accounting organizations in various companies/agencies to supplement their current end-of-year financial reporting structure to include the objectives of ABC.

B.5 SELECT A COMPUTER-BASED MODEL TO FACILITATE THE ANALYSIS PROCESS

In the selection of a computer-based model, one must ensure that the tool selected does what is expected, is sensitive to the problem at hand, and allows for the visibility needed in addressing the system as an entity, as well as any of its major components on an individual-by-individual basis. The model must enable the comparison of *many* different alternatives and aid in selecting the best among them rapidly and efficiently. The model must be *comprehensive,* allowing for the integration of many different parameters; *flexible* in structure, enabling the analyst to look at the system as a whole or any part of the system; *reliable,* in terms of repeatability of results; and *user-friendly.* So often, one selects a computer model based on the material in the advertising brochure alone, purchases the necessary equipment and software, uses the model to manipulate data, and believes in the output results without having any idea as to how the model was put together, the internal analytical relationships established, whether it is sensitive to the variation of input parameters in terms of output results, and so on. The results of a recent survey indicate that there are many computer-based tools available in the commercial marketplace and intended for use in accomplishing different levels of analysis. Each was developed on a relatively "independent" or "isolated" basis in terms of selected platform, language used, input data needs, and interface requirements. In general, the models do not "talk to each other," are not user-friendly, and are too complex for use in early system design and development.

When using a model, it is essential that the analyst become thoroughly familiar with the tool, know how it was put together, and understand what it can do. For the purposes of accomplishing a life-cycle cost analysis, it may be appropriate to select a group of models, combined as illustrated in Figure B.9 and integrated in such a manner that will enable the analyst to look not only at the cost for the system overall, but at some of the key functional areas representing potential high-cost contributors. The model(s) must be structured around the cost breakdown structure (CBS) and in such a way that will allow the analyst to look at the costs associated with each of the major functions. Further, it must be *adaptable* for use during the early stages of conceptual design as well as in the detail design and development phase.

B.6 DEVELOP A BASELINE COST PROFILE

Through the application of various estimating methods, the costs for each CBS category and for each year in the system life cycle are projected in the form of a cost profile. The worksheet format presented in Figure B.10 can serve as a vehicle for recording costs, and the profile shown in Figure B.11 can represent the anticipated cost stream.

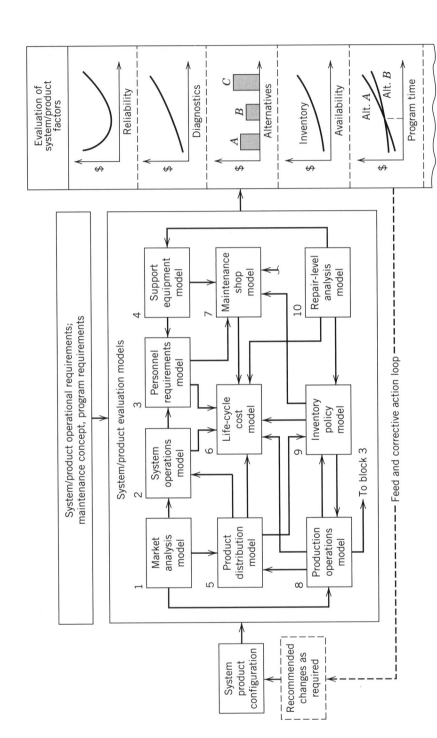

FIGURE B.9 Example models in life-cycle costing.

Program Activity	Cost Category Designation	Cost by Program Year ($)												Total cost ($)	Percent Contr. (%)
		1	2	3	4	5	6	7	8	9	10	11	12		
<u>Alternative A</u> 1. Research and development cost a. Program management b. Engineering design c. Electrical design d. Engineering data 2. 3. Others	C_R C_{RM} C_{RE} C_{RED} C_{RD}														
Total Actual Cost	C														
Total P.V. Cost (10%)	$C_{(10)}$														
<u>Alternative B</u> 1. Research and development cost a. Program management b. Engineering design Etc.	C_R C_{RM} C_{RE}														

FIGURE B.10   Cost collection worksheet.

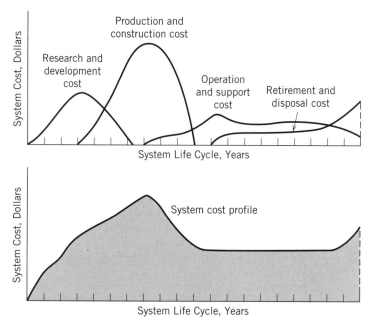

FIGURE B.11   Development of a cost profile.

In developing profiles, it may be feasible to start out with one presented in terms of *constant* dollars first (i.e., the costs for each year in the future presented in terms of today's dollars) and then develop a second profile by adding the appropriate inflationary factors for each year to reflect a *budgetary* stream. In comparing alternative profiles, the appropriate economic analysis methods must be applied in converting the various alternative cost streams to the *present value* or to the point in time when the decision is to be made in selecting a preferred approach. It is necessary to evaluate alternative profiles on the basis of some form of *equivalence*.[3]

B.7   DEVELOP A COST SUMMARY AND IDENTIFY THE HIGH-COST CONTRIBUTORS

In order to gain some insight pertaining to the costs for each major category in the CBS and to readily identify the *high-cost contributors,* it may be appropriate to view the results presented in a tabular form. In Figure B.12, the costs for each category are identified along with the percent contribution of each. Note that in this example, the high-cost areas include the initial costs associated with "facilities" and "capital equipment" and the operating and maintenance costs related to the "inspection and test" function being accomplished within the production process. For the purposes of product and/or process improvement, the "inspection and test" area should be investigated further. Through the planned life cycle, 17% of the total cost is attributed to the operation and support of this functional area of activity, and the analyst should proceed with determining some of the reasons for this high cost.

B.8   DETERMINE THE CAUSE-AND-EFFECT RELATIONSHIPS PERTAINING TO HIGH-COST AREAS

Given the presentation of costs (and the percent contribution) as shown in Figure B.12, the next step is to determine the likely "causes" for these costs. The analyst will need to revisit the CBS, the assumptions made leading to the determination of the costs, and the cost-estimating relationships utilized in the process. It is to be hoped that an *activity-based costing* (ABC) approach was used, or something of an equivalent nature, to ensure the proper traceability. The application of an Ishikawa cause-and-effect diagram, illustrated in Figure C.4 (Appendix C), may be used to assist in pinpointing the actual "causes." The problem may relate to an unreliable product requiring a lot of maintenance, an inadequate procedure or poor process, a supplier problem, or other such factors.

---

[3]The treatment of cost streams considering the "time value of money" is presented in most texts dealing with engineering economy as listed in Appendix F. Two good references for this example are (1) G. J. Thuesen and W. J. Fabrycky, *Engineering Economy,* 9th ed. (Upper Saddle River, NJ: Pearson Prentice-Hall, 2001); and B. S. Blanchard and W. J. Fabrycky, *Systems Engineering and Analysis,* 4th ed. (Upper Saddle River, NJ: Pearson Prentice-Hall, 2006).

Production Operation-Functional Flow

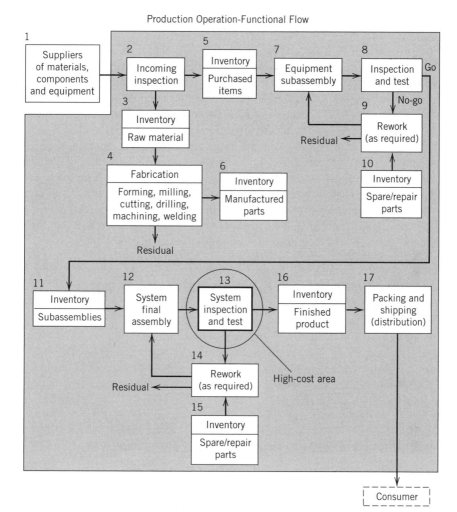

Cost Category	Cost × 1,000 ($)	% of Total
1. Architecture and design	2,248	7
2. Architecture and design	12,524	39
(a) Facilities	6,744	21
(b) Capital equipment	5,780	18
3. Future operation and maintenance	17,342	54
(a) Incoming inspection	963	3
(b) Fabrication	3,854	12
(c) Subassembly	1,927	6
(d) Final assembly	3,533	11
(e) Inspection and test	5,459	17
(f) Packing and shipping	1,606	5
Grand Total	$32,114	100%

FIGURE B.12 Life-cycle cost breakdown summary.

B.9   CONDUCT A SENSITIVITY ANALYSIS

To properly assess the results of the life-cycle cost analysis, the validity of the data presented in Figure B.12, and the associated risks, the analyst needs to conduct a *sensitivity analysis*. One may challenge the accuracy of the input data (i.e., the factors used and the assumptions made in the beginning) and determine their impact on the analysis results. This may be accomplished by identifying the critical factors at the input stage (i.e., those parameters that are suspected as having a large impact on the results), introducing variations over a designated range at the input stage, and determining the differences in output. For example, if the initially predicted reliability MTBF value is "suspect," it may be appropriate to apply variations at the input stage and determine the changes in cost at the output. The object is to identify those areas in which a small variation at the input stage will cause a large delta cost at the output. This, in turn, leads to the identification of potential high-risk areas, a necessary input to the risk management program described in Section 6.7 (Chapter 6).

B.10   CONDUCT A PARETO ANALYSIS TO IDENTIFY MAJOR PROBLEM AREAS

With the objective of implementing a program for *continuous process improvement*, the analyst may wish to rank the problem areas on the basis of relative importance, the higher-ranked problems requiring immediate attention. This may be facilitated through the conductance of a Pareto analysis and the construction of a diagram, as shown in Figure B.13.

B.11   IDENTIFY AND EVALUATE FEASIBLE ALTERNATIVES

In referring to the requirements for the communication system described in Section B.1, two potential suppliers were considered through a feasibility analysis; that is, configuration A and configuration B. Figure B.14 presents a *budgetary* profile for each of three configurations, with configuration C being eliminated for noncompliance. For the purposes of comparison on an equivalent basis, the two remaining profiles have been converted to reflect *present value* costs. Figure B.15 presents a breakdown summary of these present value costs by major CBS category and identifies the relative percent contribution of each category in terms of the total. A 10% interest rate was used in determining present value costs.

Although a review of Figure B.15 might lead one to immediately select configuration A as being preferable, prior to making such a decision the analyst needs to project the two cost streams in terms of the life cycle and determine the point in time when configuration A assumes the position of *preference*. Figure B.16 shows the results of a break-even analysis, and it appears that A is preferable after approximately 6.5 years into the future. The question arises as to whether this break-even point is reasonable in considering the type of system and its mission, the technologies being

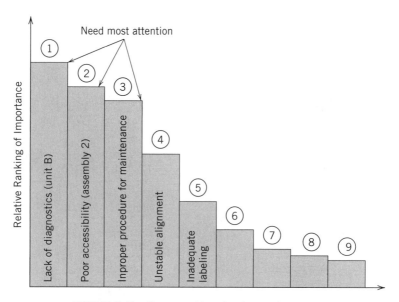

FIGURE B.13 Pareto ranking of major problem areas.

utilized, the length of the planned life cycle, and the possibilities of obsolescence. For systems in which the requirements are changing constantly and obsolescence may become a problem two to three years hence, the selection of configuration B may be preferable. On the other hand, for larger systems with longer life cycles (e.g., 10 to 15 years and greater), the selection of configuration A may be the best choice.

In this case, it is assumed that configuration A is preferable. However, when the cost profile for this alternative is converted back to a *budgetary* projection, it is realized that a further reduction of cost is necessary. This, in turn, leads the analyst

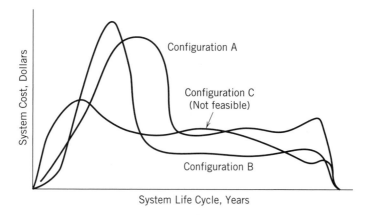

FIGURE B.14 Alternative cost profiles.

Cost Category	Configuration A		Configuration B	
	Present Cost	% of Total	Present Cost	% of Total
1. Research and development	$70,219	7.8	$53,246	4.2
(a) Management	9,374	1.1	9,252	0.8
(b) Engineering	45.552	5.0	28,731	2.3
(c) Test and evaluation	12,176	1.4	12,153	0.9
(d) Technical data	3,117	0.3	3,110	0.2
2. Production (investment)	407,114	45.3	330,885	26.1
(a) Construction	45,553	5.1	43,227	3.4
(b) Manufacturing	362,261	40.2	287,658	22.7
3. Operations and maintenance	422,217	46.7	883,629	69.4
(a) Operations	37,811	4.2	39,301	3.1
(b) Maintenance	382,106	42.5	841,108	66.3
-maintenance personnel	210,659	23.4	407,219	32.2
-spares/repair parts	103,520	11.5	228,926	18.1
-Test equipment	47,713	5.3	131,747	10.4
-Transportation	14,404	1.6	51,838	4.1
-Maintenance training	1,808	0.2	2,125	0.1
-Facilities	900	0.1	1,021	Neg.
-Field data	3,102	0.4	18,232	1.4
4. Phaseout and disposal	2,300	0.2	3,220	0.3
Grand Total	$900,250	100%	$1,267,760	100%

FIGURE B.15   Life-cycle cost breakdown (evaluation of two alternative configurations).

to Figure B.15 and the identification of potential *high-cost* contributors. Given that a large percentage of the total cost of a system is often in the area of maintenance and support, one might investigate the categories of "maintenance personnel" and "spares/repair parts," representing 23.4% and 11.5% of the total cost, respectively. The next step is to identify the applicable cause-and-effect relationships and to determine the actual causes for such high costs. This may be accomplished by being able to trace the costs back to a specific function, process, product design characteristic, or a combination thereof. The analyst also needs to refer back to the CBS and review how the costs were initially derived and the assumptions that were made at the input stage. In any event, the problem may be traced back to a specific function in which the resource consumption is high, a particular component of the system with low reliability and requiring frequent maintenance, a specific system operating function that requires a lot of highly skilled personnel, or something of an equivalent nature. Various design tools can be effectively utilized to aid in making visible these causes and to help identify areas where improvement can be made; for example, the failure mode, effects, and criticality analysis, the detailed task analysis, and so on.

As a final step, the analyst needs to conduct a sensitivity analysis to properly assess the risks associated with the selection of configuration A. Figure B.17 illustrates

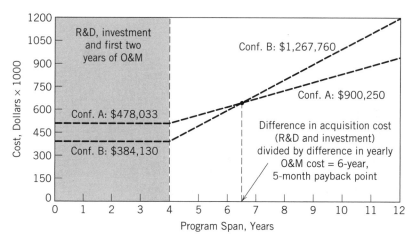

FIGURE B.16   Break-even analysis.

this approach as it applies to the "maintenance personnel" and "spares/repair parts" categories addressed earlier. The objective is to identify those areas where a small variation at the input stage will cause a large delta cost at the output. This, in turn, leads to the identification of potential high-risk areas, a necessary input to the risk management program described in Section 6.7.

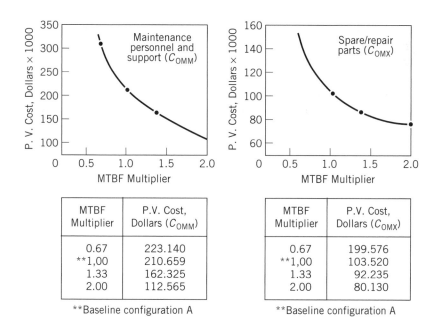

FIGURE B.17   Sensitivity analysis.

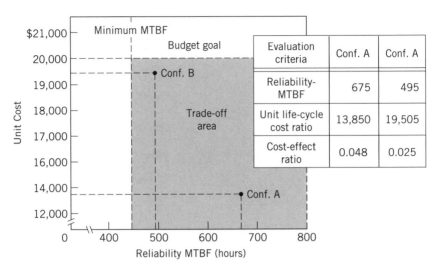

FIGURE B.18 Reliability versus unit life-cycle cost.

B.12 SELECT A PREFERRED DESIGN APPROACH

The cost issue having been addressed, it is necessary to view the results in the context of the overall cost-effectiveness balance illustrated in Figure 1.24 (Chapter 1). Although the emphasis here has been on cost, the ultimate decision-making process must consider both sides of the spectrum; that is, *cost* and *effectiveness*. For example, the two alternative communication system configurations discussed earlier must meet the reliability and cost goals described in Section B.1. In Figure B.18, the shaded area represents the allowable design trade-off "space," and the alternatives must be viewed not only in terms of cost, but in terms of reliability as well. As indicated in Section 3.4.12, the ultimate decision may be based on an overall cost-effectiveness ratio or some equivalent metric. Configuration A remains as the preferred choice.

APPENDIX C

SELECTED CASE STUDIES (SEVEN EXAMPLES)

Throughout the early stages of the system life cycle, and as an inherent part of the system engineering process, numerous applications of different tools can facilitate trade-off studies in system design. Of particular interest here are some of the tools that address the downstream aspects of system support but can be effectively utilized earlier. Seven specific applications are noted as follows and described further in Figure C.1[1]:

C.1. Failure mode, effects, and criticality analysis (FMECA)

C.2. Fault-tree analysis (FTA)

C.3. Reliability-centered maintenance (RCM)

C.4. Maintenance task analysis (MTA)

C.5. Level-of-repair analysis (LORA)

C.6. Design evaluation of aternatives

C.7. Life-cycle cost analysis (LCCA) (which supports the material in Appendix B)

Each of these case-study examples is presented with a *definition of the problem, the analysis process*, and *the analysis results*.

C.1 FAILURE MODE, EFFECTS, AND CRITICALITY ANALYSIS (FMECA)

C.1.1 Definition of the Problem

Company ABC, a manufacturer of gaskets for automobiles, was experiencing problems related to declining productivity and increased product costs. At the same time,

---

[1]Case studies C.1, C.2, and C.3 were taken in part from B. S. Blanchard, D. Verma, and E. L. Peterson, *Maintainability: A Key to Effective Serviceability and Maintenance Management* (Hoboken, NJ: John Wiley & Sons, 1995).

Analysis Tools	Description of Application
C.1 Failure mode, effects, and criticality analysis (FMECA)	Identification of potential product and/or process failures, the expected modes of failure and "causes," failure effects and mechanisms, anticipated frequency, criticality, and the steps required for compensation (i.e., the requirement for redesign and/or the accomplishment of preventive maintenance). An Ishikawa "cause-and-effect" diagram may be used to facilitate the identification of "causes," and a Pareto analysis may help in identifying those areas requiring immediate attention.
C.2 Fault-tree analysis (FTA)	A deductive approach involving the graphical enumeration and analysis of different ways in which a particular system failure can occur, and the probability of its occurrence. A separate fault tree may be developed for every critical failure mode, or undesired top-level event. Attention is focused on this top-level event and the first-tier causes associated with it. Each of these causes is next investigated for its causes, and so on. The FTA is narrower in focus than the FMECA and does not require as much input data.
C.3 Reliability-centered maintenance (RCM)	Evaluation of the system/process, in terms of the life cycle, to determine the best overall program for preventive (scheduled) maintenance. Emphasis is on the establishment of a cost-effective preventive maintenance program based on reliability information derived from the FMECA; that is, failure modes, effects, frequency, criticality, and compensation through preventive maintenance.
C.4 Maintenance task analysis (MTA)	Evaluation of those *maintenance* functions that are to be allocated to the human. Identification of maintenance functions/tasks in terms of task times and sequences, personnel quantities and skill levels, and supporting resources requirements (i.e., spares/repair parts and associated inventories, tools and test equipment, facilities, transportation and handling requirements, technical data, training, and computer software). Identification of high resource-consumption areas.
C.5 Level-of-repair analysis (LORA)	Evaluation of maintenance,policies in terms of levels of repair. That is, should a component be repaired in the event of a failure or discarded and, given the "repair" option, should the repair be accomplished at the intermediate level of maintenance, at the supplier's factory, or at some other level? Decision factors include economic, technical, social, environmental, and political considerations. The emphasis here is based on life-cycle cost factors.
C.6 Design evaluation of alternatives	Evaluation of alternative design configurations using multiple criteria. Weighting factors are established to specify levels of importance.
C.7 Life-cycle cost analysis (LCCA) (Supplement to Appendix B)	Determination of the system/product/process life-cycle cost (design and development, production and/or construction, system utilization, maintenance and support, and retirement/disposal costs); high-cost contributors; cause-and-effect relationships; potential areas of risk; and identification of areas for improvement (i.e., cost reduction).

FIGURE C.1   Design analysis methods (case study applications)

competition was increasing and the company was losing its share of the market. As a result, the company decided to implement a *continuous process improvement program* with the objective of identifying potential problem areas and their impact and criticality on both internal company operations and the product being delivered to the customer. To aid in facilitating this objective, the company's manufacturing operations were evaluated using the failure mode, effects, and criticality analysis (FMECA).

C.1.2 The Analysis Process

An initial step included the identification of the major functions performed in the overall gasket manufacturing process by completing a functional flow diagram in accordance with the procedures described in Section 2.7 (Chapter 2). In this instance, there were 13 major functions that were subject to evaluation. For each function, required input factors and expected outputs were identified, along with the appropriate metrics. This led to the initial selection of 1 of the 13 functions, based on a perception by company personnel as to the area causing the most problems. Given this selection, the sequence of steps conveyed in Figure C.2 was followed in completing an FMECA of the selected function.

Figure C.3 represents the function, or portion of the overall manufacturing process, that was selected for evaluation. Note that although the emphasis is on the manufacturing process and its impact on the gasket, one must also consider the impact of a faulty gasket on the automobile. Thus, the FMECA needs to address both the *process* and the *product.*

As shown in Figure C.2, the approach selected for conducting the FMECA was in accordance with the practices followed in the automotive industry.[2] This included the following:

1. Identifying the different failure modes; that is, the manner in which a system element fails to accomplish its function.

2. Determining the cause(s) of failure; that is, the factor(s) responsible for the occurrence of each failure. An Ishikawa *cause-and-effect,* or *fishbone,* diagram, as illustrated in Figure C.4, was utilized to help establish the relationships between failures and their possible causes.[3]

3. Determining the effects of failure; that is, the effects on subsequent functions/processes, on the next higher-level functional entity, and on the overall system.

---

[2]Three references were used, including (1) *Potential Failure Mode and Effects Analysis;* Instruction Manual, Ford Motor Company, 1988; (2) *Failure Mode and Effects Analysis.* Instruction Manual, Saturn Quality System, Saturn Corporation, 1990; and (3) *Potential Failure Mode and Effects Analysis (FMEA),* Reference Manual FMEA-1, developed by FMEA teams at Ford Motor Company, General Motors, Chrysler, Goodyear, Bosch, and Kelsey-Hayes, under the auspices of the American Society of Quality Control (ASQC).

[3]K. Ishikawa, *Introduction to Quality Control* (London: Chapman and Hall, 1991).

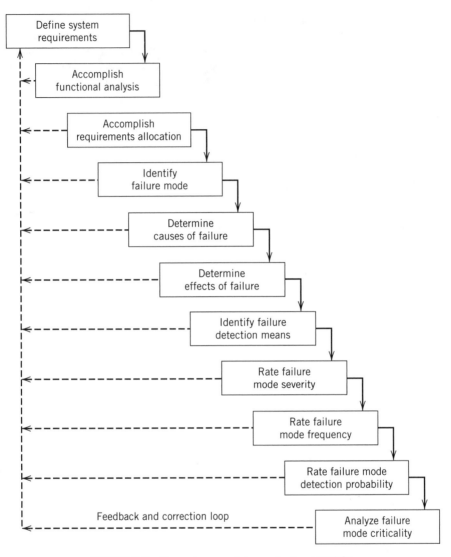

FIGURE C.2   General approach to conducting a FMECA.

4. Identifying failure detection means; that is, the current controls, design features, or verification procedures that will result in the detection of potential failure modes.

5. Determining the severity of a failure mode; that is, the seriousness of the effect or impact of a particular failure mode. The degree of severity was converted quantitatively on a scale of 1 to 10, with *minor* effects being 1, *low* effects being 2 to 3, *moderate* effects being 4 to 6, *high* effects being 7 to 8, and *very*

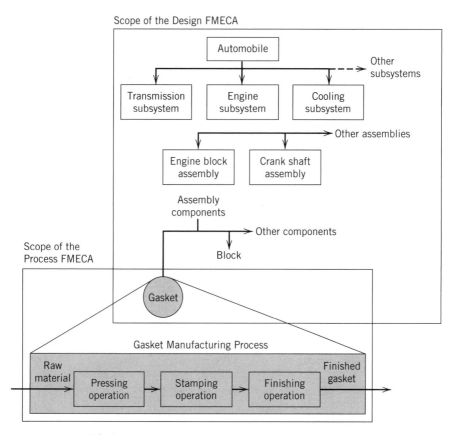

FIGURE C.3   Design and process FMECA focus and scope.

high effects being 9 to 10. The level of severity was related to issues pertaining to safety and the degree of customer dissatisfaction.

6. Determining the frequency of occurrence; that is, the frequency of occurrence of each individual failure mode or the probability of failure. A scale of 1 to 10 was applied with *remote* (failure is unlikely) being 1, *low* (relatively few failures) being 2 to 3, *moderate* (occasional failures) being 4 to 6, *high* (repeated failures) being 7 to 8, and *very high* (failure is almost inevitable) being 9 to 10. These rating factors were based on the number of failures per segment of operating time.

7. Determining the probability that a failure will be detected; that is, the probability that the design features/aids and/or verification procedures will detect potential failure modes in time to prevent a system-level failure. For a process application, this refers to the probability that a set of process controls currently in place will be in a position to detect and isolate a failure before it is transferred to the subsequent processes or to the ultimate product output. This

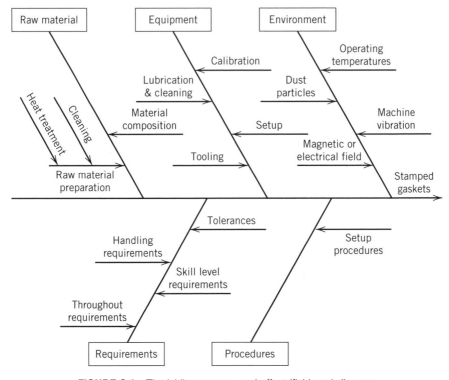

FIGURE C.4   The Ishikawa cause-and-effect (fishbone) diagram.

probability is once again rated on a scale of 1 to 10, with *very high* being 1 to 2, *high* being 3 to 4, *moderate* being 5 to 6, *low* being 7 to 8, *very low* being 9, and *absolute certainty of nondetection* being 10.

8. Analyzing failure mode criticality; that is, a function of severity (item 5), the frequency of occurrence of a failure mode (item 6), and the probability that it will be detected in time to preclude its impact at the system level (item 7). This resulted in the determination of the *risk priority number* (RPN) as a metric for evaluation. RPN can be expressed as:

$$\text{RPN} = (\text{severity rating})(\text{frequency rating})(\text{probability of detection rating})$$
$$(\text{C.1})$$

The RPN reflects failure-mode criticality. On inspection, one can see that a failure mode with a high frequency of occurrence, with significant impact on system performance, and that is difficult to detect is likely to have a very high RPN.

9. Identifying critical areas and recommending modifications for improvement; that is, the iterative process of identifying areas with high RPNs,

evaluating the causes, and initiating recommendations for process/product improvement.

Figure C.5 shows a partial example of the format used for recording the results of the FMECA. The information was derived from the functional flow diagram and expanded to include the results from the steps presented in Figure C.2. Figure C.6 lists the resulting RPNs in order of priority (relative to requiring attention), and Figure C.7 presents the results in the form of a Pareto analysis.

C.1.3 The Analysis Results

After having completed the FMECA on the function identified in Figure C.3, Company ABC proceeded to evaluate each of its other 12 major functions/processes in a similar manner, utilizing a *team* approach. The activity was very beneficial overall, the individuals participating in the effort learned more about their own activities, and numerous changes were initiated for purposes of improvement.

C.2 FAULT-TREE ANALYSIS (FTA)

C.2.1 Definition of the Problem

During the very early stages of the system design process, and in the absence of the information required to complete a FMECA (discussed in Section C.1), a fault-tree analysis (FTA) was conducted to gain insight into critical aspects of system design. A fault-tree analysis is a deductive approach involving the graphical enumeration and analysis of the different ways in which a particular system failure can occur and the probability of its occurrence. A separate fault tree is developed for every critical failure mode or undesired top-level event. The emphasis is on this top-level event and the first-tier causes associated with it. Each of these causes is next investigated for *its* causes, and so on. This top-down hierarchy, illustrated in Figure C.8, and the associated probabilities, is called a *fault tree*. Figure C.9 presents some of the symbology used in the development of such a structure.

C.2.2 The Analysis Process

One of the outputs from an FTA is the probability of occurrence of the top-level event or failure. If the probability factor is unacceptable, the causal hierarchy developed provides engineers with insight into aspects of the system to which redesign efforts may be directed or for which compensatory provisions may be provided. The logic used in developing and analyzing a fault tree has its foundations in Boolean algebra. Axioms from Boolean algebra are used to collapse the initial version of the fault tree to an equivalent reduced tree with the objective of deriving *minimum cut sets.* Minimum cut sets are unique combinations of basic failure events that can cause

Process Failure Mode and Effects Analysis

Reference Number	Process Description	Potential Failure Mode	Potential Cause of Failure	Potential Effect(s) of Failure at Federal Mongul	Potential Effect(s) of Failure at Customer	Current Controls	Occurrence	Beverly FM	Beverly C	Detection	RPN	Recommended Action(s) and Status	Responsible Activity
8.1.2	Sense for oversized queue	A) Change in free spread	1) Sensor fails	a) Up process jams b) Height variations	c) Bearing loose when installed in engine	a) Machine shops b) 1 Pc/5 min c) 2 Pcs/half hr	1 1 1	1 7	7	1 3 5	1 21 35		
			1) Sensor dirty	a) Up process jams b) Height variations	c) Bearing loose when installed in engine	a) Machine shops b) 1 Pc/5 min c) 2 Pcs/half hr	1 1 1	1 7	7	1 3 5	1 21 35		
			1) Improper setup	a) Up process jams b) Height variations	c) Bearing loose when installed in engine	a) Machine shops b) 1 Pc/5 min c) 2 Pcs/half hr	1 1 1	1 7	7	1 3 5	1 21 35		
8.1.3	Sense for undersized queue	A) Up mislocated	1) Sensor fails	a) Up process jams	b) Fillet ride	a) 100% visual b) 5 Pcs/half hr	2 2	1	3	1 7	2 42		
			2) Sensor dirty	a) Up process jams	b) Fillet ride	a) 100% visual b) 5 Pcs/half hr	2 2	1	3	1 7	2 42		
			3) Improper setup	a) Up process jams	b) Fillet ride	a) 100% visual b) 5 Pcs/half hr	2 2	1	3	1 7	2 42		
		B) Facing/back damage	1) Sensor fails 2) Sensor dirty 3) Improper setup		a) Rejected at assembly a) Rejected at assembly a) Rejected at assembly	a) 100% visual a) 100% visual a) 100% visual	3 3 3		5 5 5	4 4 4	60 60 60		
8.1.4	Load block	A) Up mislocated	1) "Hold down" not set properly	a) Up smashed in broache b) Up process jams		a) 5 Pcs/half hr b) 100% visual	3 3	7 1		7 7	147 21		
			2) Loose load block	a) Up smashed in broache b) Up process jams	c) Fillet ride	a) 5 Pcs/half hr b) 100% visual c) 5 Pcs/half hr	3 2	7	3	7 7	63 96		
					c) Fillet ride	b) 100% visual c) 5 Pcs/half hr	2	1	5	7 7	14 42		
		B) Facing/back damage	1) Misaligned pusher		a) Rejected at assembly	a) 100% visual	4		4	4	80		

FIGURE C.5   Sample FMECA worksheet.

Causes	Risk Priority Numbers (RPNs)
Chip breaker angle ground incorrectly	273
Hold-down not set correctly	210
Undersize sensor fails	200
Undersize sensor dirty	200
Undersize sensor not positioned properly	200
Loose load block	161
Sharp die edge	120
Improper projection angle/resharpening of punch	108
Oversize sensor fails	105
Oversize sensor dirty	105
Oversize sensor not positioned properly	105
Improper sharpening of insert	93
Misaligned pusher	80
Worn tooling	72
Adapter reground to wrong dimension	60
Insert loose	60
Slivers in adapter	60
Insert off location	60
Worn/loose insert	60
Burrs from punch process caught	40
Ram stroke too long	36
Ram stroke too short	21
Broken/loose punch	12
Setscrew fault	12
Insufficient stroke by ram/punch	12
Broken pressure spring	10
Total	2475

FIGURE C.6   Risk priority numbers (RPNs).

the undesired top-level event to occur. These minimum cut sets are necessary to evaluate a fault tree from a qualitative and quantitative perspective. The basic steps in conducting an FTA are as follows:

1. *Identify the top-level event.* It is essential that the analyst be quite specific in defining this event. For example, it may be delineated as the "system catches fire," rather than the "system fails." Further, the top-level event should be clearly observable and unambiguously definable and measurable. A generic and nonspecific definition is likely to result in a broad-based fault tree with a scope that is too wide and lacking in focus.

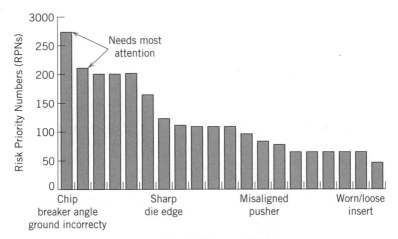

FIGURE C.7   Partial Pareto analysis.

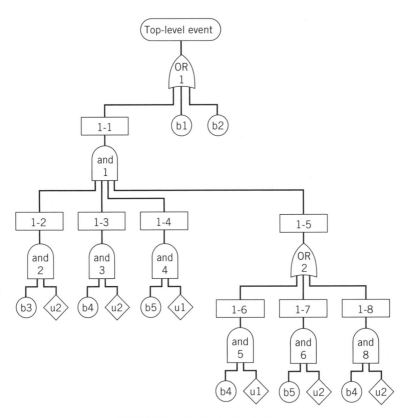

FIGURE C.8   An illustrative fault tree.

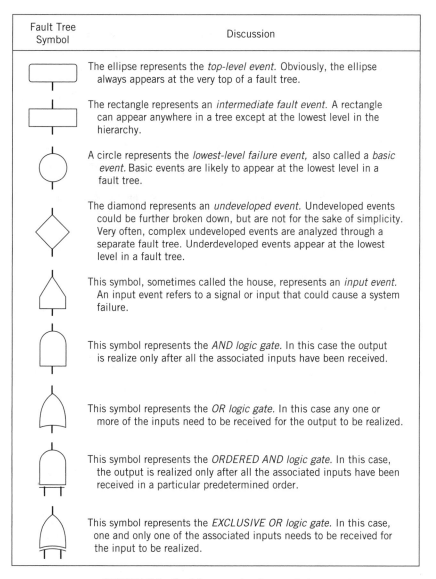

Fault Tree Symbol	Discussion
	The ellipse represents the *top-level event.* Obviously, the ellipse always appears at the very top of a fault tree.
	The rectangle represents an *intermediate fault event.* A rectangle can appear anywhere in a tree except at the lowest level in the hierarchy.
	A circle represents the *lowest-level failure event,* also called a *basic event.* Basic events are likely to appear at the lowest level in a fault tree.
	The diamond represents an *undeveloped event.* Undeveloped events could be further broken down, but are not for the sake of simplicity. Very often, complex undeveloped events are analyzed through a separate fault tree. Underdeveloped events appear at the lowest level in a fault tree.
	This symbol, sometimes called the house, represents an *input event.* An input event refers to a signal or input that could cause a system failure.
	This symbol represents the *AND logic gate.* In this case the output is realize only after all the associated inputs have been received.
	This symbol represents the *OR logic gate.* In this case any one or more of the inputs need to be received for the output to be realized.
	This symbol represents the *ORDERED AND logic gate.* In this case, the output is realized only after all the associated inputs have been received in a particular predetermined order.
	This symbol represents the *EXCLUSIVE OR logic gate.* In this case, one and only one of the associated inputs needs to be received for the input to be realized.

FIGURE C.9 Fault-tree constructive symbology.

2. *Develop the fault tree.* Once the top-level event has been satisfactorily defined, the next step is to construct the initial causal hierarchy in the form of a fault tree. Once again, a technique such as Ishikawa's cause-and-effect diagram can be beneficial (refer to Figure C.4). In developing the fault tree, all hidden failures must be considered and incorporated.

For the sake of consistency and communication, a standard symbology to develop the fault tree is recommended. Figure C.9 depicts and defines the

symbology to comprehensively represent the causal hierarchy and intercon-
nects associated with a particular top-level event. In Figure C.8, the symbols
OR1 and OR2 represent the two OR logic gates, "and 1" through "and 8"
represent eight AND logic gates, 1-1 through 1-8 represent eight intermediate
fault events, b1 through b5 represent five basic events, and u1 and u2 represent
two undeveloped failure events. In constructing a fault tree, it is important to
break every branch down to a reasonable and consistent level of detail.

3. *Analyze the fault tree.* The third step in conducting the FTA is to analyze the
   initial fault tree developed. A comprehensive analysis of a fault tree involves
   both a quantitative and a qualitative perspective. The important steps in completing the analysis of a fault tree are as follows:
   (a) *Delineate the minimum cut sets.* As part of the analysis process, the mini-
       mum cut sets in the initial fault tree are first delineated. These are necessary
       to evaluate a fault tree from a qualitative and/or quantitative perspective.
       The objective of this step is to reduce the initial tree to a simpler equiv-
       alent reduced fault tree. The minimum cut sets can be derived using two
       different approaches. The first approach involves a graphical analysis of
       the initial tree, an enumeration of all the cut sets, and the subsequent de-
       lineation of the minimal cut sets. The second approach, on the other hand,
       involves translating the graphical fault tree into an equivalent Boolean ex-
       pression. This Boolean expression is then reduced to a simpler equivalent
       expression by eliminating all the redundancies. For example, the fault tree
       depicted in Figure C.8 can be translated into a simpler and equivalent fault
       tree, through Boolean reduction, as depicted in Figure C.10.
   (b) *Determine the reliability of the top-level event.* This is accomplished by
       first determining the probabilities of all relevant input events, and the sub-
       sequent consolidation of these probabilities in accordance with the under-
       lying logic of the tree. The reliability of the top-level event is computed by
       taking the product of the reliabilities of the individual minimum cut sets.

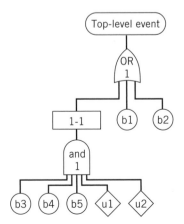

FIGURE C.10   A reduced equivalent fault tree (refer to Figure C.8).

(c) *Review analysis output.* If the derived top-level probability is unacceptable initiate necessary redesign or compensation efforts. The development of the fault tree and subsequent delineation of minimum cut sets provides engineers and analysts with the kind of foundation needed for making sound decisions.

C.2.3 The Analysis Results

An FTA can be effectively applied in the early phases of design to specific areas where potential problems are suspected. It is narrow in focus and easier to accomplish than an FMECA, requiring less input data to complete. For large and complex systems, which are highly software-intensive and where there are many interfaces, the use of the FTA is often preferred in lieu of the FMECA. The FTA is most beneficial if conducted, not in isolation, but as part of an overall system analysis process.[4]

C.3 RELIABILITY-CENTERED MAINTENANCE (RCM)

C.3.1 Definition of the Problem

Reliability-centered maintenance (RCM) is a systematic approach to develop a focused, effective, and cost-efficient preventive maintenance program and control plan for a system or product. This technique is best initiated during the early system design process and evolves as the system is developed, produced, and deployed. However, the technique can also be used to evaluate preventive maintenance programs for existing systems, with the objective of continuous product/process improvement.

The RCM technique was developed in the 1960s primarily through the efforts of the commercial airline industry.[5] The approach is through a structured decision tree that leads the analyst through a "tailored" logic in order to delineate the most applicable preventive maintenance tasks (their nature and frequency). The overall process involved in implementing the RCM technique is illustrated in Figure C.11. Note that the functional analysis and the FMECA are necessary inputs to the RCM, and that there are trade-offs resulting in a balance between preventive maintenance

---

[4]Reliability Information Analysis Center (RIAC), *Fault Tree Analysis Guide*, 6000 Flanagan Rd., Suite 3, Utica, NY 13502, 1991. An excellent "how-to" source for the application of FTA depicting numerous case studies.

[5]A maintenance steering group (MSG) was formed in the 1960s that undertook the development of this technique. The result was a document titled *747 Maintenance Steering Group Handbook: Maintenance Evaluation and Program Development (MSG-1),* published in 1968. This effort, focused on a particular aircraft, was next generalized and published in 1970 as *Airline/Manufacturer Maintenance Program Planning Document-MSG2.* The MSG-2 approach was further developed and published in 1978 as *Reliability Centered Maintenance,* Report Number A066-579, prepared by United Airlines, and in 1980 as *Airline/Manufacturer Maintenance Program Planning Document-MSG3.* The MSG-3 report has been revised and is currently available as *Airline/Manufacturer Maintenance Program Development Document (MSG-3), 1993.* MSG3 activity continues with the results of analyses from the C-5 program. These reports are available from the Air Transport Association.

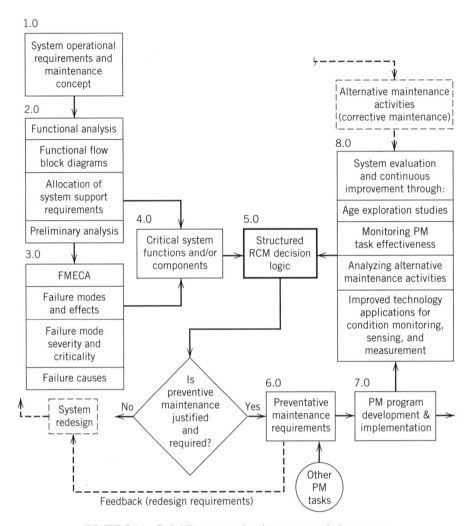

FIGURE C.11   Reliability-centered maintenance analysis process.

and the accomplishment of corrective maintenance. Figure C.12 presents a simplified RCM decision logic, where system safety is a prime consideration along with performance and cost.

C.3.2   The Analysis Process

Three major steps in accomplishing an RCM analysis are as follows:

1. *Identify the critical system functions and/or components.* For example, these might be airplane wings, car engine, printer head, video head, and so on. Criticality in terms of this analysis is a function of the failure frequency, the failure effect severity, and the probability of detection of the relevant failure modes. The concept

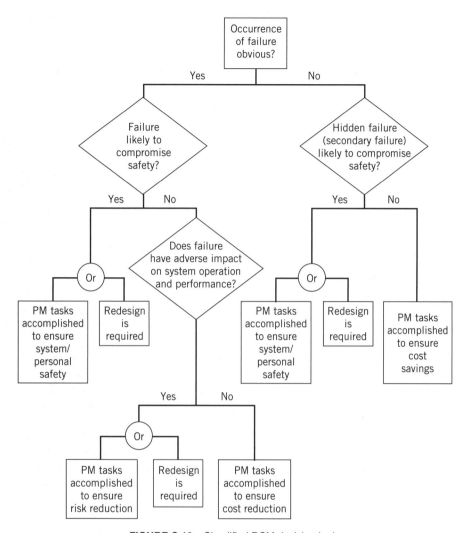

FIGURE C.12 Simplified RCM decision logic.

of criticality is discussed in more detail in Section C.1. This step is facilitated through outputs from the system functional analysis (see Section 2.7) and the failure mode, effects, and criticality analysis (FMECA). This is also depicted in Figure C.11, Blocks 1.0 to 4.0.

2. *Apply the RCM decision logic and preventive maintenance (PM) program development approach.* The critical system elements are subjected to the tailored RCM decision logic. The objective here is to better understand the nature of failures associated with the critical system functions or components. In each case, and whenever feasible, this knowledge is translated into a set of preventive maintenance tasks, or a set of redesign requirements. A simplified illustrative RCM decision logic is depicted

in Figure C.12. Numerous decision logics, with slight variations to the original MSG-3 logic and tailored to better address certain types of systems, have been developed and are currently being utilized.[6]

These slight variations notwithstanding (as illustrated in Figure C.12), the first concern is whether a *failure is evident or hidden.* A failure can become evident through the aid of certain color-coded visual gauges and/or alarms. It may also become evident if it has a perceptible impact on system operation and performance. On the other hand, a failure may not be evident (i.e., hidden) in the absence of an appropriate alarm, and even less so if it does not have an immediate or direct impact on system performance. For example, a leaking engine gasket is not likely to reflect an immediate and evident change in an automobile's operation, but it may in time and, after most of the engine oil has leaked, cause engine seizure. In the event that a failure is not immediately evident, it may be necessary to either initiate a specific fault-finding task as part of the overall PM program or design in an alarm that signals a failure (or pending failure).

The next concern is whether the failure is likely to compromise personal safety or system functionality. Queries exist in the decision logic to clarify this and other likely impacts of failures. This step in the overall process can be facilitated by the results of the FMECA (Section C.1). The objective is to better understand the basic nature of the failure being studied. Is the failure likely to compromise the system or personnel safety? Does it have an operational or economic impact? For example, a failure of an aircraft wing may be safety-related, whereas a certain failure in the case of an automobile engine may result in increased oil consumption without any operational degradation and will therefore have an economic impact. In another case, a failed printer head may result in a complete loss of printing capability and may be said to have an operational impact, and so on.

Once the failure has been identified as a certain type, it is then subjected to another set of questions. However, in order to answer this next set of questions adequately, the analyst must thoroughly understand the nature of the failure from a *physics-of-failure* perspective. For example, in the event of a crack in an airplane wing, how fast is this crack likely to propagate? How long before such a crack causes a functional failure?

These questions have an underlying objective of delineating a feasible set of compensatory provisions or preventive maintenance tasks. Is a lubrication or servicing task applicable and effective, and, if so, what is the most cost-effective and efficient frequency? Will a periodic check help preclude a failure, and at what frequency? Periodic inspections or checkouts are likely to be most applicable in situations where a failure is unlikely to occur immediately, but is likely to develop at a certain rate over a period of time. The frequency of inspections can vary from very infrequently to continuously, as in the case of condition monitoring. Some of the more specific

---

[6]RCM decision logics, with some variations, have also been proposed in (1) MIL-STD-2173(AS), *Reliability-Centered Maintenance Requirements for Naval Aircraft. Weapons Systems, and Support Equipment*; (2) AMC-P-750-2, *Guide to Reliability-Centered Maintenance;* (3) John Moubray, *Reliability-Centered Maintenance,* 2d ed. (New York: Industrial Press, 1997); and (4) A. M. Smith, *Reliability-Centered Maintenance* (New York, McGraw-Hill, Inc., 1993).

queries are presented in Figure C.12. In each case, the analyst must not only respond with a yes or no, but should also give specific reasons for each response. Why would lubrication either make, or not make, any difference? Why would periodic inspection be a *value-added* task? It may be that the component's wear-out characteristics have a predictable trend, in which case inspections at predetermined intervals could preclude corrective maintenance. Would it be effective to discard and replace certain system elements in order to upgrade the overall inherent reliability? And, if so, at what intervals or after how many hours of system operation (e.g., changing the engine oil after 3,000 miles of driving)? Further, in each case a trade-off study, in terms of the benefit/cost and overall impact on the system, needs to be accomplished to determine the trade-offs between performing a task and not performing it.

In the event that a set of applicable and effective preventive maintenance requirements are delineated, they are input to the preventive maintenance program development process and subsequently implemented, as shown in Figure C.11, blocks 5.0 to 7.0. If no feasible and cost-effective provisions or preventive maintenance tasks can be identified, a redesign effort may have to be initiated.

3. *Accomplish PM program implementation and evaluation.* Very often, the PM program initially delineated and implemented is likely to have failed to consider certain aspects of the system, delineated a very conservative set of PM tasks, or both. Continuous monitoring and evaluation of preventive maintenance tasks along with all other (corrective) maintenance actions is imperative in order to realize a cost-effective preventive maintenance program. This is depicted in Figure C.11, block 8.0. Further, given the continuously improving technology applications in the field of condition monitoring, sensing, and measurement, PM tasks need to be reevaluated and modified whenever necessary.

Often, when the RCM technique is conducted in the early phases of the system design and development process, decisions are made in the absence of ample data. These decisions may have to be verified and modified, whenever justified, as part of the overall PM evaluation and continuous improvement program. Age exploration studies are often conducted to facilitate this process. Tests are conducted on samples of unique system elements or components with the objective of better understanding their reliability and wear-out characteristics under actual operating conditions. Such studies can aid the evaluation of applicable PM tasks and help delineate any dominant failure modes associated with the component being monitored and/or any correlation between component age and reliability characteristics. If any significant correlation between age and reliability is noticed and verified, the associated PM tasks and their frequency may be modified and adapted for greater effectiveness. In addition, redesign efforts may be initiated to account for some, if any, of the dominant component failure modes.

C.3.3 The Analysis Results

Quite often in the early design process, as system components are being selected, the issue of maintenance is ignored altogether. If maintenance is addressed, however, the designer may tend to specify components requiring some preventive maintenance (usually recommended by the manufacturer). If this is done, the perception is that such PM recommendations are based on actual knowledge of the component in terms of its physical characteristics, expected modes of failure, and so on. It is also believed that the more preventive maintenance required, the better the reliability. In any event, there is often a tendency to overspecify the need for PM because of the reliability issue, particularly if the component *physics-of-failure* characteristics are not known and the designer assumes a conservative approach, just in case.

Experience indicates that although the accomplishment of some selective preventive maintenance is essential, the overspecification of PM activities can actually cause a degradation of system reliability and can be quite costly. The objective is to specify the correct amount of PM, to the depth required, and at the proper frequency; that is, *not too much or too little*. Further, as systems age, the required amount of PM may shift from one level to another. The application of RCM methods on a continuing basis is highly recommended, particularly in evaluating systems from a life-cycle cost perspective.

C.4 MAINTENANCE TASK ANALYSIS (MTA)

C.4.1 Definition of the Problem

Company DEF has been manufacturing Product 12345 for the past few years. The costs have been higher than anticipated, and international competition has been increasing. As a result, company management has decided to conduct an evaluation of the overall production capability, identify "high-cost" contributors through the accomplishment of a life-cycle cost analysis, and identify possible problem areas where improvement can be realized. One area for possible improvement is the manufacturing test function where frequent failures have occurred during Product 12345 test. By reducing maintenance costs, it is likely that one can reduce the overall cost of the product and improve the company's competitive position in the marketplace. With the objective of identifying some specifics, a detailed maintenance task analysis of the manufacturing test function is accomplished. Specific recommendations for improvement are being solicited.

C.4.2 The Analysis Process

In response, a detailed maintenance task analysis is performed, using the format included in Appendix E of B. S. Blanchard, *Logistics Engineering and Management,* 6th ed. (Upper Saddle River, NJ: Prentice-Hall, 2004). The format, as adapted for the purpose of this evaluation, includes the following general steps:

1. Review of historical information covering the performance of the manufacturing test capability indicated the frequent loss of power during the final testing of Product 12345. From this, a typical "symptom of failure" was identified, and a sample logic troubleshooting flow diagram was developed, as shown in Figure C.13.

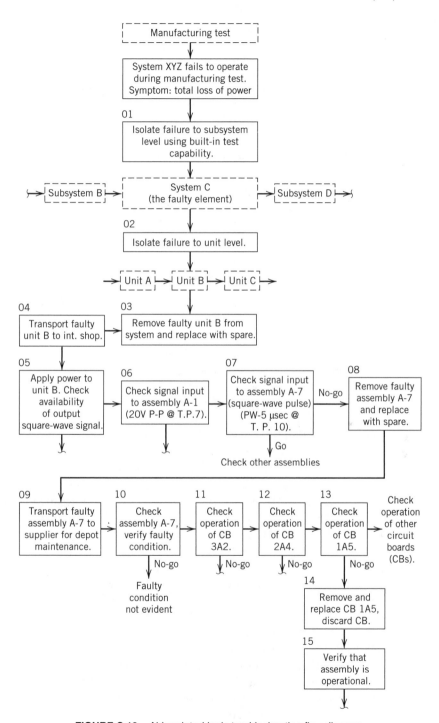

FIGURE C.13 Abbreviated logic troubleshooting flow diagram.

2. The applicable "go/no-go" functions, identified in Figure C.13, are converted to the task analysis format in Figures C.14 and C.15. The functions are analyzed on the basis of determining task requirements (task durations, parallel-series relationships, sequences), personnel quantity and skill-level requirements, spare/repair part requirements, test and support equipment requirements, special facility requirements, technical data requirements, and so on. The intent in Figure C.14 is to lay out the applicable maintenance tasks required, determine the anticipated frequency of occurrence, and identify the logistic support resources that are likely to be necessary for the performance of the required maintenance. This information, in turn, can be evaluated on the basis of cost.

3. Given the preliminary results of the analysis in terms of the layout of expected maintenance functions/tasks, the next step is to evaluate the information presented in Figures C.14 and C.15 and suggest possible areas where improvement can be made.

C.4.3 The Analysis Results

A review of the information presented in Figures C.14 and C.15 suggests that the following seven areas be investigated further:

1. With the extensive resources required for the repair of assembly A-7 (e.g., the variety of special test and support equipment, the necessity for a "clean room" facility for maintenance, the extensive amount of time required for the removal and replacement of CB-1A5, etc.), it may be feasible to identify assembly A-7 as being nonrepairable. In other words, the analyst should investigate the feasibility of whether the assemblies of unit B should be classified as "repairable" or "discard at failure."

2. For tasks 01 and 02, a "built-in test" capability exists at the organizational level for fault isolation to the subsystem. However, fault isolation to the unit requires a special system tester (0-2310B), and it takes 25 minutes of testing plus a highly skilled (supervisory skill) individual to accomplish the function. In essence, one should investigate the feasibility of extending the built-in test down to the unit level and eliminate the need for the special system tester and the high-skill-level individual.

3. The physical removal of unit B from the system and its replacement takes 15 minutes, which seems rather extensive. Although perhaps not a major item, it would be worthwhile investigating whether the removal/replacement time can be reduced (to less than 5 minutes, for example).

4. In tasks 10 to 15, a special clean-room facility is required for maintenance. Assuming that the various assemblies of unit B are repaired (versus being classified as "discard at failure"), then it would be worthwhile to investigate changing the design of these assemblies so that a clean-room environment is not required for maintenance. In other words, can the expensive maintenance facility requirement be eliminated?

1. System: XYZ
2. Item name/part no.: Manufacturing Test/A4321
3. Next higher assy.: Assembly and test
4. Description of requirement: During manufacturing and test of Product 12345 (Serial No. 654), System XYZ failed to operate. The symptom of failure was "loss of total power output." Requirement: Troubleshoot and repair system.

5. Req. No.: 01
6. Requirement: Diag./Repair
7. Req. Freq.: 0.00450
8. Maint. Level: Org/Inter.
9. Ma. Cont. No.: A12B100

10. Task Number	11. Task Description	13. Total elap. time	14. Task Freq.	15. B	16. I	17. S	18. Total
01	Isolate failure to subsystem level (Subsystem C is faulty)	5	0.00450	5	-	-	5
02	Isolate failure to unit level (Unit B is faulty)	25		-	25	-	25
03	Remove Unit B from system and replace with a spare Unit B (2nd cycle)	15		15	-	-	15
				-	-	25	25
04	Transport faulty unit to int. shop	30		30	-	-	30
05	Apply power to faulty unit. Check for output squarewave signal (3rd cycle)	20		-	20	-	20
06	Check signal input to Assembly A-1 (20v P-P @ T.P.7)	15		-	15	-	15
07	Check signal input to Assembly A-7 (squarewave, PW-5μsec @ T.P. 2) (4th cycle)	20		-	20	-	20
08	Remove faulty A-7 & replace	10		10	-	-	10
09	Transport faulty Assembly A-7 to supplier for depot maintenance (14 calendar days in transit)			-	-	-	-
10	Check A-7 & verify faulty condition (5th cycle)	25		-	-	25	25
11	Check operation of CB-3A2	15		-	-	15	15
12	Check operation of CB-2A4 (6th cycle)	10		-	-	10	10
13	Check operation of CB-1A5 (7th cycle)	20		-	-	20	20
14	Remove and replace faulty CB-1A5	40		-	40	-	40
	Discard faulty circuit board						
15	Verify that assembly is operational and return to inventory	15		-	-	15	15
Total		**265**	**0.00450**	**60**	**120**	**110**	**290**

12. Elapsed time - minutes (scale: 2 4 6 8 10 12 14 16 18 20 22 24 26 28 30 32 34 36 38). Personnel Man-min (columns 15–18).

FIGURE C.14 Maintenance task analysis (Part 1).

461

		Replacement Parts				Test & Support/Handling Equipment				
7. Task No.	8. Qty per Assy.	9. Part nomenclature / 11. Part number	10. Rep. Freq.		12. Qty.	13. Item part nomenclature / 15. Item part number	14. Use time (min)	16. Description of Facility Requirements	17. Special Technical Data Instructions
01	-	- / -	-		1	Built-in test equip. A123456	5	-	Organizational Maintenance
02	1	- / -	-		1	Special system tester 0-2310B	25	-	
03	-	Unit B / B180265X	0.01866		1	Standard tool kit STK-100-B	15	-	
04	-	- / -	-		1	Standard cart (M-10)	30	-	Intermediate maintenance
05	-	- / -	-		1	Special system tester I-8891011-A	20	-	
06	-	- / -	-		1	Special system tester I-8891011-A	15	-	
07	-	- / -	-		1	Special system tester I-8891011-A	20	-	
08	1	Assembly A-7 / MO-2378A	0.00995		1	Special extractor tool EX20003-4	10	-	Refer to special removal instructions
09	-	- / -	-		1	Container, special handling T-300A	14 days	-	Normal trans. environment
10	-	- / -	-		1	Special system tester I-8891011-B	25	Clean room environment	Supplier (depot) maintenance
11	-	- / -	-		1	C.B. test set D-2252-A	15		
12	-	- / -	-		1	C.B. test set D-2252-A	10		
13	-	- / -	-		1	C.B. test set D-2252-A	20		
14	1	CB-1A5 / GDA-221056C	0.00450		1	Special extractor tool EX45112-63 Standard tool kit STK-200	40		
15	-	- / -	-		1	Special system tester I-8891011-B	15		Return operating assy. to inventory

Header information:
1. Item name/Part No.: Manufacturing Test/A4321
2. Req No.: 01
3. Requirement: Diagnostic Troubleshooting and repair
4. Req. Freq.: 0.00450
5. Maint. level: Organization, Intermediate, Depot
6. Ma. Cont. No.: A12B100

FIGURE C.15 Maintenance task analysis (Part 2).

5. There is an apparent requirement for a number of new "special" test equipment/tool items; that is, special system tester 0-2310B, special system tester I-8891011-A, special system tester I-8891011-B, CB test set D-2252-A, special extractor tool EX20003-4, and special extractor tool EX45112-63. Usually, these *special* items are limited as to general application for other systems and are expensive to acquire and maintain. Initially, one should investigate whether these items can be eliminated; if test equipment/tools are required, can *standard* items be utilized (in lieu of special items)? Moreover, if the various special testers are required, can they be integrated into a "single" requirement? In other words, can a single item be designed to replace the three special testers and the CB test set? Reducing the overall requirements for special test and support equipment is a major objective.

6. For task 09, there is a special handling container for the transportation of assembly A-7. This may impose a problem in terms of the availability of the container at the time and place of need. It would be preferable if normal packaging and handling methods could be utilized.

7. For task 14, the removal and replacement of CB-1A5 takes 40 minutes and requires a highly skilled individual to accomplish the maintenance task. Assuming that assembly A-7 is repairable, it would be appropriate to simplify the circuit board removal/replacement procedure by incorporating plug-in components, or at least simplify the task to allow a person with a basic skill level to accomplish it.

C.5   LEVEL-OF-REPAIR ANALYSIS (LORA)

C.5.1   Definition of the Problem

In the design of system components, one of the decision factors relates to the question: Should the component be designed to be repairable, or should it be designed to be discarded in the event of failure? If it is designed to be repairable, at what level of maintenance should the repair be accomplished? Although these questions can be applied to any component of the system (e.g., equipment, unit, assembly, module, and element of software), this case study applies to the design of assembly A-1. This assembly is one of 15 assemblies in unit B of system XYZ. The objective is to evaluate design alternatives for the assembly on the basis of economic criteria, as shown in Figure C.16.

C.5.2   The Analysis Process

The accomplishment of a level-of-repair analysis requires that the item being evaluated be presented in terms of a system operational requirement, a maintenance concept, and a program plan. In this instance, it is assumed that system XYZ is installed in an aircraft. When a maintenance action is required, there is a built-in test capability within the aircraft that allows one to isolate the fault to unit A, unit B, or unit C. The applicable unit is removed, replaced with a spare, and the faulty item is transported to the intermediate-level maintenance shop for corrective maintenance. In the

Evaluation Criteria	Repair at Intermediate Cost ($)	Repair at Depot Cost ($)	Discard at Failure Cost ($)	Description and Justification
1. Estimated acquisition cost for assembly A-1 (to include design and development, production cost)	1,700/Assembly or 102,000 (47.8%)	1,700/Assembly or 102,000 (54.7%)	1,600/Assembly or 96,000 (19.5%)	Acquisition cost is based on 60 systems. Assembly design and production cost are less in the discard case (simplified configuration).
2. Maintenance labor cost	12,240 (5.7%)	18,360 (9.8%)	Not applicable	Based on 452,600 hours of operation and a maintenance rate of 0.00045, the estimated quantity of maintenance actions is 204. When repair is accomplished, one (1) technician is assigned on a full-time basis. The Mct is 3 hours. The labor rate is $20/hour for intermediate and $30/hour for depot.
3. Supply support—spare assemblies	8,500 (4%)	17,000 (9.1%)	326,400 (66.4%)	For intermediate maintenance, 5 spare assemblies are required to compensate for turnaround time, the maintenance queue, etc. 10 spares are required for depot maintenance. 100% spares are required for the discard case.
4. Supply support—spare components	10,200 (4.8%)	10,200 (5.5%)	Not applicable	Assume $50 per maintenance action.
5. Supply support—inventory maintenance	3,740 (1.8%)	5,440 (2.9%)	65,280 (13.3%)	Assume 20% of the inventory value (spare assemblies and spare components).
6. Special test and support equipment	60,000 (28.1%)	12,000 (6.4%)	Not applicable	Special test equipment is required in the repair case. The acquisition cost is $12,000 per installation. There are five (5) installations at intermediate and one(1) at depot.
7. Transportation and handling	Negligible	12,240 (6.6%)	Not applicable	Transportation costs at the intermediate level are negligible. For depot maintenance, assume 408 one-way trips at $150/100 pounds. One assembly weighs 20 pounds.
8. Maintenance training	4,500 (2.1%)	900 (0.5%)	Not applicable	Assume 10 students for 3 days at $150 per student day for intermediate, and 2 students for 3 days at $150 per student day for depot.
9. Maintenance facilities	5,612 (2.6%)	1,918 (1%)	Not applicable	Assume $1.00 per direct maintenance manhour for intermediate, and $1.50 per direct manhour for depot. Also, assume an initial fixed cost of $1,000 per installation.
10. Technical data	6,100 (2.9%)	6,100 (3.3%)	Not applicable	For repair case, assume $1,000 for the cost of preparation of maintenance instructions. Also, assume $25 per maintenance action for maintenance data.
11. Disposal	408)0.2%)	408 (0.2%)	4,080 (0.8%)	Assume $20 per assembly and $2 per component as the cost of disposal.
Total Estimated Cost	$213,300	$186,566	$491,760	

FIGURE C.16 Repair versus discard evaluation (assembly A-1).

maintenance shop, fault isolation is accomplished within the unit to the assembly level. The faulty assembly is removed, replaced with a spare, and the unit is checked out and returned to the inventory as an operational spare. The basic question pertains to the disposition of the assembly.

In approaching this problem, the first step is to accomplish a level-of-repair analysis on assembly A-1 as an individual entity. Subsequently, the results of this part of the analysis have to be viewed in the context of the whole; that is, the results of similar analyses involving assemblies A-2, A-3, ..., and A-15, and the applicable assemblies of unit A and unit C. There is usually a feedback effect between the individual assembly analysis, the unit-level analysis, and the overall maintenance concept for the system as a whole.

For completing the level-of-repair analysis on assembly A-1, the following information is provided:

1. System XYZ is installed in each of 60 aircraft, which are distributed equally at five operating sites over an eight-year time period. System utilization is on the average of 4 hours per day, and the total operating time for all systems is 452,600 hours.

2. As stated earlier, system XYZ includes three units: unit A, unit B, and unit C. unit B includes 15 assemblies, one of which is assembly A-1. The estimated acquisition cost for assembly A-1 (including design and development cost and production cost) is $1,700 each if the assembly is designed to be repairable, and $1,600 each if the assembly is designed to be discarded at failure. The design for repairability considers the incorporation of diagnostic provisions, accessibility, internal labeling, and so on, which is apt to be more expensive in terms of design and production costs.

3. The estimated failure rate (or corrective maintenance rate) of assembly A-1 is 0.00045 failure per hour of system operation. When failures occur, repair is accomplished by a single technician who is assigned for the duration of the allocated active maintenance time. The estimated corrective maintenance downtime ($\overline{\text{Mct}}$) is three hours. The loaded labor rate is $20 per labor hour for intermediate-level maintenance and $30 per labor hour for depot-level maintenance.

4. Supply support includes three categories of cost: the cost of spare assemblies in inventory, the cost of spare components to enable the repair of faulty assemblies, and the cost of inventory management and maintenance. Assume that 5 spare assemblies will be required in inventory when maintenance is accomplished at the intermediate level, and that 10 spare assemblies will be required when maintenance is accomplished at the depot level. For component spares, assume that the average cost of material consumed per maintenance action is $50. The estimated cost of inventory maintenance is assumed to be 20 of the inventory value (the summation of the costs for assembly and component spares).

5. When assembly repair is accomplished, special test and support equipment is required for fault diagnosis and assembly checkout. The cost per test station is $12,000, which includes acquisition cost and amortized maintenance cost. This cost is that part of the total cost that is attributed to the maintenance requirement

for assembly A-1, and there are five test stations required for intermediate-level maintenance.

6. Transportation and handling cost is considered as being negligible when maintenance is accomplished at the intermediate level. However, assembly maintenance accomplished at the depot level will involve an extensive amount of transportation. For depot maintenance, assume $150 per 100 pounds per one-way trip (independent of distance), and that the packaged assembly weighs 20 pounds.

7. The allocation for assembly A-1 relative to maintenance facility cost is categorized in terms of an initial fixed cost and a sustaining recurring cost proportional to facility utilization requirements. The initial fixed cost is $1,000 per installation, and the assumed usage cost allocation is $1.00 per direct maintenance labor hour at the intermediate level and $1.50 per direct labor hour at the depot level.

8. Technical data and maintenance software requirements include the maintenance instructions to be included in the technical manuals to support assembly repair activities, and the failure reporting and maintenance data covering each maintenance action in the field. Assume that the cost for preparing and distributing maintenance instructions (and supporting computer software) is $1,000, and that the cost for field maintenance data is $25 per maintenance action.

9. There will be some initial formal training costs associated with maintenance personnel in considering the assembly repair option. Assume 30 student-days of formal training for the intermediate level of maintenance (for the five sites in total) and 6 student-days for depot-level maintenance. The cost of training is $150 per student-day. The requirement for replenishment training as a result of attrition or turnover is considered as being negligible.

10. As a result of maintenance, there will be a requirement for disposal and/or the recycling of material. The assumed disposal cost is $20 per assembly and $2 per component.

The objective is to evaluate assembly A-1 based on the information provided. Should assembly A-1 be designed for (1) repair at the intermediate level of maintenance, (2) repair at the depot level of maintenance, or (3) discard at failure?

C.5.3 The Analysis Results

Figure C.16 presents a worksheet with the results from the evaluation of assembly A-1. Based on the information shown, it is recommended that the assembly be *repaired at the depot level of maintenance.*

Prior to making a final decision, however, one should review the data in Figure C.16 in terms of "high-cost" contributors and the sensitivities of various input factors. Some of the initial assumptions may have a great impact on the analysis results and, perhaps, should be challenged. The analyst may also wish to review the source of prediction data covering reliability, maintainability, and some of the input cost factors.

Assembly Number	Repair Policy			Decision
	Repair at Intermediate	Repair at Depot	Discard at Failure	
A-1	$213,300	$186,566	$491,760	Repair—depot
A-2	130,800	82,622	75,440	Discard
A-3	215,611	210,420	382,452	Repair—depot
A-4	141,633	162,912	238,601	Repair—intermediate
A-5	132,319	98,122	121,112	Repair—depot
A-6	112,189	96,938	89,226	Discard
A-7	125,611	142,206	157,982	Repair—intermediate
A-8	99,812	131,413	145,662	Repair—intermediate
A-9	128,460	79,007	66,080	Discard
A-10	167,400	141,788	314,560	Repair—depot
A-11	185,850	142,372	136,740	Discard
A-12	135,611	122,453	111,502	Discard
A-13	105,667	113,775	133,492	Repair—intermediate
A-14	111,523	89,411	99,223	Repair—depot
A-15	142,119	120,813	115,723	Discard
Policy Cost	$2,147,905	$1,920,808	$2,679,555	Repair—depot

FIGURE C.17   Summary of repair-level decisions.

Given that the repair policy decision for assembly A-1 is verified in terms of its evaluation in an "isolated" sense (i.e., a decision has been made relative to the results of the individual analysis in Figure C.16), then it is essential that this decision be reviewed in context with other assemblies of system XYZ and with the maintenance concept. Figure C.17 reflects the results of individual level-of-repair analyses accomplished for each of the major assemblies in Unit B. The same approach used for assembly A-1 is used for the evaluation of assemblies A-2 through A-15.

As shown in Figure C.17, there are two major choices: (1) Adopt the individual repair policy for each assembly (i.e., a "mixed" overall policy) and (2) adopt a uniform overall policy for *all* assemblies based on the lowest total policy cost (i.e., repair at depot). Both options must be reviewed in terms of the feedback effects that occur, life-cycle cost implications, and associated risks.

Figure C.18 illustrates the basic process that has been discussed herein. There are many candidate items that can be evaluated in terms of repair-versus-discard decisions. Quite often, such decisions will be made based on noneconomic criteria. It may not be technically feasible to repair an item at the intermediate level. Safety criteria and/or the need for a specialized repair facility dictates that repair must be accomplished at the depot level. The proprietary aspects of a product dictate that an item must be repaired at the producer's facility (i.e., depot). The approach used in this example deals with those components for which economic evaluation is feasible. As shown in the figure, there are some decisions that may initially be clear-cut, and there are other decisions where a more in-depth analysis is required.

C.6   DESIGN EVALUATION OF ALTERNATIVES

C.6.1   Definition of the Problem

Company DEF is responsible for the design and development of a major system, which, in turn, comprises a number of large subsystems. Subsystem XYZ is to be procured from an outside supplier, and there are three different configurations being evaluated for selection. Each of the configurations represents an existing design, with some redesign and additional development necessary to be compatible with the requirements for the new system. The evaluation criteria include various parameters, such as performance, operability, effectiveness, design characteristics, schedule, and cost. Both qualitative and quantitative considerations are covered in the evaluation process.

C.6.2   The Analysis Process

The analyst commences with the development of a list of evaluation parameters, as depicted in Figure C.19. In this instance, there is no single parameter (or figure of merit) that is appropriate by itself, but there are 11 factors that must be considered on an integrated basis. Given the evaluation parameters, the next step is to determine the level of importance of each. Quantitative weighting factors from 0 to 100 are assigned to each parameter in accordance with the degree of importance. The Delphi method, or an equivalent evaluation technique, may be used to establish the weighting factors. The sum of all weighting factors is 100.

For each of the 11 parameters identified in Figure C.19, the analyst may wish to develop a special checklist including criteria against which to evaluate the three proposed configurations. For instance, the parameter "performance" may be described in terms of degrees of desirability; that is, "highly desirable," "desirable," or "less desirable." Although each configuration must comply with a minimum set of requirements, one may be more desirable than the next when looking at the proposed performance characteristics. In other words, the analyst should break down each evaluation parameter into "levels of goodness."

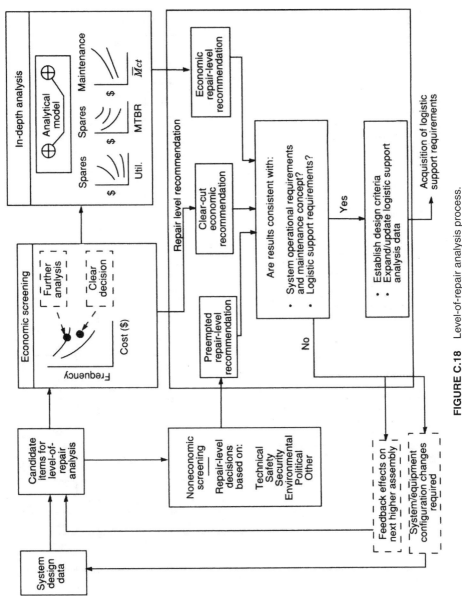

FIGURE C.18 Level-of-repair analysis process.

Item	Evaluation parameter	Weighting factor	Configuration A		Configuration B		Configuration C	
			Base rate	Score	Base rate	Score	Base rate	Score
1	Performance—input, output, accuracy, range, compatibility	14	6	84	9	126	3	42
2	Operability—simplicity and ease of operation	4	10	40	7	28	4	16
3	Effectiveness—Ao, MTBM, Mct, Mpt, MDT, MLH/OH	12	5	60	8	96	7	84
4	Design characteristics— reliability, maintainability, human factors, supportability, producibility, interchange-ability	9	8	72	6	54	3	27
5	Design data—design drawings, specifications, logistics data, operating and maintenance procedures	2	6	12	8	16	5	10
6	Test aids—common and standard test equipment, calibration standards, maintenance and diagnostic computer programs	3	5	15	8	24	3	9
7	Facilities and utilities— space, weight, volume, environment, power, heat, water, air conditioning	5	7	35	8	40	4	20
8	Spare/repair parts—part type and quantity, standard parts, procurement time	6	9	54	7	42	5	30
9	Flexibility/growth potential— for reconfiguration, design change acceptability	3	4	12	8	24	6	18
10	Schedule—research and development, production	17	7	119	8	136	9	153
11	Cost—life cycle (R & D, investment, O & M)	25	10	250	9	225	5	125
Subtotal				753		811		534
Derating factor (development risk)				113 15%		81 10%		197 20%
Grand Total		100		640		730		427

FIGURE C.19   Evaluation summary (three alternatives).

Each of the three proposed configurations of subsystem XYZ is evaluated independently, using the special checklist criteria. Base rating values from 0 to 10 are applied according to the degree of compatibility with the desired goals. If a "highly desirable" evaluation is realized, a rating of 10 is assigned.

The base-rate values are multiplied by the weighting factors to obtain a score. The total score is then determined by adding the individual scores for each configuration. Because some redesign is required in each instance, a special derating factor is applied to cover the risk associated with the failure to meet a given requirement. The resultant values from the evaluation are summarized in Figure C.19.

C.6.3 The Analysis Results

In Figure C.19, configuration B represents the preferred approach, based on the highest total score of 730 points. This configuration is recommended on the basis of its inherent features relating to performance, operability, effectiveness, design characteristics, design data, and so on.

C.7 LIFE-CYCLE COST ANALYSIS (LCCA)

C.7.1 Definition of the Problem

A large metropolitan area has a need for a new communication system network capability (i.e., system XYZ) that will enable day-to-day active communication between each and all of the following nodes: (1) a centralize city operational terminal located in the city center; (2) three remote ground district operational facilities located in the city's suburban areas; (3) 50 ground vehicles patrolling the city and within a 30-mile range; (4) five helicopters flying at low altitude and within a 50-mile range; (5) three low-flying aircraft with a 200-mile range; and (6) a centralize maintenance facility located in the city's outskirts. The proposed network needs to enable "live" two-way voice and data communication, 24 hours per day, and throughout all of its branches and to any one of the stated nodes as requested.

In response to this new system requirement, need and feasibility analyses have been completed, a solicitation for proposal has been initiated, and two prospective suppliers have responded, each with a different design approach. The objective is to evaluate each of the two supplier proposals, on the basis of system life-cycle cost (LCC), and to select a preferred approach (i.e., configuration A or configuration B).

C.7.2 The Analysis Process

The life-cycle cost analysis (LCCA) process is discussed in detail in Appendix B. The objective herein is to further illustrate this process, covering a different system in more of an overview context.

The first major step in accomplishing an LCCA is to establish a good baseline description of system operational requirements, the maintenance concept, primary

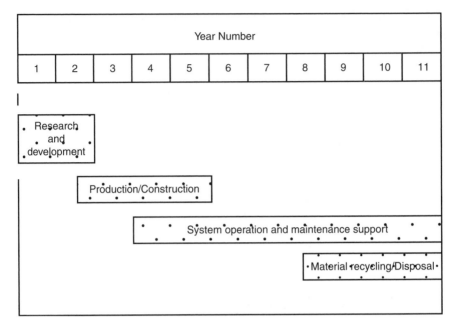

FIGURE C.20 System XYZ life-cycle plan.

TPM requirements, and a top-level functional analysis, and to present these require-ments in the context of a proposed life-cycle framework (refer to Appendix B, Sec-tion B.1). The proposed life-cycle plan for the communication network (i.e., system XYZ) is illustrated in Figure C.20.

Given the planned program phases, the next step is to describe the basic activities in each phase of the system life cycle, develop a cost breakdown structure (CBS), estimate the costs for each program activity, and categorize the activity costs per each year within the applicable category of the CBS (refer to Appendix B, Sections B.2–B.4). The proposed CBS for the communication network in Figure C.21.

Cost profiles are then developed for each configuration A and configuration B, incorporating the effects of inflation from year to year (refer to Appendix B, Section B.6). At the same time, present value (PV) costs are calculated to allow for the eval-uation of comparable alternatives on the basis of economic equivalence. A 6% cost of capital was assumed for this LCCA effort. The equivalent profiles are shown in Figure C.22.

Based on the results shown in Figure C.22, it appears as though configuration A is the preferred approach, since the present value (PV) cost of $5,927,885 is less than that for the other configuration. The question is, how much better is configuration A, and at what point in time does this configuration assume a point of preference? It should be noted that, on the basis of acquisition cost only (i.e., categories C_r and C_p), it appears that configuration B would be preferred ($4,417,404 for B and $4,509,271 for A). However, based on the overall LCC, configuration A is preferred.

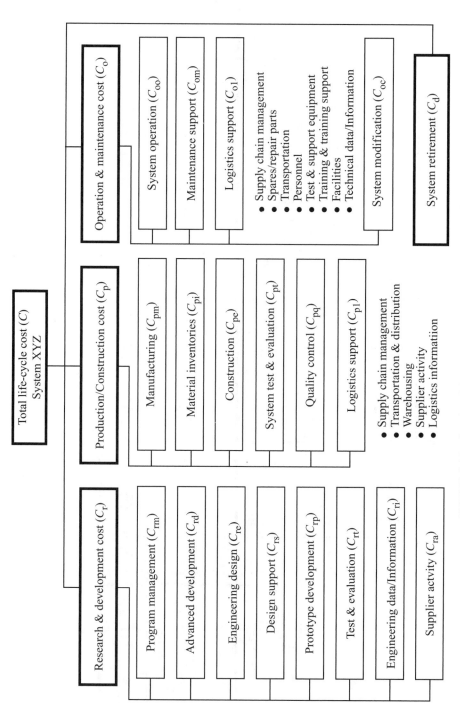

FIGURE C.21 Cost breakdown structure (CBS) for system XYZ.

The hierarchy shown in the figure:

Total life-cycle cost (C) System XYZ

- Research & development cost (C_r)
  - Program management (C_{rm})
  - Advanced development (C_{rd})
  - Engineering design (C_{re})
  - Design support (C_{rs})
  - Prototype development (C_{rp})
  - Test & evaluation (C_{rt})
  - Engineering data/Information (C_{ri})
  - Supplier actvity (C_{ra})

- Production/Construction cost (C_p)
  - Manufacturing (C_{pm})
  - Material inventories (C_{pi})
  - Construction (C_{pe})
  - System test & evaluation (C_{pt})
  - Quality control (C_{pq})
  - Logistics support (C_{pl})
    - Supply chain management
    - Transportation & distribution
    - Warehousing
    - Supplier activity
    - Logistics informatiion

- Operation & maintenance cost (C_o)
  - System operation (C_{oo})
  - Maintenance support (C_{om})
  - Logistics support (C_{ol})
    - Supply chain management
    - Spares/repair parts
    - Transportation
    - Personnel
    - Test & support equipment
    - Training & training support
    - Facilities
    - Technical data/Information
  - System modification (C_{oc})

- System retirement (C_d)

473

Cost Category	1	2	3	4	5	6	7	8	9	10	11	Total ($)
Configuration A												
Research & development (C_r)	615,725	621,112										1,236,837
Production/Construction (C_p)		364,871	935,441	985,911	986,211							3,272,434
Operation & maintenance (C_o)				179,203	207,098	448,248	465,660	483,945	503,122	523,297	544,466	3,355,039
System retirement (C_d)									27,121	41,234	45,786	114,141
Total cost ($)	615,725	985,983	935,441	1,165,114	1,193,309	448,248	465,660	483,945	530,243	564,531	590,252	7,978,451
Present value cost − 6% ($)	580,875	877,525	785,396	922,887	891,760	316,015	309,770	303,627	313,851	315,234	310,945	5,927,885
Configuration B												
Research & development (C_r)	545,040	561,223										1,106,263
Production/Construction (C_p)		379,119	961,226	982,817	987,979							3,311,141
Operation & maintenance (C_o)				192,199	225,268	456,648	472,236	592,717	613,005	625,428	650,342	3,827,843
System retirement (C_d)								20,145	35,336	45,455	50,816	151,752
Total cost ($)	545,040	940,342	961,226	1,175,016	1,213,247	456,648	472,236	612,862	648,341	670,883	701,158	8,396,999
Present value cost − 6% ($)	514,191	836,904	807,045	930,730	906,659	321,937	314,089	384,510	383,621	374,621	369,370	6,143,809

Life-cycle Year

FIGURE C.22 Life-cycle cost profile for system XYZ.

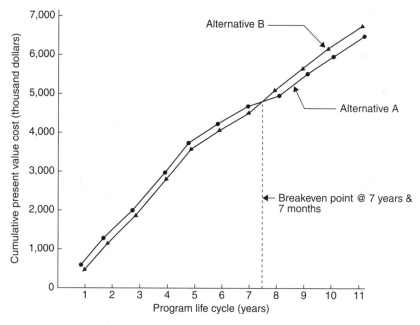

FIGURE C.23 Breakeven analysis for system XYZ.

Relative to the time of preference, (i.e., when *A* assumes the point of preference—see Appendix B, Section B.11 and Figure B.16), the analyst conducted a breakeven analysis as illustrated in Figure C.23. From the figure, it can be seen that configuration A assumes a favorable position at about the 7-year, 7-month point in the projected life cycle. It was decided in this instance that this was early enough for the selection of configuration A.

Having initially selected configuration *A* as being preferred over the alternative, the next step is to further evaluate the costs that make up the $7,978,451 for this configuration, identify the high-cost contributors, determine the cause-and-effect relationships, and re-evaluate the system XYZ design to determine whether improvements can be implemented that will result in an overall reduction in LCC. A breakout of the costs for this configuration is presented in Figure C.24.

Referring to the figure, for example, it should be noted that the costs associated with logistics support activities (i.e., C_{pl} and C_{ol}) make up about 21.38% of the total. Within this spectrum, the categories of spares/repair parts and transportation, represent the high-cost contributors (4.57% and 3.73% respectiyely) under category C_{ol} Additionally, transportation and distribution costs within category C_{pl} are also relatively high (2.75%). Through a reevaluation of the basic design configuration, the extensive requirements for spares/repair parts could perhaps be reduced through some form of reliability improvement, particularly for critical items with relatively high failure rates. For transportation, it may be possible to repackage elements of the system such that internal transportation attributes in the design can be improved, or

Configuration A

Cost Category	Cost ($) (Undiscounted)	Percent (%)
1. Research & development (C_r)	1,236,660	15.50
(a) Program management (C_{rm})	79,785	1.00
(b) Advanced development (C_{rd})	99,731	1.25
(c) Engineering design (C_{re})	276,852	3.47
(d) Design support (C_{rs})	193,876	2.43
(e) Prototype development (C_{rp})	89,359	1.12
(f) Test & evaluation (C_{rt})	116,485	1.46
(g) Engineering data/Information (C_{ri})	75,795	0.95
(h) Supplier activity (C_{ra})	304,777	3.82
2. Production/Construction (C_p)	3,272,762	41.02
(a) Manufacturing (C_{pm})	1,716,166	21.51
(b) Material invetories (C_{pi})	453,176	5.68
(c) Construction (C_{pc})	95,741	1.20
(d) System test & evaluation (C_{pt})	228,184	2.86
(e) Quality control (C_{pq})	76,593	0.96
(f) Logistics support (C_{pl})	702,902	8.81
(1) Supply chain management	39,892	0.50
(2) Transportation & distribution	219,408	2.75
(3) Warehousing	168,345	2.11
(4) Supplier activity	263,289	3.30
(5) Logistics information	11,968	0.15
3. Operation & maintenance (C_o)	3,354,939	42.05
(a) System operation (C_{oo})	1,458,461	18.28
(b) Maintenance support (C_{om})	768,325	9.63
(c) Logistics support (C_{ol})	1,002,891	12.57
(1) Supply chain management	79,785	1.00
(2) Spares/repair parts	364,615	4.57
(3) Transportation	297,596	3.73
(4) Personnel	153,984	1.93
(5) Test & support equipment	46,275	0.58
(6) Training & training support	24,733	0.31
(7) Facilities	20,744	0.26
(8) Technical data/Information	15,159	0.19
(d) System modifications (C_{oc})	125,262	1.57
4. System retirement (C_d)	114,092	1.43
GRAND TOTAL	7,978,451	100.00

FIGURE C.24   Cost breakdown structure (CBS) summary.

to select alternative modes of transportation that will still meet the specified TPM requirements for the system overall, but at a lesser overall cost.

C.7.3   The Analysis Results

Through implementation of this process on an iterative basis, experience has indicated that significant system design improvements can often be realized. It should be noted that by improving one area of concern, the result could lead to an improvement in another area. For example, if improvement can be made in the spares/repair parts area (within category C_{ol}), this also may result in an overall reduction of the maintenance support cost (category C_{om}) as well. There are numerous interactions that could occur throughout the analysis process, and care must be exercised to ensure that improvement in any given area will not result in a significant degradation in another.

For system XYZ configuration A was selected initially, and then design improvements were made, through the iterative process of analysis, to further reduce the projected life-cycle costs associated with the selected configuration.

APPENDIX D

DESIGN REVIEW CHECKLIST

Periodically, throughout the system design and development process, it is appropriate to conduct an informal review to (1) determine whether the necessary tasks have been completed in a given program to ensure optimum results overall, and (2) assess whether the appropriate characteristics have been considered and incorporated into the system/product design configuration. Questions, presented in the form of a checklist, have been developed to reflect certain features. System-level requirements and detailed design considerations are included. Not all questions are applicable in all reviews; however, the answer to those questions that are applicable should be yes in order to reflect desirable results. For some questions, it may be necessary to pursue a more in-depth study of the subject through a review of selected references prior to arriving at a decision. These questions directly support the abbreviated checklist presented in Figure 5.4 (Chapter 5).

D.1 GENERAL REQUIREMENTS

1.0 Feasibility Analysis

1.1   Has the need for the system/product been defined and justified?

1.2   Has the overall technical design approach for the system been justified through a feasibility analysis?

1.3   In the conductance of the feasibility analysis, have *all* appropriate technology applications been considered and prioritized in the development of a technical design approach for the system?

1.4   Have *existing* technologies been selected where feasible, as compared with the selection of new state-of-the-art technologies?

1.5   In the evaluation of alternative technology applications, have life-cycle cost considerations been employed? Have peculiar support requirements been identified?

1.6   Will the technologies selected be available in time to meet program schedule requirements?

1.7   Are there multiple sources of supply for each of the technologies selected for application (i.e., more than one supplier)?

1.8   Have research-and-development activities been defined for areas where deficiencies exist?

1.9   Have areas of risk and uncertainty been identified?

2.0 Operational Requirements

2.1   Has the mission for the system been defined (i.e., primary and secondary missions with applicable scenarios or profiles)?

2.2   Have system performance requirements been defined?

2.3   Have the technical performance measures (TPMs) for the system/program been identified, described, and prioritized? Are they measurable?

2.4   Has the system/product life cycle, and the major activities within, been adequately defined (i.e., design and development, production and/or construction, distribution, operational use, sustaining support, retirement and disposal)?

2.5   Has the planned operational deployment, or the geographical distribution of system components, been defined (i.e., customer requirements, quantity of items per "user" location, distribution schedule)?

2.6   Have system utilization requirements been defined? These may include projected hours of system operations, or quantity of operational cycles in a given time period. A "dynamic" operational scenario is desired.

2.7   Has the projected operational environment been adequately described in terms of temperature cycles and extremes, humidity, vibration and shock, storage, transportation and handling?

2.8   Have system-of-system (SOS) requirements been addressed and defined where applicable?

2.9   Have the operational interfaces with other systems within the same SOS configuration been identified and defined?

2.10 Have "interoperability" requirements been defined for other systems being utilized in the same user's environment (where the newly developed system will be operational)?

3.0 Maintenance Concept

3.1   Have the anticipated levels of maintenance been identified and defined?

3.2   Have the basic maintenance functions been identified for each level?

3.3   Have the organizational responsibilities for system maintenance and support been assigned (i.e., user support, contractor support, supplier support, third-party maintenance)?

3.4   Have level-of-repair policies been established (i.e., repair versus discard)? Have the criteria for level-of-repair decisions been adequately defined?

3.5   Have the requirements for "standardization" been established (as they apply to the overall system support capability)?

3.6   Have criteria been established for personnel quantities and/or skills at each level of maintenance?

3.7   Have criteria been established for test and support equipment at each level of maintenance? Built-in versus external test equipment? Diagnostic requirements?

3.8   Have software requirements for system and/or component testing been defined? Test language requirements?

3.9   Have criteria been established for maintenance facilities?

3.10   Have criteria been established for packaging, transportation, and handling?

3.11   Have the appropriate effectiveness factors been established for design for the overall support capability (i.e., spare part demand rates, inventory locations and levels, test equipment availability and utilization, maintenance shop queue and process time, facility utilization, turnaround time, and so on)?

3.12   Have the maintenance and support environments been defined in terms of temperature cycle and extremes, humidity, vibration and shock, transportation and handling, and storage?

4.0 Effectiveness Factors

4.1   Have the appropriate system effectiveness and cost-effectiveness figures-of-merit (FOMs) been defined for the system/product (i.e., availability, dependability, capability, readiness, life-cycle cost, design to cost)?

4.2   Have applicable quantitative factors for reliability, maintainability, human factors, and supportability been specified (i.e., MTBM, MTBR, MTBF, λ, fpt, MDT, $\overline{M}$, ADT, LDT, $\overline{M}ct$, $\overline{M}pt$, MLH/OH, Cost/OH, Cost/MA, TAT)? For an explanation of these terms, refer to Chapter 3.

4.3   Are the effectiveness factors that have been specified directly traceable to either the system operational requirements (mission scenario) or the maintenance concept?

4.4   Can each of the effectiveness factors identified for the system be measured? Have test and evaluation provisions been incorporated in the Test and Evaluation Master Plan (TEMP) for the purposes of verification?

4.5   In the event that two or more effectiveness measures are applicable, are the measures properly weighted to indicate degree of significance or relative level of importance?

4.6   Are all significant effectiveness FOMs properly integrated into the TPM evaluation and reporting capability?

5.0 Functional Analysis and Allocation

5.1   Has the system/product been adequately defined in *functional* terms utilizing the functional block diagram approach?

5.2   Have all major system operational functions and maintenance functions been defined?

5.3   Does the functional analysis evolve directly from the system operational requirements and the maintenance concept? Are the functions directly traceable to top system-level requirements (i.e., the mission scenario)?

5.4   Do the maintenance functions evolve directly from the operational functions?

5.5   Is the functional analysis presented in enough detail to allow for the proper development of the reliability block diagram, fault-tree analysis (FTA), failure mode, effects, and criticality analysis (FMECA), maintainability prediction, maintenance analysis, detailed operator task analysis, operational sequence diagrams, safety hazard analysis, and supportability analysis (SA)?

5.6   Is the functional analysis presented in enough detail for development of the system specification?

5.7   Have the functional interfaces been properly defined for all systems within the same overall SOS configuration?

5.8   Have the appropriate system-level requirements been allocated to the depth necessary for adequate design definition (i.e., subsystem level and below)? This may include the allocation of reliability requirements, maintainability requirements, supportability factors, and cost parameters.

5.9   In the allocation of factors from the system to the subsystem, unit, and below, are the parameters traceable from one level to the next? Are the parameters meaningful in terms of being good measures for the level of the system specified?

6.0 System Specification

6.1   Has a program/project specification tree been developed (showing governing specifications presented in a hierarchal manner)?

6.2   Has a system specification been prepared (i.e., type A specification)?

6.3   Have the appropriate development, procurement, process, and material specifications been prepared (i.e., types B, C, D, and E)?

6.4   Does the system specification include operating requirements, interoperability requirements, the maintenance concept, a functional definition of the system, and effectiveness requirements (i.e., reliability, maintainability, human factors, safety, supportability, economic, and quality factors)?

6.5   Are the various specifications that are applicable compatible with each other? Have conflicting specification requirements been eliminated? If not, has precedence been established?

7.0 Supplier Requirements[1]

7.1   Have the criteria and procedures for the initial identification, evaluation, and selection of component suppliers been established?

7.2   Have all component suppliers been identified? Has an on-site evaluation been conducted for each potential supplier?

7.3   Have supplier specifications been prepared and properly applied through the appropriate contractual arrangements?

7.4   Have individual supplier program plans been prepared and implemented?

7.5   Are supplier design data, databases and documentation compatible with the requirements for the overall program?

7.6   Have the appropriate configuration management procedures been imposed on supplier design and development activities?

7.7   Have the appropriate quality control procedures been established for the ongoing monitoring and control of supplier activities? Has a supplier rating system been implemented for the purposes of evaluation?

8.0 System Engineering Management Plan (SEMP)

8.1   Has the System Engineering Management Plan (SEMP) been developed?

8.2   Does the SEMP address the overall system life cycle and its phases/activities?

8.3   Does the plan adequately describe the system engineering process?

8.4   Does the plan convey the proper integration of the different engineering specialties involved in the system/product design process?

8.5   Does the SEMP properly integrate other plans such as the reliability program plan, maintainability program plan, human-factors program plan, safety and security engineering plan, integrated and logistic support (ILS) plan, logistics and supportability analysis plan, configuration management plan, test and evaluation master plan (TEMP), and so on?

8.6   Have major system trade-off studies been adequately documented, and are they appropriately referenced in the SEMP?

8.7   Does the SEMP adequately support the system specification?

8.8   Are program tasks, organizational structure and responsibilities, work breakdown structure (WBS), schedules, cost projections, and program monitoring and control functions included?

8.9   Have a personnel development plan and an organizational training plan been included?

8.10   Have supplier program requirements been covered?

---

[1] Also refer to Appendix E.

8.11   Have formal design reviews been covered? Has a formal system evalua-
        tion and corrective-action procedure been described and implemented?

Additional questions pertaining to the SEMP are included in Section 6.2 (Chapter 6).

D.2 DESIGN FEATURES

1.0 Accessibility

1.1   Are key system components directly accessible for the performance of
        both operator and maintenance tasks?

1.2   Is access easily attained?

1.3   Are access requirements compatible with the frequency of mainte-
        nance (or criticality of need)? Accessibility for items requiring fre-
        quent maintenance should be greater than for items requiring infrequent
        maintenance.

1.4   Are access doors provided where appropriate? Are hinged doors utilized?
        Can access doors that are hinged be supported in the open position?

1.5   Are access openings adequate in size and optimally located for the access
        required?

1.6   Are access door fasteners minimized?

1.7   Are access door fasteners of the quick-release variety?

1.8   Can access be attained without the use of tools?

1.9   If tools are required to gain access, is the number of tools held to a mini-
        mum? Are the tools of the standard variety?

1.10   Are access provisions between modules and components adequate?

1.11   Are access doors and openings labeled in terms of items that are accessi-
        ble from within?

2.0 Adjustments and Alignments

2.1   Have adjustment/alignment requirements been minimized, if not elimi-
        nated?

2.2   Are adjustment requirements and frequencies known where applicable?

2.3   Are adjustment points accessible?

2.4   Are adjustment-point locations compatible with the maintenance level at
        which the adjustment is made?

2.5   Have adjustment/alignment interaction effects been eliminated?

2.6   Are factory adjustments specified?

2.7   Are adjustment points adequately labeled?

2.8   Can adjustments/alignments be made without the requirement for special tools?

3.0   Cables and Connectors

3.1   Are cables fabricated in removable sections?

3.2   Are cables routed to avoid sharp bends?

3.3   Are cables routed to avoid pinching?

3.4   Is cable labeling adequate?

3.5   Is cable clamping adequate?

3.6   Are the connectors used of the quick-disconnect variety?

3.7   Are connectors that are mounted on surfaces far enough apart so that they can be firmly grasped for connecting and disconnecting?

3.8   Are connectors and receptables labeled?

3.9   Are connectors and receptables keyed?

3.10   Are connectors standardized?

3.11   Do the connectors incorporate provisions for moisture prevention?

4.0   Calibration

4.1   Have calibration requirements been minimized, if not eliminated?

4.2   Are calibration requirements known where applicable?

4.3   Are calibration frequencies and tolerances known?

4.4   Have the facilities for calibration been identified?

4.5   Are the necessary standards available for calibration?

4.6   Have calibration procedures been prepared?

4.7   Is traceability to the National Institute of Standards and Technology (NIST) possible?

4.8   Are calibration requirements compatible with the maintenance concept and the supportability analysis (SA)?

5.0   Data Requirements

5.1   Has the design been properly defined through good data and documentation (i.e., layouts, drawings, functional diagrams, and materials and parts lists)?

5.2   Are system components adequately covered through good up-to-date design data?

5.3   Have the results of significant design trade-off studies been properly recorded through good documentation?

5.4   Have all data requirements been defined for each applicable program?

5.5   Have all supplier data requirements been defined and properly integrated into the overall data requirements for the program?

5.6   Have the procedures for data collection, distribution, and processing been developed and described?

5.7   Is standardization employed, where appropriate, in data formatting, processing, and reporting?

5.8   Are the data properly controlled in accordance with approved configuration management procedures?

6.0   Disposability

6.1   Has the equipment been designed for disposability (e.g., selection of materials, packaging)?

6.2   Have procedures been prepared to cover system/equipment/component disposal?

6.3   Can the components or materials used in system/equipment design be recycled for use in other products?

6.4   If component/material recycling is not feasible, can decomposition be accomplished?

6.5   Can recycling and/or decomposition be accomplishing using existing logistic support resources?

6.6   Are recycling and/or decomposition methods and results consistent with environmental, ecological, safety, political, and social requirements?

6.7   Is the method(s) used for recycling and/or decomposition economically feasible?

7.0   Ecological Requirements

7.1   Has an environmental impact study been completed (to determine if the system will have an adverse impact on the environment)?

7.2   Are the required standards associated with air quality, water quality, noise level(s), solid-waste processing, and so on, being maintained in spite of the introduction, operation, and sustaining support of the system/product?

7.3   Have potentially degrading ecological effects been identified? Is corrective action being taken to eliminate problems in this area?

7.4   Have the appropriate handling and transportation methods/procedures been described for the processing of solid waste?

8.0   Economic Feasibility

8.1   Has the system/product been justified in terms of total *life-cycle* revenues and costs?

8.2   Are all cost elements considered?

8.3   Are all cost categories adequately defined?

8.4   Are cost estimates relevant?

8.5   Are variable and fixed costs separately identifiable?

8.6   Are escalation factors specified and employed where applicable?

8.7   Are learning curves specified and employed where applicable?

8.8   Is the project economically feasible, considering all possible alternatives?

8.9   Is *activity-based costing* (ABC) used in determining costs?

9.0 Environmental Requirements

9.1   Has system/product design considered all possible phases of activity from an environmental standpoint; for example, environmental requirement during system handling, operation/utilization, transportation, storage, and maintenance?

9.2   Has system/product design considered the following: temperature, humidity, vibration, shock, pressure, wind, salt spray, sand, and dust? Have the ranges and extreme conditions been specified and properly addressed in design? Have the proper environmental profiles been addressed?

9.3   Is the design compatible with air and water quality standards?

9.4   Have provisions been made to specify and control noise, illumination, temperature, and humidity in areas where personnel are required to perform operating and maintenance tasks?

10.0 Facility Requirements

10.1   Have facility requirements (space, volume, capital equipment, utilities, etc.) necessary for system operation been defined?

10.2   Have facility requirements (space, volume, capital equipment, utilities, etc.) necessary for system maintenance at each level been defined?

10.3   Have operational and maintenance facility requirements been minimized to the greatest extent possible?

10.4   Have environmental system requirements (e.g., temperature, humidity, and dust control) associated with operational and maintenance facilities been identified?

10.5   Have storage or shelf-space requirements for spare/repair parts been defined?

10.6   Have storage environments been defined?

10.7   Are the designated facility and storage requirements compatible with the supportability analysis and human-factors data?

11.0 Fasteners

11.1   Are quick-release fasteners used on doors and access panels?

11.2   Is the total number of fasteners minimized?

11.3   Is the number of different types of fasteners held to a minimum? This relates to standardization.

11.4   Have fasteners been selected based on the requirement for standard tools rather than special tools?

12.0 Handling

12.1   For heavy items, are hoist lugs (lifting eyes) or base-fitting provisions for forklift-truck application incorporated? Hoist lugs should be provided on all items weighing more than 150 pounds.

12.2   Are hoist and base-lifting points identified relative to lifting capacity?

12.3   Are weight labels provided?

12.4   Are packages, units, components, or other items weighing over more than 10 pounds provided with handles? Are the proper-sized handles used, and are they located in the right positions? Are the handles optimally located from the weight-distribution standpoint? Handles should be located over the centers of gravity.

12.5   Are packages, units, or other items weighing more than 40 pounds provided with two handles (for a two-person carrying capability)?

12.6   Can normal packing materials be used for shipping? If not, are special containers, cases, or covers provided to protect component-vulnerable areas from damage during handling?

13.0 Human Factors

13.1   Has a system analysis been accomplished to verify optimum human–machine interfaces? Are automated and manual functions adequately identified?

13.2   Are the identified automated/manual functions consistent with the results of the overall system-level functional analysis?

13.3   Have operational sequence diagrams (OSDs) been prepared where appropriate?

13.4   Has a detailed *operator* task analysis been accomplished to verify task sequence, task complexities, personnel skills, and so on?

13.5   Has a detailed *maintenance* task analysis been accomplished to verify maintenance task sequences, task complexities, personnel skills, and so on?

13.6   Is the detailed maintenance task analysis compatible with reliability data, maintainability data, and supportability analysis (SA) data?

13.7   Are the detailed operator and maintenance task analyses compatible with system/product operating and maintenance procedures (e.g., task sequences, depth of explanatory material based on task complexity)?

13.8   For human-interface functions, is the system/product design optimum in considering anthropometric factors, human sensory factors, psychological factors, and physiological factors? For manual tasks, does the design reflect "ease of operation" by low-skilled personnel? Is the design such that potential human error rates are minimized?

13.9   Has a detailed training plan for operator and maintenance personnel been prepared? Have training facility, equipment, material, software, and data requirements been identified?

13.10   Is the human-factors effort compatible with safety and security engineering requirements?

13.11   Has an approach been established for personnel test and evaluation?

14.0   Interchangeability

14.1   Are equipment, modules, and/or components that perform similar operations electrically, functionally, and physically interchangeable?

14.2   Can replacements of like items be made without adjustments and/or alignments?

15.0   Maintainability

15.1   Is the system/product maintainable in terms of troubleshooting and diagnostic provisions, accessibility, ease of replacement, handling capabilities, accuracy of test and verification, and economics in the performance of maintenance (corrective and preventive)? Actually, many of the other items in this checklist may be appropriately included under maintainability, depending on the organization involved in the design.

15.2   Have maintainability requirements for the system/equipment been adequately defined? Are they compatible with system performance, reliability, supportability, and effectiveness factors?

15.3   Have maintainability requirements been allocated to the appropriate level (e.g., MTBM, MDT, MLH/OH, $\overline{M}$ct, $\overline{M}$pt, \$/MA to the unit assembly, subassembly, and/or other appropriate component of the system)? For an explanation of these terms, please refer to Chapter 3.

15.4   Have anticipated system/product corrective and preventive maintenance requirements been identified through a detailed maintenance engineering analysis? Have the proper trade-off studies been conducted to attain the proper balance between corrective and preventive maintenance? Too much preventive maintenance can be costly and can significantly impact corrective maintenance requirements. Are the results compatible with supportability analysis (SA) data?

15.5   Has a level-of-repair analysis (LORA) been completed? Are the results consistent with the maintenance concept and the supportability analysis (SA)?

15.6   Have maintainability predictions been accomplished to assess the design in terms of the specified requirements? Do the predictions indicate compliance with the requirements?

15.7   Have maintainability demonstrations been conducted? Do the results indicate compliance with the requirements? Are these requirements included in the TEMP?

16.0 Mobility

16.1   Can the equipment/component be easily transported? Moved from one location to another?

16.2   Can the system component be moved utilizing common and standard support/handling equipment? The use of special handling equipment should be avoided.

17.0 Operability

17.1   Is the system designed for ease of operation?

17.2   Can the system be operated effectively by individuals with basic skills and with a minimum of special training?

17.3   Can system operation be accomplished with a minimum of error?

18.0 Packaging and Mounting

18.1   Is the packaging design attractive from the standpoint of consumer appeal (e.g., color, shape, size)?

18.2   Is functional packaging incorporated to the maximum extent possible? Interaction effects between packages should be minimized. It should be possible to limit maintenance to the removal of one module (the one containing the failed part) when a failure occurs and not require the removal of two, three, or four modules in order to resolve the problem.

18.3   Is the packaging design compatible with level-of-repair analysis decisions? Repairable items are designed to include maintenance provisions such as test points, accessibility, and plug-in components. Items classified as "discard at failure" should be encapsulated and relatively low in cost. Maintenance provisions for a disposable module are not required.

18.4   Are disposable modules incorporated to the maximum extent practical? It is highly desirable to reduce overall support through a no-maintenance-design concept as long as the items being discarded are relatively high in reliability and low in cost.

18.5   Are plug-in modules and components utilized to the maximum extent possible (unless the use of plug-in components significantly degrades the equipment reliability)?

18.6   Are accesses between modules adequate to allow for hand grasping?

18.7   Are modules and components mounted so that the removal of any single item for maintenance will not require the removal of other items? Component stacking should be avoided where possible.

18.8   In areas where module stacking is necessary because of limited space, are the modules mounted in such a way that access priority has been assigned in accordance with the predicted removal and replacement frequency? Items that require frequent maintenance should be more accessible.

18.9   Are modules and components, not of a plug-in variety, mounted with four fasteners or fewer? Modules should be securely mounted, but the number of fasteners should be held to a minimum.

18.10   Are shock-mounting provisions incorporated where shock and vibration requirements are excessive?

18.11   Are provisions incorporated to preclude installation of the wrong module?

18.12   Are plug-in modules and components removable without the use of tools? If tools are required, they should be of the standard variety.

18.13   Are guides (slides or pins) provided to facilitate module installation?

18.14   Are modules and components labeled?

18.15   Are module and component labels located on top or immediately adjacent to the items and in plain sight?

18.16   Are the labels permanently affixed and unlikely to come off during a maintenance action or as a result of environment? Is the information on the label adequate? Disposable modules should be so labeled. In equipment racks, are the heavier items mounted at the bottom of the rack? Unit weight should decrease with the increase in installation height.

18.17   Are operator panels optimally positioned? For personnel in the standing position, panels should be located between 40 and 70 inches above the floor. Critical or precise controls should be between 48 and 64 inches above the floor. For personnel in the sitting position, panels should be located 30 inches above the floor.

18.18   Are drawers in equipment racks mounted on roll-out slides?

19.0 Panel Displays and Controls

19.1   Are controls standardized?

19.2   Are controls sequentially positioned?

19.3   Is control spacing adequate?

19.4   Is control labeling adequate?

19.5   Have the proper control/display relationships been incorporated (based on good human-factors criteria)?

19.6   Are the proper types of panel switches used?

19.7   Is the control panel lighting adequate?

19.8   Are the controls placed according to frequency and/or criticality of use?

20.0 Personnel and Training

20.1   Have operational and maintenance personnel requirements (quantity and skill levels) been defined?

20.2   Are operational and maintenance personnel requirements minimized to the greatest extent possible?

20.3   Are operational and maintenance personnel requirements compatible with supportability analysis and with human-factors data? Personnel quantities and skill levels should "track" both sources.

20.4   Are the planned personnel skill levels at each location compatible with the complexity of the operational and maintenance tasks specified?

20.5   Has maximum consideration been given to the use of existing personnel skills for new equipment?

20.6   Have personnel attrition rates been established?

20.7   Have personnel effectiveness factors been determined (actual time that work is accomplished per the total time allowed for work accomplishment)?

20.8   Have operational and maintenance training requirements been specified? This includes consideration of both initial training and replenishment training throughout the life cycle.

20.9   Have specific training programs been planned? The type of training, frequency of training, duration of training, and student entry requirements should be identified.

20.10   Are the planned training programs compatible with the personnel skill-level requirements specified for the performance of operational and maintenance tasks?

20.11   Have training equipment requirements been defined? Have the applicable training equipment been acquired?

20.12   Have maintenance provisions for training equipment been planned?

20.13   Have training data requirements been defined?

20.14   Are the planned operating and maintenance procedures (designated for support of the system throughout its life cycle) utilized to the maximum extent possible in the training program(s)?

21.0 Producibility

21.1   Does the design lend itself to economic production? Can simplified fabrication and assembly techniques be employed?

21.2   Has the design stabilized (minimum change)? If not, are changes properly controlled through good configuration management methods?

21.3   Is the design such that rework requirements are minimized? Are spoilage factors held to a minimum?

21.4   Has the design been verified through prototype testing, environmental qualification, reliability qualification, maintainability demonstration, and the like?

21.5   Is the design such that many models of the same item can be produced with identical results? Are fabrication steps, manufacturing processes, and assembly methods adequately controlled through good quality assurance procedures?

21.6   Has adequate consideration been given to the application of just-in-time (JIT), Taguchi, material requirements planning (MRP), enterprise resource planning (ERP), and related methods in the production process?

21.7   Are production drawings, computer-aided design (CAD), computer-aided manufacturing (CAM), computer-aided support (CAS) data, materials lists, and so on, adequate for production needs?

21.8   Can currently available facilities, standard tools, and existing personnel be used for fabrication, assembly, manufacturing and test operations?

21.9   Is the design such that automated manufacturing processes (e.g., CAM, numerical control techniques) can be applied for high-volume repetitive functions?

21.10   Is the design definition such that two or more suppliers can produce the system/product from a given set of data with identical results?

22.0 Reconfigurability

22.1   Is the design configuration such that it can be readily upgraded for improved capability?

22.2   Have preplanned product improvements been considered in the initial design of the system?

22.3   Can modifications for performance enhancement be incorporated at minimum cost?

23.0 Reliability

23.1   Is the design simple? Have the number of component parts been kept to a minimum?

23.2   Are standard high-reliability parts being utilized?

23.3   Are item failure rates known? Has the mean life been determined?

23.4   Have parts been selected to meet reliability requirements?

23.5   Have parts with excessive failure rates been identified (unreliable parts)?

23.6   Have adequate derating factors been established and adhered to where appropriate?

23.7   Have the shelf life and wear-out characteristics of parts been determined?

23.8   Have all critical-useful-life items been eliminated from the design? If not, have they been identified with inspection/replacement requirements specified? Has a critical-useful-life analysis been accomplished?

23.9   Have critical parts that require special procurement methods, testing, and handling provisions been identified?

23.10   Has the need for the selection of "matching" parts been eliminated?

23.11   Have fail-safe provisions been incorporated where possible (protection against secondary failures resulting from primary failures)?

23.12   Has the use of "adjustable" components been minimized?

23.13   Have safety factors and safety margins been used in the application of parts?

23.14   Have component failure modes and effects been identified? Has a failure mode and effects analysis (FMEA), a failure mode, effects, and criticality analysis (FMECA), and/or a fault-tree analysis (FTA) been accomplished?

23.15   Has a stress-strength analysis been accomplished?

23.16   Have cooling provisions been incorporated in design "hot spot" areas? Is cooling directed toward the most critical items?

23.17   Has redundancy been incorporated in the design where needed to meet specified reliability requirements?

23.18   Are the best available methods for reducing the adverse effects of operational and maintenance environments on critical components being incorporated?

23.19   Have the risks associated with critical-item failures been identified and accepted? Is corrective action in design being taken?

23.20   Have reliability requirements for spares and repair parts been considered?

23.21   Have reliability predictions been accomplished? Have reliability testing requirements been defined? Test requirements in design? Test requirements in production/construction? Have they been covered in the TEMP?

23.22   Has a reliability failure analysis and corrective action capability been installed?

24.0 Safety

24.1   Has an integrated safety engineering plan been prepared and implemented?

24.2   Has a hazard analysis been accomplished to identify potential hazardous conditions? Is the hazard analysis compatible with the reliability FMECA/FMEA and FTA (where applicable)?

24.3   Have system/product hazards related to heat, cold, thermal change, barometric change, humidity change, shock, vibration, light, mold, bacteria, corrosion, rodents, fungi, odors, chemicals, oils, greases, handling and transportation, and so on, been eliminated?

24.4   Have fail-safe provisions been incorporated in the design?

24.5   Have protruding devices been eliminated, or are they suitably protected?

24.6   Have provisions been incorporated for protection against high voltages? Are all external metal parts adequately grounded?

24.7   Are sharp metal edges, access openings, and corners protected with rubber, fillets, fiber, or plastic coating?

24.8   Are electrical circuit interlocks employed?

24.9   Are standoffs or handles provided to protect system components from damage during the performance of shop maintenance?

24.10   Are tools that are used near high-voltage areas adequately insulated at the handle or at other parts of the tool that maintenance personnel are likely to touch?

24.11   Are the environments such that personnel safety is ensured? Are noise levels with a safe range? Is illumination adequate? Is the air clean? Are the temperatures at a proper level? Are the requirements of the Occupational Health and Safety Administration (OSHA) being maintained?

24.12   Has the proper protective clothing been identified for areas where the environment could be detrimental to human safety? Radiation, intense cold or heat, gas, loud noise, and so on, are examples of detrimental factors.

24.13   Are safety equipment requirements identified for areas where ordinance devices (and the like) are activated?

25.0 Selection of Parts/Materials

25.1   Have appropriate standards been consulted for the selection of components and materials?

25.2   Have all component parts and materials selected for the design been adequately evaluated prior to their procurement and application? Evaluation should consider performance parameters, reliability, maintainability, supportability, human factors, quality, and cost.

25.3   Have supplier sources for component-part and material procurement been established?

25.4   Are the established supplier sources reliable in terms of quality level, ability to deliver on time, and willingness to accept component-warranty provisions? There is an ongoing concern in regard to control specifications, process variations, stresses, tolerances, item interchangeability, and so on.

25.5   Have alternative supplier sources been identified for use in the event that the prime source fails to deliver?

26.0 Servicing and Lubrication

26.1   Have servicing requirements been held to a minimum?

26.2   Where servicing is indicated, are the specific requirements identified?

26.3   Are procurement sources for servicing materials known?

26.4   Are servicing points accessible?

26.5   Have personnel and equipment requirements for servicing been identified? This includes handling equipment, vehicles, carts, and so on.

26.6   Does the design include servicing indicators?

27.0 Societal Requirements

27.1   Does the system/product satisfy societal needs?

27.2   Have the societal effects of introducing the system/product into the inventory been evaluated (to determine the impact of system operation and support on community life)?

27.3   Have all adverse societal effects caused by the introduction, operation, and support of the system/product been minimized, if not eliminated?

28.0 Software

28.1   Have all system software requirements for operating and maintenance functions been identified? Have these requirements been developed through the system-level functional analysis (i.e., is there traceability indicated)?

28.2   Is the software complete in terms of scope and depth of coverage?

28.3   Is the software compatible relative to the equipment with which it interfaces? Is operating software compatible with maintenance software? With other elements of the system?

28.4   Are the language requirements for operating software and maintenance software compatible?

28.5   Is all software adequately covered through good documentation (i.e., logic functional flows and coded programs)?

28.6   Has the software been adequately tested and verified for accuracy (performance), reliability, and maintainability?

29.0 Standardization

29.1   Are standard commercial off-the shelf (COTS) components and parts incorporated in the design to the maximum extent possible (except for items not compatible with effectiveness factors)? Maximum standardization is desirable.

29.2   Are the same items and/or parts used in similar applications?

29.3   Are identifying equipment labels and nomenclature assignments standardized to the maximum extent possible?

29.4   Are equipment-control-panel positions and layouts (from panel to panel) the same or similar when a number of panels are incorporated and provide comparable functions?

30.0 Storage

30.1   Can the equipment be stored for extended periods of time without excessive degradation (beyond specification limits)?

30.2   Have scheduled maintenance requirements for stored equipment been defined?

30.3   Have scheduled maintenance requirements for stored equipment been eliminated or minimized?

30.4   Have the required maintenance resources necessary to service stored equipment been identified?

30.5   Have storage environments been defined?

31.0 Supportability

31.1   Have spare/repair part requirements been minimized to the greatest extent possible? Is the number of different part types used throughout the design minimized?

31.2   Are the types and quantity of spare/repair parts compatible with the system maintenance concept, the supportability analysis (SA), and level-of-repair analysis data?

31.3   Are the types and quantity of spare/repair parts designated for a given location appropriate for the estimated demand at that location? Too many or too few spares can be costly.

31.4   Have the distribution channels and inventory points for spare/repair parts been established?

31.5   Are spare/repair part provisioning factors (e.g., replacement frequencies) directly traceable to reliability and maintainability predictions?

31.6   Are the specified logistics pipeline times compatible with effective supply support? Long pipeline times place a tremendous burden on logistic support.

31.7 Have spare/repair parts been identified and provisioned for preoperational support activities (e.g., interim producer or supplier support, test programs)?

31.8 Have test and acceptance procedures been developed for spare/repair parts? Spare/repair parts should be processed, produced, and accepted on a similar basis as their equivalent components in the prime equipment.

31.9 Have the consequences (risks) of stock-out been defined in terms of effect on mission requirements and cost?

31.10 Has an inventory safety stock level been defined?

31.11 Has a provisioning or procurement cycle been defined (procurement or order frequency)? Have economic order quantity (EOQ) factors been determined?

31.12 Has a supply-availability requirement been established (the probability of having a spare available when required)?

31.13 Have the test and support equipment requirements been defined for each level of maintenance?

31.14 Have standard test and support equipment items been selected? Newly designed equipment should not be necessary unless standard equipment is unavailable.

31.15 Are the selected test and support equipment items compatible with the prime equipment? Does the test equipment do the job?

31.16 Are the test and support equipment requirements compatible with maintenance concept, supportability analysis (SA), and level-of-repair analysis data?

31.17 Have test and support equipment requirements (in terms of both variety and quantity) been minimized to the greatest extent possible?

31.18 Are the reliability and maintainability features in the test and support equipment compatible with those equivalent features in the prime equipment? It is not practical to select an item of support equipment that is not as reliable as the item it supports.

31.19 Have logistic support requirements for the selected test and support equipment been defined? This includes maintenance tasks, calibration equipment, spare/repair parts, personnel and training, data, and facilities.

31.20 Is the test and support equipment selection process based on cost-effectiveness considerations (i.e., life-cycle cost)?

31.21 Have test and maintenance software requirements been adequately defined?

31.22 Have operational and maintenance personnel requirements (quantity and skill levels) been defined?

31.23 Are operational and maintenance personnel requirements minimized to the greatest extent possible?

31.24 Are operational and maintenance personnel requirements compatible with supportability analysis (SA) and with human-factors data? Personnel quantities and skill levels should track both sources.

31.25 Are the planned personnel skill levels at each location compatible with the complexity of the operational and maintenance tasks specified?

31.26 Has maximum consideration been given to the use of existing personnel skills for the new system?

31.27 Have personnel attrition rates been established?

31.28 Have personnel effectiveness factors been determined (actual time that work is accomplished per the total time allowed for work accomplishment)?

31.29 Have operational and maintenance training requirements been specified? This includes consideration of both initial training and replenishment training throughout the life cycle.

31.30 Have specific training programs been planned? The type of training, frequency of training, duration of training, and student entry requirements should be identified.

31.31 Are the planned training programs compatible with the personnel skill-level requirements specified for the performance of operational and maintenance tasks?

31.32 Have training equipment requirements been defined? Has the needed training equipment be procured?

31.33 Have maintenance provisions for training equipment been planned?

31.34 Have training data requirements been defined?

31.35 Are the planned operating and maintenance procedures (designated for support of the system throughout its life cycle) utilized to the maximum extent possible in the training program(s)?

32.0 Support Equipment Requirements

Refer to questions 31.13 through 31.21, category 31.0, "Supportability," for coverage of test and support equipment.

33.0 Survivability

33.1 Has an integrated survivability engineering plan been prepared and implemented?

33.2 Have survivability measures been established, and are they appropriately integrated with other system TPMs?

33.3 Has a test and evaluation approach been defined for the verification of system survivability? Is this covered in the TEMP?

34.0 Testability

34.1   Have self-test provisions been incorporated where appropriate?

34.2   Is reliability degradation due to the incorporation of built-in test (BIT) minimized? The BIT capability should not significantly impact the reliability of the overall system.

34.3   Is the extent or depth of self-testing compatible with the level-of-repair analysis?

34.4   Are self-test provisions automatic?

34.5   Have direct fault indicators been provided (a fault light, an audio signal, or a means of determining that a malfunction positively exists)? Are continuous condition monitoring provisions incorporated where appropriate?

34.6   Are test points provided to enable checkout and fault isolation beyond the level of self-test? Test points for fault isolation within an assembly should not be incorporated if the assembly is to be discarded at failure. Test point provisions must be compatible with the level of repair analysis.

34.7   Are test points accessible? Accessibility should be compatible with the extent of maintenance performed. Test points on the operator's front panel are not required for a depot maintenance action.

34.8   Are test points functionally and conveniently grouped to allow for sequential testing (following a signal flow), testing of similar functions, or frequency of use when access is limited?

34.9   Are test points provided for a direct test of all replaceable items?

34.10   Are test points adequately labeled? Each test point should be identified with a unique number, and the proper signal or expected measured output should be specified on a label located adjacent to the test point.

34.11   Are test points adequately illuminated to allow the technician to see the test point number and labeled signal value?

34.12   Can every component malfunction (degradation beyond specification tolerance limits) that could possibly occur be detected through a no-go indication at the system level? Are false alarm rates minimized? This is a measure of test thoroughness.

34.13   Will the prescribed maintenance software provide adequate diagnostic information?

35.0 Transportability

35.1   Have transportation and handling requirements been defined?

35.2   Have transportation requirements been considered in the equipment design? This includes consideration of temperature ranges, vibration and shock, humidity, and so on. Has the possibility of equipment degradation been minimized if transported by air, ground vehicle, ship, or rail?

35.3   Can the equipment be easily disassembled, packed, transported from one location to another, reassembled, and operated with a minimum of performance and reliability degradation?

35.4   Have container requirements been defined?

35.5   Have the requirements for ground handling equipment been defined?

35.6   Was the selection of handling equipment based on cost-effectiveness considerations?

36.0 Quality

36.1   Has a total quality management (TQM) plan been prepared and implemented? Does it include coverage of customer, contractor (producer), and supplier activities and interfaces?

36.2   Are formal quality (i.e., TQM) training programs being conducted within the customer/contractor/supplier organization?

36.3   Are statistical process control (SPC) and Six Sigma methods/techniques being implemented where appropriate?

36.4   Are quality control requirements specified and imposed on all suppliers?

The preceding questions are representative of what may be considered in conducting a program/design review. They should not be considered as being all-inclusive by any means. In fact, it is appropriate for the reviewer to develop a list covering applicable issues tailored to the program in question.

APPENDIX E

SUPPLIER EVALUATION CHECKLIST

This checklist, to be applied in the evaluation of suppliers, is tailored and is a supplemental version of the in-depth design review checklist presented in Appendix D. Not all of the questions are applicable in all situations; however, the answer to those questions that are applicable should be yes in order to reflect the desired results.

E.1 GENERAL CRITERIA

1.1 Has a technical performance specification been prepared covering the product being acquired? Is this specification "supportive" of and "traceable" from the system specification?

1.2 Is the product a commercial off-the-shelf (COTS) item requiring no adaptation, modification, and/or rework for installation?

1.3 Has the COTS item been assessed in terms of effectiveness and life-cycle cost?

1.4 If the product is a COTS item and requires some modification for installation,

    1. Has the degree of modification been clearly defined and minimized to the extent possible?

    2. Has the impact of the modification been assessed in terms of effectiveness and life-cycle cost? Has the life-cycle cost been minimized to the extent possible?

    3. Can the modification be accomplished easily and with a minimum of interaction effects?

    4. Have common and standard parts, reusable software, recycleable material, and so on, been incorporated in the modification/interface package or kit?

1.5   Have alternative sources of supply for the same product been identified?

1.6   If new design is required, has it been justified to the extent that the COTS, modified COTS, and comparable options are not feasible?

E.2   PRODUCT DESIGN CHARACTERISTICS

2.1   Technical Performance Parameters

1. Does the product fully comply with the functional performance specification (i.e., development, product, process, and/or material specification as applicable)?

2. Has the applicable mission scenario (or operational/utilization profile) been defined for the product?

3. Are the product's design characteristics responsive to the prioritized technical performance measures (TPMs)? Does the design reflect the most important features?

4. Were the design characteristics derived through the use of a quality function deployment (QFD) (or equivalent) approach?

5. Are the performance requirements easily traceable from those specified for the system level?

6. Are the performance requirements measurable? Can they be verified or validated?

2.2   Technology Applications

1. Does the design utilize state-of-the-art and commercially available technologies?

2. Do the technologies utilized have a life cycle that is at least equivalent to the product life cycle?

3. Have "short-life" technologies been eliminated? If not, have such applications been minimized?

4. Has an "open-architecture" approach been utilized in the design such that new technologies can be inserted without causing a redesign of other elements of the product?

5. Have alternative sources for each of the technologies being utilized been identified?

6. Have the technologies being utilized reached a point of maturity/stability relative to their applications?

2.3   Physical Characteristics

1. Is the product both functionally and physically interchangeable?

2. Can the product be physically removed and replaced with a like item without requiring any subsequent adjustments or alignments? If not, have such interaction effects been minimized?

3. Does the product design comply with the physical requirements in the technical specification (i.e., size, shape, and weight)?

2.4 Effectiveness Factors

1. Have the appropriate effectiveness factors been defined and included in the technical specification (i.e., TPMs applicable to the product being acquired)?

2. Can the effectiveness requirements be traced back to comparable requirements specified at the system level?

3. Has the supplier provided a measure of reliability for the product (e.g., R, failure rate, and/or MTBF)? Is this figure of merit based on actual field experience?

4. Have the applicable reliability requirements been considered in the product design?

5. Has the supplier provided a measure of maintainability for the product (e.g., MTBM, MLH/OH, $\overline{M}$ct, $\overline{M}$pt, MDT, and/or equivalent)? Is this figure or merit based on actual field experience? Refer to Chapter 3 for an explanation of these acronyms.

6. Have the applicable maintainability requirements been considered in the product design?

7. Have the applicable human-factors requirements been considered in the product design?

8. Have the applicable safety and security requirements been considered in the product design?

9. Have the applicable supportability/serviceability requirements been considered in the product design?

10. Have the applicable quality requirements been considered in the product design?

2.5 Producibility Factors

1. Has the product been designed for producibility?

2. Is the design data/documentation such that any other supplier with comparable facilities/equipment, capabilities, and experience can manufacture the product in accordance with the specification?

2.6 Disposability Factors

1. Has the product been designed for disposability?

2. Has the supplier developed the appropriate planning documentation and procedures covering the disposal and/or recycling of the product?

2.7 Environmental Factors

1. Has the product been designed with ecological and environmental requirements in mind?

2. Has the supplier prepared an environmental impact statement for the introduction of the product?

2.8 Economic Factors

1. Has the product been designed with economic considerations in mind?

   2. Has the supplier conducted a life-cycle cost analysis for the product? Are the results realistic? Refer to Appendix B.

E.3 PRODUCT MAINTENANCE AND SUPPORT INFRASTRUCTURE

3.1   Maintenance and Support Requirements

   1. Does the supplier have an established maintenance and support infrastructure in place?
   2. Has the supplier defined the maintenance concept/plan for the product?
   3. Have the appropriate supportability "metrics" been established for the product and included in the maintenance concept/plan (i.e., response time, turnaround time, maintenance process time, test equipment reliability and maintainability factors, facility utilization, spare parts demand rates and inventory levels, transportation rates and times, etc.)?
   4. Does the maintenance concept/plan facilitate or allow for the required degree of *responsiveness* on the part of the supplier?
   5. Have the preventive maintenance requirements been established for the product (if any)? Have these requirements been justified through a reliability-centered maintenance (RCM) approach?
   6. Have the product maintenance and support resource requirements been defined (i.e., spares, repair parts, and associated inventories; personnel quantities, skill levels, and training; test and support equipment; facilities; packaging, transportation and handling; technical data; and computer resources)? Have these requirements been adequately justified through a maintenance engineering analysis (MEA), a supportability analysis (SA), or equivalent?

3.2   Data/Documentation

   1. Does the supplier have a computerized maintenance management data capability in place? Is this capability being effectively utilized for the purposes of *continuous product/process improvement*? Does it provide visibility relative to how well the product is performing in the field?
   2. Does the supplier have in place a reliability data collection, analysis, feedback, and corrective-action process? Are product failures properly recorded and are they traceable to the cause?
   3. Is the supplier monitoring and measuring the effectiveness of its preventive maintenance program? Where applicable, have the preventive maintenance requirements been revised to reflect a more cost-effective approach?

3.3   Warranty/Guarantee Provisions

   1. Have product warranties/guarantees been established?
   2. Have the established warranty provisions been adequately defined through some form of a contractual mechanism?

3. Are the warranty provisions consistent with the defined maintenance concept?

3.4 Customer Service

1. Does the supplier have an established customer service capability in place?
2. Will the supplier provide assistance in the installation and checkout of the product at the producer's site and/or the user's site (if required)?
3. Will the supplier provide on-site field service support if required?
4. Does the supplier provide operator and maintenance training at the producer's site and/or the user's site when necessary? Is this training available "on call"? Will it be available throughout the product life cycle?
5. In support of training activities, will the supplier provide the necessary data, training manuals, software, aids, equipment, simulators, and so on? Will the supplier provide updates/revisions to the training material as applicable?
6. Does the supplier have a program for measuring training effectiveness?

3.5 Economic Factors

1. Is the product support infrastructure cost-effective?
2. Have the requirements been based on life-cycle cost objectives?

E.4 SUPPLIER QUALIFICATIONS

4.1 Planning/Procedures

1. Does the supplier have a standard policies and procedures manual/guide?
2. Are the appropriate management procedures properly documented and followed on a day-to-day basis?
3. Are the procedures/processes periodically reviewed, evaluated, and revised as necessary for the purposes of *continuous process improvement?*
4. Has the supplier identified the activities and tasks that are essential in the successful accomplishment of system engineering requirements?

4.2 Organizational Factors

1. Has the supplier's organization been adequately defined in terms of activities, responsibilities, interface requirements, and so on?
2. Does the organizational structure support the overall program objectives for the system? Is it compatible with the producer's organizational structure?
3. Has the supplier identified the organizational element responsible for the accomplishment of system engineering tasks (as applicable)?

4.3 Available Personnel and Resources

    1. Does the supplier have the available personnel and associated resources to assign to the task(s) being contracted? Will these personnel/ resources be available for the duration of the program?

    2. Do the personnel assigned have the proper background, experience, and training to do the job effectively?

4.4   Design Approach

    1. Has the supplier implemented the system engineering process in the design of its products?

    2. Has an effective design database been established, and is it compatible with the system-level database established by the producer (prime contractor)?

    3. Does the supplier have in place a configuration management program, along with a disciplined change-control process? Has a configuration "baseline" approach been implemented in the development and growth of the product?

    4. Has the supplier's design process been enhanced through the use of such tools as computer-aided design (CAD), simulation, rapid prototyping, EC applications, and so on?

4.5   Manufacturing Capability

    1. Does the supplier have a well-defined manufacturing process in place?

    2. Does the process incorporate the latest technologies and computer-aided methods (i.e., robotics, the use of CAD or computer-integrated manufacturing (CIM) technology, etc.)?

    3. Is the process flexible, and does it support an "agile" and/or "lean" manufacturing approach?

    4. Does the supplier utilize materials requirements planning (MRP), capacity planning (CP), shop floor control (SFC), just-in-time (JIT), master production scheduling (MPS), enterprise resource planning (ERP), statistical process control (SPC), Six Sigma, and other such methods in the manufacturing process?

    5. Has the supplier implemented a formal quality program in accordance with ISO-9000 and ISO-14,000 (or equivalent)? Is the supplier ISO-9000-certified? Does the supplier have a formal procedure in place for correcting deficiencies?

    6. Has the supplier implemented a total productive maintenance (TPM) program within its manufacturing plant? Has a TPM measure of effectiveness been established (i.e., OEE/overall equipment effectiveness)?

4.6   Test and Evaluation Approach

    1. Has the supplier developed an integrated test and evaluation plan for the product?

    2. Have the requirements for testing been derived in a logical manner, and are they compatible with the identified technical performance measures (TPMs) for the system, and as allocated for the product?

3. Does the supplier have the proper facilities and resources to support all product testing requirements (i.e., people, facilities, equipment, data)?

4. Does the supplier have in place a data collection, analysis, and reporting capability covering all testing activities?

5. Does the supplier have a plan for "retesting" if required?

4.7 Management Controls

1. Has the supplier incorporated the necessary controls for monitoring, reporting, providing feedback, and initiating corrective action in regard to technical performance measurement, cost measurement, and scheduling?

2. Has the supplier implemented a configuration management capability?

3. Has the supplier implemented an integrated data management capability?

4. Has the supplier developed a risk management plan?

4.8 Experience Factors

1. Has the supplier had experience in designing, testing, manufacturing, handling, delivering, and supporting this product before?

2. Has the supplier utilized experiences from other projects to help respond to the requirements for this program; that is, the transfer of "lessons learned"?

4.9 Past Performance

1. Has the supplier successfully completed similar projects in the past?

2. Has the supplier been responsive to all of the requirements for past projects?

3. Has the supplier been successful in delivering products in a timely manner and within cost?

4. Has the supplier delivered reliable and high-quality products?

5. Has the supplier been responsive in initiating any corrective action that has been required to correct deficiencies?

6. Has the supplier stood behind all product warranties/guarantees?

7. Does the supplier's organization reflect stability, growth, and high quality?

8. Is the supplier's business posture good?

9. Does the supplier enjoy an excellent reputation?

4.10 Maturity

1. Has the supplier established a process for benchmarking?

2. Has the supplier implemented an organizational assessment program (i.e., Systems Engineering Capability Model (SECM), Capability Maturity Model Integration (CMMI), or equivalent)?

4.11 Economic Factors

1. Has the supplier implemented a life-cycle cost-analysis approach for all of its functions, products, processes, and so on?

2. Has the supplier implemented an activity-based costing (ABC) approach with the objective of acquiring full visibility relative to the high-cost contributors and cause-and-effect relationships, and leading to the implementation of improvements for cost-reduction purposes?

APPENDIX F

SELECTED BIBLIOGRAPHY

When addressing the subject of *system engineering*, one should become familiar not only with the available literature in the field but with some of the subject areas that are directly aligned with system engineering. System engineering, by its nature, is highly *interdisciplinary,* and acquiring knowledge in closely related areas is essential if one is to progress and successfully accomplish the objectives specified herein. With this in mind, this bibliography has been developed to cover selected references in each of the following areas:

F.1. Systems, System Analysis, and System Engineering

F.2. Concurrent and Simultaneous Engineering

F.3. Software and Computer-Aided Systems

F.4. Reliability Engineering

F.5. Maintainability Engineering and Maintenance

F.6. Human Factors, Safety, and Security Engineering

F.7. Logistics, Supply Chain Management, and Supportability

F.8. Production, Manufacturing, Quality Control and Assurance

F.9. Operations Research and Operations Management

F.10. Engineering Economy and Life-Cycle Cost Analysis

F.11. Management and Supporting Areas

It should be emphasized that, in preparing a bibliography, one is often guilty by virtue of what has been left out. Nevertheless, an attempt has been made to provide some additional guidance in the field.

F.1 SYSTEMS, SYSTEMS ANALYSIS, AND SYSTEMS ENGINEERING

1. ANSI/GEIA ElA-632. *Processes for Engineering a System.* Electronic Industries Alliance (EIA). Arlington, VA, September 2003.
2. Blanchard, B. S. *System Engineering Management,* 3rd ed. John Wiley & Sons, Hoboken, NJ, 2004.

3. Blanchard, B. S., and W.J. Fabrycky. *Systems Engineering and Analysis,* 4th ed. Prentice Hall, Upper Saddle River, NJ, 2006.

4. Boyd, D.W. *Systems Analysis and Modeling: A Macro-Micro Approach with Multidisciplinary Applications.* Academic Press, San Diego, CA, 2001.

5. Buede. D.M. *The Engineering Design of Systems: Models and Methods.* John Wiley & Sons, Hoboken, NJ, 2000.

6. DAU. *Systems Engineering Fundamentals.* Defense Acquisition University (DAU) Press, Fort Belvoir, VA 22060-5565, December 2000.

7. Eisner, H. *Essentials of Project and Systems Engineering Management*, 2nd ed. John Wiley & Sons, Hoboken, NJ, 2002.

8. GEIA EIA-731-1. *Systems Engineering Capability Maturity Model (SECM).* Electronic Industries Alliance (EIA), Arlington, VA, August 2002.

9. GEIA ElA-731-2. *Systems Engineering Capability Model Appraisal Method.* Electronic Industries Alliance (EIA), Arlington, VA, August 2002.

10. Grady, J.O. *System Engineering Deployment.* CRC Press, Boca Raton, FL, 2000.

11. IEEE 1220-1998. *Standard for Application and Management of the Systems Engineering Process.* Institute of Electrical and Electronics Engineers (IEEE), 345 East 47th St., New York, NY 10017, 1998.

12. IEEE 1233-1998. *IEEE Guide for Developing System Requirements Specifications.* Institute of Electrical and Electronics Engineers (IEEE), 345 East 47th St., New York, NY 10017, 1998 (ISBN 0738103373).

13. INCOSE-The International Council on Systems Engineering. *Systems Engineering.* Quarterly Journal, published by John Wiley & Sons, Hoboken, NJ.

14. INCOSE-The International Council on Systems Engineering, *INSIGHT*. Quarterly Newsletter, INCOSE, 2150 N. 107th St., Suite 205, Seattle, WA 98133-9009.

15. INCOSE-TP-2003-002-03.1. *Systems Engineering Handbook.* International Council on Systems Engineering (INCOSE), 2150N. 107th St., Suite 205, Seattle, WA WA 98133-9009, Version 3.1, August 2007.

16. ISO/IEC-15288. *Systems Engineering—Systems Life-Cycle Processes.* ISO, Geneva, Switzerland, 2002.

17. ISO/IEC-19760. *A Guide for the Application of ISO/IEC-15288 System Life-Cycle Processes.* 2004.

18. Kossiakof F, A., and W. Sweet. *Systems Engineering: Principles and Practice.* John Wiley & Sons, Hoboken, NJ, 2003.

19. Maier, M.W., and E. Rechtin. *The Art of Systems Architecting,* 2nd ed. CRC Press, Boca Raton, FL, 2000.

20. Martin, J.N. *Systems Engineering Guidebook: A Process for Developing Systems and Products.* CRC Press, Boca Raton, FL, 1996.

21. Pugh, S. *Total Design: Integrated Methods for Successful Product Engineering.* Addison-Wesley Publishing, Reading, MA, 1991 (ISBN 100201416395).

22. Sage, A. P. *Systems Engineering.* John Wiley & Sons, Hoboken, NJ, 1992.

23. Sage, A. P., and W. B. Rouse (eds.). *Handbook of Systems Engineering and Management.* John Wiley & Sons, Hoboken, NJ, 1999 (ISBN 0471154059).

24. Sage, A. P., and J. E. Armstrong. *Introduction to Systems Engineering.* John Wiley & Sons, Hoboken, NJ, 2000.

25. Sandquist, G. M. *Introduction to System Science.* Prentice Hall, Upper Saddle River, NJ, 1885.

26. Wasson, C. S. *System Analysis, Design, and Development: Concepts, Principles, and Practice.* John Wiley & Sons, Hoboken, NJ, 2006.

27. Young, R. R. *Effective Requirements Practices.* Addison-Wesley Publishing, Reading, MA, 2001.

F.2 CONCURRENT AND SIMULTANEOUS ENGINEERING

1. Anderson. D. M. *Design for Manufacturability and Concurrent Engineering.* CIM Press, 2003.

2. Hartley, J. R. *Concurrent Engineering: Shortening Lead Times, Raising Quality, and Lowering Costs.* Productivity Press, Portland, OR, 1998.

3. Prasad, B. *Concurrent Engineering Fundamentals: Integrated Product Development.* Prentice Hall, Upper Saddle River, NJ, 1997.

4. Shina, S. G. (ed.). *Successful Implementation of Concurrent Engineering Products and Processes.* John Wiley & Sons, Hoboken, NJ, 1994.

F.3 SOFTWARE AND COMPUTER-AIDED SYSTEMS

1. Boehm, B. W. *Software Engineering Economics.* Prentice Hall, Upper Saddle River, NJ, 1981.

2. *Crosstalk—The Journal of Defense Software Engineering.* Published by the Software Technology Support Center (STSC), 00-ALC/MAS, 6022 Fir Ave., Building 1238, Hill AFB, UT 84056-5820 (monthly).

3. IEEE/ElA-12207. *Information Technology—Software Life-Cycle Processes.* Defense Automated Printing Services, Building 4/D, 700 Robins Ave., Philadelphia, PA 19111-5094.

4. ISO/IEC 15939. *Software Engineering—Software Measurement Process.* 2002.

5. Jones, C. *Software Assessments, Benchmarks, and Best Practices.* Addison Wesley Longman, Reading, MA, 2000.

6. Leach, R. *Introduction to Software Engineering.* CRC Press, Boca Raton, FL, 1999.

7. McConnell, S. *Software Project Survival Guide.* Microsoft Press, Redmond, WA, 1998.

8. Moore, J. W. *The Roadmap to Software Engineering: A Standards-Based Guide*. Wiley IEEE Computer Society Press, Los Alamitos, CA, 2006.

9. Pressman, R.S. *Software Engineering: A Practitioner's Approach*, 6th ed. McGraw-Hill, New York, NY, 2005.

10. Sage, A. P., and J. D. Palmer. *Software Systems Engineering*. John Wiley & Sons, Hoboken, NJ, 1990.

11. Sage, A. P. *Systems Management for Information Technology and Software Engineering*. John Wiley & Sons, Hoboken, NJ, 1995.

12. Shih, R. H. *Parametric Modeling with Autodesk Inverter 2008*. Software Development Corp., Mission, KS, 2007 (ISBN 9781585033713).

13. Wiegers, K. E. *Software Requirements,* 2nd ed. Microsoft Press, Redmond, WA, 2003.

F.4 RELIABILITY ENGINEERING

1. *Annual Reliability and Maintainability Symposium (RAMS)*. Proceedings, Sponsored by 10 Technical Societies, Scien-Tech Associates, Inc., P.O. Box 2097, Banner Elk, NC 28604-2097.

2. Barlow. R. E. *Engineering Reliability (Statistics and Applied Probability)*. Society of Industrial & Applied Mathematics, New York, NY, 1998.

3. Dhillon, B. S. *Reliability, Quality, and Safety for Engineers*. CRC Press, P.O. Box 409267, Atlanta, GA 30384, 2005.

4. Dodson, B., and D. Nolan. *Reliability Engineering Handbook*. CRC Press, Boca Raton, FL, 1999.

5. Ebeling, C. E. *An Introduction to Reliability and Maintainability Engineering*. McGraw-Hill, New York, NY, 1997 (ISBN 0070188521).

6. Institute of Electrical and Electronics Engineers (IEEE). *IEEE Transactions on Reliability*. Published quarterly, IEEE, P.O. Box 1331, Piscataway, NJ 08854-1331.

7. IEEE 1332-1998. *IEEE Standard Reliability Program for Development and Production of Electronic Systems and Equipment*. Institute of Electrical and Electronics Engineers (IEEE), 345 East 47th St., New York, NY 10017, 1998.

8. IEEE 1413-1998. *IEEE Standard Methodology for Reliability Predictions and Assessment for Electronic Systems and Equipment*. Institute of Electrical and Electronics Engineers (IEEE), 345 East 47th St., New York, NY 10017, 1998.

9. Ireson, W. G. (ed.). *Handbook of Reliability Engineering and Management,* 2nd ed. McGraw-Hill, New York, 1996.

10. Knezevic, J. *Reliability, Maintainability, and Supportability: A Probabilistic Approach*. McGraw-Hill, London, UK, 1993.

11. Musa, J. D. *Software Reliability Engineering: More Reliable Software, Faster Development and Testing*. McGraw-Hill, New York, 1998.

12. Nachlas. J. A. *Reliability Engineering: Probabilistic Models and Maintenance Methods*. CRC Press, P.O. Box 409267. Atlanta, GA 30384, 2005.

13. Nikolaidis, E., D. M. Ghiocel, and S. Singhal. *Engineering Design Reliability Handbook*. CRC Press, P.O. Box 409267, Atlanta, GA 30384, 2005.

14. O'Connor, P.D.T., D. Newton, and R. Bromley. *Practical Reliability Engineering*, 4th ed. John Wiley & Sons, Hoboken, NJ, 2002.

15. RIAC. *Handbook of 217Plus Reliability Prediction Models*. Reliability Information Analysis Center (RIAC), 6000 Flanagan Road, Suite 3, Utica, NY 13502, 2006.

16. RIAC. *System Reliability Toolkit: A Practical Guide for Understanding and Implementing a Program for System Reliability*. Reliability Information Analysis Center (RIAC), 6000 Flanagan Road, Suite 3, Utica, NY 13502, 2005.

17. Smith, D. J. *Reliability, Maintainability, and Risk*, 6th ed., Butterworth-Heinemann, Woburn, MA, 2001.

F.5 MAINTAINABILITY ENGINEERING AND MAINTENANCE

1. Blanchard, B. S., D. Verma, and E. L. Peterson. *Maintainability: A Key to Effective Serviceability and Maintenance Management*. John Wiley & Sons, Hoboken, NJ, 1995.

2. Dhillon, B. S. *Engineering Maintenance*. CRC Press, P.O. Box 409267, Atlanta GA 30384, 2002.

3. Dhillon, B. S. *Maintainability, Maintenance, and Reliability for Engineering*. CRC Press, P.O. Box 409267, Atlanta, GA 30384, 2006.

4. Knezevic, J. *System Maintainability: Analysis, Engineering, and Management*. Chapman and Hall, London, 1997.

5. Maintenance Steering Group 3 Task Force. *Operator/Manufacturer Schedule Maintenance Development*. Air Transport Association (ATA) of America, Washington, DC, April 2001 (http://www.airlines.org).

6. Moubray, J. *Reliability-Centered Maintenance*, 2nd ed. Industrial Press, Boca Raton, FL, 1997.

7. Nakajima, S. *Introduction to TPM: Total Productive Maintenance*. Productivity Press, Portland, OR, 1994.

8. Nyman, D., and J. Levitt. *Maintenance Planning Scheduling, and Coordination*. Industrial Press, Boca Raton, FL, 2002.

9. Reliability Information Analysis Center (RIAC). *Maintainability Toolkit: A Practical Guide for Designing and Developing Maintainable Products and Systems*. RIAC, 6000 Flanagan Rd., Suite 3, Utica, NY 13502-1348, 2000.

10. Willmott, P., and D. McCarthy, *Total Productive Maintenance: A Route to World-Class Performance*. Butterworth-Heinemann, Woburn, MA, 2001.

11. Wireman, T. *Total Productive Maintenance*. 2nd ed. Industrial Press, Boca Raton, FL, 2003.

F.6 HUMAN FACTORS, SAFETY, AND SECURITY ENGINEERING

1. Bahr, N.J. *System Safety Engineering and Risk Assessment: A Practical Approach.* Taylor & Francis, New York, 1997.

2. Booer, H. R. *Handbook of Human Systems Integration.* Wiley-Interscience, Hoboken, NJ, 2003.

3. Chapanis, A. *Human Factors in Systems Engineering.* John Wiley & Sons, Hoboken, NJ, 1996.

4. GEIA HEB 1. *Human Engineering—Principles and Practices.* Government Electronics and Information Technology Association (GEIA), June 2002.

5. Kroemer, K. H. E., H. J. Kroemer, and K. E. Kroemer-Elbert. *Ergonomics*, 2nd ed. Prentice Hall, Upper Saddle River, NJ, 2001.

6. Roland, H. E., and B. Moriarity. *System Safety Engineering and Management*, 2nd ed. John Wiley & Sons, Hoboken, NJ, 1990.

7. Salvendy, G. (ed.). *Handbook of Human Factors and Ergonomics*, 3rd ed. John Wiley & Sons, Hoboken, NJ, 2006.

8. Sanders, M. S., and E. J. McCormick. *Human Factors in Engineering and Design.* 7th ed. McGraw-Hill, New York, 1993 (ISBN 007054901X).

9. Wickens, C. D., J. Lee, Y. D. Liu, and S. Gordon-Becker. *An Introduction to Human Factors Engineering.* Prentice Hall, Upper Saddler River, NJ, 2003.

10. Warkentin, M., and R Vaughn. *Enterprise Information Systems Assurance and System Security: Managerial and Technical Issues.* IGI Publishing, Hershey, PA, 2006.

11. Woodson, W. E., P. Tillman, and B. Tillman. *Human Factors Design Handbook*, 2nd ed. McGraw-Hill, New York, 1992.

F.7 LOGISTICS, SUPPLY CHAIN MANAGEMENT, AND SUPPORTABILITY

1. Ayers, J. B. *Handbook of Supply Chain Management.* Saint Lucie Press, 2000.

2. Ballou, R. H. *Business Logistics Management: Planning, Organizing, and Controlling the Supply Chain*, 5th ed. Prentice Hall, Upper Saddle River, NJ, 2003.

3. Blanchard, B. S. *Logistics Engineering and Management*, 6th ed. Prentice Hall, Upper Saddle River, NJ, 2004.

4. Bowersox, D. J. *Supply Chain Logistical Management*, 2nd ed. McGraw-Hill, New York, NY, 2005.

5. Chopra, S., and P. Meindl. *Supply Chain Management*, 2nd ed. Prentice Hall, Upper Saddle River, NJ, 2003.

6. Council of Supply Chain Management Professionals (CSCMP). *Logistics Comment*. CSCMP Newsletter, 2805 Butterfield Rd., Suite 200, Oak Brook, IL 60523.

7. Council of Supply Chain Management Professionals (CSCMP). *Annual Conference Proceedings*. CSCML, 2805 Butterfield Rd., Suite 200, Oak Brook, IL 60523.

8. Council of Supply Chain Management Professionals (CSCMP). *Journal of Business Logistics*. CSCMP, 805 Butterfield Rd., Suite 200, Oak Brook, IL 60523.

9. Coyle, J. J. E. J. Bardi, and R. A. Novack. *Transportation*, 5th ed. South-Western Publisher, St. Paul, MN, 2000.

10. Coyle, J. J., E. J. Bardi, and C. J. Langley. *The Management of Business Logistics*, 7th ed. South-Western Publisher, Mason, OH, 2003.

11. DOD. *Designing and Assessing Supportability in DOD Weapon Systems: A Guide to Increased Reliability and Reduced Logistics Footprint*. Department of Defense, Washington, DC, 2003.

12. Frazelle, E. H. *Supply Chain Strategy: The Logistics of Supply Chain Management*. McGraw-Hill, New York, 2002.

13. Frohne, P.T. *Quantitative Measurements for Logistics*. McGraw-Hill and SOLE Press, New York, 2008.

14. Jones, J. V. *Integrated Logistics support Handbook*, 3rd ed. McGraw-Hill Professional and SOLE Press, New York, 2006 (ISBN 0071471685).

15. Jones, J. V. *Supportability Engineering Handbook: Implementation, Measurement, and Management*. McGraw-Hill and SOLE Press, New York, 2007.

16. Langford, J. W. *Logistics: Principles and Applications*, 2nd ed. McGraw-Hill Professional and SOLE Press, New York, 2007.

17. MIL-HDBK-502. *Department of Defense Handbook on Acquisition Logistics*. Department of Defense, Washington, DC, 1997.

18. MIL-PRF-49506. Performance Specification. *Logistics Management Information*. Department of Defense, Washington, DC (latest edition).

19. Reliability Information Analysis Center (RIAC). *Supportability Toolkit*. Prepared by B. S. Blanchard and J. W. Langford for RIAC, 6000 Flanagan Rd., Suite 3, Utica NY 13502-1348, 2005.

20. SOLE—The International Society of Logistics. *Logistics Spectrum*. Quarterly Journal, SOLE, 8100 Professional Place, Suite 111, Hyattsville, MD 20785.

21. SOLE—The International Society of Logistics. *Annual Symposium Proceedings*. SOLE, 8100 Professional Place, Suite 111, Hyattsville, MD 20785.

22. *Supply Chain Management Review*. Published by Reed Business Information, Reed Elsevier, 225 Wyman St., Waltham, MA 02451 (published bimonthly).

23. Taylor, G.D., editor. *Logistics Engineering Handbook*. CRC Press/Taylor Francis Group, Boca Raton, FL, 2008.

24. U.S. Air Force. *Air Force Journal of Logistics*. Quarterly Journal, Superintendent of Documents, U.S. Government Printing Office, Washington, D.C. 20402.

F.8 PRODUCTION, MANUFACTURING, QUALITY CONTROL, AND ASSURANCE

1. Breyfogle, F. W., J. M. Cupello, and B. Meadows. *Managing Six Sigma: A Practical Guide to Understanding, Assessing, and Implementing the Strategy that Yields Bottom-Line Success*. John Wiley & Sons, Hoboken, NJ, 2000.

2. Cohen, L. *Quality Function Deployment: How to Make QFD Work for You*. Addison-Wesley, Reading, MA, 1995.

3. Evans, J. R., and J. W. Dean. *Total Quality Management, Organization, and Strategy*. 3rd ed. Thomson/South-Western, Mason, OH, 2002.

4. Fowlkes, W. Y., and C. M. Creveling. *Engineering Methods for Robust Design: Using Taguchi Methods in Technology and Product*. Addison-Wesley, Reading, MA, 1995.

5. Fryman, M. *Quality and Process Improvement*. Delmar Learning Publishers, New York, NY, 2001.

6. Gaal, A. *ISO 9001:2000 for Small Business: Implementing Process Approach in Quality Management*. St. Lucie Press, Boca Raton, FL, 2001.

7. George, M., D. Rowlands, and W. Kastle, *What Is Lean Six Sigma*. McGraw Hill, New York, 2004.

8. Gradel, T. E., and B. R. Allenby. *Industrial Ecology*, 2nd ed. Prentice Hall, Upper Saddle River, NJ, 2003.

9. Juran, J. M., and G. A. Blanton (eds.). *Juran's Quality Control Handbook*, 5th ed. McGraw-Hill, New York, 1998.

10. Hoyle. D. *ISO 9000 Quality Systems Development Handbook: A Systems Engineering Approach*. Butterworth-Heinemann, New York, 1998.

11. ISO 9000. *Quality Management Systems—Fundamentals and Vocabulary.* International Organization for Standards(ISO), December 2000.

12. ISO 9001. *Quality Management Systems—Requirements*. International Organization for Standards (ISO), December 2000.

13. ISO 9004. *Quality Management Systems—Guidelines for Performance Improvements*. International Organization for Standards (ISO), December 2000.

14. ISO 14001. *Environmental Management Systems—Specific Guidance for Use*. International Organization for Standards (ISO), September 1996.

15. *Journal of Industrial Ecology*, published by the International Society for Industrial Ecology (ISIE), MIT Press, Cambridge, MA.

16. Lowenthal, J. N. *Six Sigma Project Management: A Pocket Guide*. ASQ Quality Press, Milwaukee, WI, 2001.

17. Madu, C. N. (ed.). *Handbook of Environmentally Conscious Manufacturing*. Kluwer Academic Publishers, New York, NY, 2001.

18. McMahon, C., and J. Browne. *CADCAM: Principles, Practice and Manufacturing Management*, 2nd ed. Addison-Wesley, Reading, MA, 1998 (ISBN 0201178192).

19. Montgomery, D.C. *Introduction to Statistical Quality Control*, 5th ed. John Wiley & Sons, Hoboken, NJ, 2005.

20. Moody, J. A., W. L. Chapman, F. D. Van Voorhees, and A.T. Bahill. *Metrics and Case Studies for Evaluating Engineering Designs*. Prentice Hall, Upper Saddle River, NJ, 1997.

21. Reliability Information Analysis Center (RIAC). *Quality Toolkit*. RIAC, 6000 Flanagan Rd, Suite 3, Utica, NY 13502-1348.

22. Revelle, J. B., J. W. Moran, and C. Cox. *The QFD Handbook*. John Wiley & Sons, Hoboken, NJ, 1997.

23. Revelle, J. B. *Manufacturing Handbook of Best Practices: An Innovation, Productivity, and Quality Focus*. St. Lucie Press, Boca Raton, FL, 2001.

24. Roy, R. K. *Design of Experiments Using the Taguchi Approach: 16 Steps to Product and Process Improvement*. John Wiley & Sons, Hoboken, NJ, 2001.

25. Smith, G. M. *Statistical Process Control and Quality Improvement*, 4th ed. Prentice Hall, Upper Saddle River, NJ, 2001.

26. Swamidass, P. M. (ed.). *Innovations in Competitive Manufacturing*. American Management Association (AMACOM), Boston, MA, 2002.

F.9 OPERATIONS RESEARCH AND OPERATIONS MANAGEMENT

1. Fabrycky, W. J., P. M. Ghare, and P. E. Torgersen. *Applied Operations Research and Management Science*. Prentice Hall, Upper Saddle River, NJ, 1984.

2. Hall, R.W. *Queuing Methods for Service and Manufacturing*. Prentice Hall, Upper Saddle River, NJ, 2001.

3. Hillier, F. S., and G. V. Lieberman. *Introduction to Operations Research*, 8th ed. McGraw-Hill, New York, NY, 2005.

4. Krajewski, L. J., and L. P. Ritzman. *Operations Management : Strategy and Analysis*, 5th ed. Addison-Wesley, Reading, MA, 1998.

5. Russell, R. S., and B. W. Taylor. *Operations Management*, 4th ed. Prentice Hall, Upper Saddle River, NJ, 2002.

6. Stevenson, W.J. O*perations Management,* 8th ed. McGraw-Hill, Boston, 2005.

7. Taha, H. A. *Operations Research: An Introduction*, 8th ed. Prentice Hall, Upper Saddle River, NJ, 2006.

F.10 ENGINEERING ECONOMY AND LIFE-CYCLE COST ANALYSIS

1. Blank, L., and A. Tarquin. *Engineering Economy*, 5th ed. McGraw-Hill, New York, 2002.
2. Conkins, G. *Activity-Based Cost Management: An Executive's Guide.* John Wiley & Sons, Hoboken, NJ, 2001.
3. Sullivan, W. G., E. M. Wicks, and J. Luxhoj. *Engineering Economy,* 12th ed. Pearson Prentice Hall, Upper Saddle River, NJ, 2002.
4. Fisher, G. H. *Cost Considerations in Systems Analysis.* American Elsevier Publishing Co., New York, 1971.
5. Hicks. D. T. *Activity-Based Costing: Making it Work for Small and Mid-Sized Companies*, 2nd ed. John Wiley & Sons, Holbken, NJ, 1999.
6. Thuesen G. J., and W. J. Fabrycky. *Engineering Economy*, 9th ed. Prentice Hall, Upper Saddle River, NJ, 2001.

F.11 MANAGEMENT AND SUPPORTING AREAS

1. ANSI/GEIA EIA 649-A. *National Consensus Standard for Configuration Management.* Electronic Industries Alliance (EIA), Arlington, VA April 2004.
2. Camp, R. C. *Business Process Benchmarking: Finding and Implementing Best Practices.* Quality Press, Milwaukee, WI, 1993.
3. Chaffey, D. *E-Business and E-Commerce Management*, 3rd ed. Pearson Prentice Hall, Upper Saddle River, NJ, 2006.
4. Chapman, C., and S. Ward. *Project Risk Management: Processes, Techniques, and Insights.* John Wiley & Sons, Hoboken, NJ, 2003.
5. Cleland, D. I. *Project Management: Strategic Design and Implementation,* 3rd ed. McGraw-Hill, New York, 1998.
6. Corbitt, R. A. *Standard Handbook of Environmental Engineering.* McGraw-Hill, New York, 1999.
7. D0D 5000.2-R. *Mandatory Procedures for Major Defense Acquisition Programs (MDAPS) and Majar Automated Information System (MAIS) Acquisition Programs.* Office of the Secretary of Defence, The Pentagon, Washington, DC 20301, April 5, 2002.
8. Gibson, J. L., J. H. Donnelly, J. M. Ivancevich, and R. Konopaske. *Organizations: Behavior, Structure, and Processes*, 12th ed. McGraw-Hill/Irwin, New York, 2005.
9. Gordon, J. R., and S. R. Gordon. *Information Systems: A Management Approach*, 2nd ed. Dryden Press, New York, 1998.
10. Haimes, Y. Y. *Risk Modeling, Assessment, and Management*, 2nd ed. John Wiley & Sons, Hoboken, NJ, 2004.

11. Kerzner, H. *Proiect Management: A Systems Approach to Planning, Scheduling, and Controlling*, 9th ed. John Wiley & Sons, Hoboken, NJ, 2005.

12. Lewis, J. P. *Fundamentals of Project Management,* 2nd ed. AMACOM, New York, 2002.

13. Magrab, E. B. *Integrated Product and Process Design and Development: The Product Realization Process.* CRC Press, Boca Raton, FL, 1997.

14. Smith, P. G., and G. M. Merritt. *Proactive Risk Management: Controlling Uncertainty in Product Development.* Productivity Press, New York, 2002.

15. Thomas, G. C., and M. R. Smith. *Using Structured Benchmarking to Fast-Track CMM Process Improvement.* IEEE Computer Society Press, Vol. 18, Issue 5, 2001.

16. Zairi, M., and P. Leonard. *Practical Benchmarking: The Complete Guide.* Chapman and Hall, 1994.

INDEX